Mathematics and Biosciences in Interaction

Managing Editor

Wolfgang Alt
Division of Theoretical Biology
Botanical Institute
University of Bonn
Kirschallee 1
D-53115 Bonn
Germany
e-mail: wolf.alt@uni-bonn.de

Editorial Board

Fred Adler (Dept. Mathematics, Salt Lake City)
Mark Chaplain (Dept. Math. & Computer Sciences, Dundee)
Andreas Deutsch (Div. Theoretical Biology, Bonn)
Andreas Dress (Math. Institut, Bielefeld)
David Krakauer (Dept. of Zoology, Oxford)
Robert T. Tranquillo (Dept. Chem. Engineering, Minneapolis)

Mathematics and Biosciences in Interaction is devoted to the publication of advanced textbooks, monographs, and multi-authored volumes on mathematical concepts in the biological sciences. It concentrates on truly interdisciplinary research presenting currently important biological fields and relevant methods of mathematical modelling and analysis. Emphasis will be put on mathematical concepts and methods being developed and refined in close relation to problems and results relevant for experimental bioscientists.

The series aims at publishing not only monographs by individual authors presenting their own results, but welcomes, in particular, volumes arising from collaborations, joint research programs or workshops. These can feature concepts and open problems as a result of such collaborative work, possibly illustrated with computer software providing statistical analyses, simulations or visualizations.

The envisaged readership includes researchers and advanced students in applied mathematics – numerical analysis as well as statistics, genetics, cell biology, neurobiology, bioinformatics, biophysics, bio(medical) engineering, biotechnology, evolution and behavioral sciences, theoretical biology, systems theory.

DYNAMICS of CELL and TISSUE MOTION

Wolfgang Alt
Andreas Deutsch
Graham Dunn
Editors

Birkhäuser
Basel · Boston · Berlin

Editors:

Wolfgang Alt and Andreas Deutsch
Division of Theoretical Biology
Botanical Institute
University of Bonn
Kirschallee 1
53115 Bonn
Germany
e-mail: wolf.alt@uni-bonn.de
       deutsch@io.bota.uni-bonn.de

Graham A. Dunn
MCR Muscle and Cell Motility Unit
The Randall Institute
King's College London
26–29 Drury Lane
London WC2B 5RL
U.K.
e-mail: gad@helios.rai.kcl.ac.uk

1991 Mathematics Subject Classification 92C99

Library of Congress Cataloging-in-Publication Data

Dynamics of cell and tissue motion / Wolfgang Alt, Andreas Deutsch,
  Graham Dunn. editors.
      p.   cm. -- (mathematics and biosciences in interaction)
    Includes bibliographical references and index.
      ISBN 3-7643-5781-9 (alk. paper). -- ISBN 0-8176-5781-9 (alk. paper)
    1. Calls--Motility. 2. Biophysics. 3. Dynamics. I. Alt. W.
  (Wolfgang) II. Deutsch, Andreas, 1960-   . III. Dunn, Graham,
  1944-   . IV. Series.
  QH647.D96   1997
  571.6'34--dc21

Deutsche Bibliothek Cataloging-in-Publication Data

Dynamics of cell and tissue motion / Wolfgang Alt ... ed. - Basel
Boston ; Berlin : Birkhäuser, 1997
        (Mathematics and biosciences in interaction)
        ISBN 3-7643-5781-9 (Basel ...)
        ISBN 0-8176-5781-9 (Boston)

This work is subject to copyright. All rights are reserved, whether the whole or part of the material is concerned, specifically the rights of translation, reprinting, re-use of illustrations, broadcasting, reproduction on microfilms or in other ways, and storage in data banks. For any kind of use whatsoever, permission from the copyright owner must be obtained.

© 1997 Birkhäuser Verlag, P.O. Box 133, CH-4010 Basel, Switzerland
Printed on acid-free paper produced of chlorine-free pulp. TCF ∞
Printed in Germany
ISBN 3-7643-5781-9
ISBN 0-8176-5781-9

9 8 7 6 5 4 3 2 1

Preface

Six years after the first International Workshop on "Biological Motion" took place at Königswinter near Bonn in March 1989 (see Alt & Hoffmann 1990) a subsequent Workshop, organized under the same DFG Research Program (SFB 256: Nonlinear Partial Differential Equations), assembled in March 1995 to concentrate on an important sub-topic of the first meeting - "Cell and Tissue Motion". About 45 scientists with specialities ranging from biology and medicine to physics and applied mathematics, from ten European countries and from North America, gathered at the hospitable conference centre of the "Landjugendakademie" in Bonn-Röttgen. The presentations of physico-chemical theories and mathematical models in plenary talks, as well as their discussion in several work sessions, concentrated on eight particular themes: Animal Tissues, Plant Tissues, Proteins and Forces of Motion, Cytoplasm and Fiber Dynamics, Cell Shape and Motion Analysis, Locomotion Models, Swarm Models and, especially, *Dictyostelium* Aggregation. During the whole conference the atmosphere was open minded and very stimulating, not only in the formal sessions but also down in the "Schloßkeller", at the bowling alley, and on the communal afternoon hike through the Kottenforst.

We were glad and thankful that so many biologists and mathematicians participated in the meeting and especially that so many young scientists attended including students, coworkers and colleagues from Universities and Research Institutes near Bonn. Moreover, we express our special thanks to the "senior" participants, Prof. Günter Gerisch (München), Prof. Andrej Grebecki (Warszawa), Prof. Zigmunt Hejnowicz (Katowice) and, in particular, to the emeritus biology Professors of the "Rheinische Friedrich-Wilhelms-Universität Bonn": Prof. Karl-Ernst Wohlfarth-Bottermann (formerly at the Institute of Cytology and Micromorphology) and Prof. Andreas Sievers (Botanical Institute), who have both provided essential support to the local organizing group "Abteilung Theoretische Biologie" since the time of its foundation in 1986. Finally, we acknowledge that the preparation and realization of the Workshop would not have been so easy and effective without the kind help of Marianne and Anke Thiedemann, the secretaries of the SFB 256, as well as Edith Geigant, Boris Hinz, Volker Lendowski and Beate Pfistner, who also helped with the production of this book.

In a final plenary discussion, we identified a series of "hot topics" that either had been treated in single sessions or else had emerged as guiding principles during the meeting. These we collected into four or five groups, and we appointed one or two coordinators for each group who kindly undertook the task of contacting possible authors to contribute articles for the forthcoming book, of having the articles reviewed by independent referees, of ensuring that the articles were revised accordingly and, finally, of writing an Introduction as well as a concluding Discussion section for each Chapter. After more than one year of mutual correspondence, exchange of ideas and careful revision of the text, formulas and figures, we are now happy to present the outcome of these activities in the finished book. We have tried in selecting the material presented not only to reflect the course of the discussions at the Workshop but also, where appropriate, to cover more recent results and point to important problems of current and future research.

To receive additional information about single contributions, including available simulation software, the reader may contact the authors or one of the editors, preferably in Bonn (see the List of Addresses at the end of the book). Furthermore, we (the editors) offer to provide English translations of the German quotations heading some of the Chapter Introductions. Obviously, this selection of remarks by scientists from the first half of the century is biased but might show the kind of inspiration that biologists during this period had and, possibly, still have on biophysical and biomathematical thought.

Bonn, February 1997

Wolfgang Alt and Andreas Deutsch (Bonn)
Graham Dunn (London)

Contents

Preface v

General Introduction xi

Chapter I
Motile Dynamics at the Cellular Level
– Cytoplasmic Motion and Cell Shape –
Coordinators: Graham Dunn and Wolfgang Alt

Introduction 3

I.1 Rudolf Winklbauer, Andreas Selchow, Beate Boller
     and Jürgen Bereiter-Hahn
     Embryonic Mesoderm Cells and Larval Keratocytes
     from *Xenopus*: Structure and Motility of Single Cells 7

I.2 Boris Hinz and Oana Brosteanu
     Periodicity in Shape Changes of Human Epidermal Keratinocytes ... 15

I.3 Michael G. Vicker and Wei Xiang
     Self-organized F-Actin Autowaves Govern Pseudopodium Projection
     and the Non-random Locomotion of *Dictyostelium* Amoebae 21

I.4 (Box) Oana Brosteanu, Peter J. Plath and Michael G. Vicker
     Mathematical Analysis of Cell Shape 29

I.5 Graham Dunn, Igor Weber and Daniel Zicha
     Protrusion, Retraction and the Efficiency of Cell Locomotion 33

I.6 Marcin Inkielman and Jan Doroszewski
     Microscopic Image Classification Based on Descriptor Analysis 47

I.7 Xiaoyi He and Micah Dembo
     A Dynamical Model of Cell Division 55

I.8 Saša Svetina, Bojan Božič and Boštjan Žekš
     Shape Behavior of Closed Layered Membranes and Cytokinesis 67

I.9 Wolfgang Alt and Robert T. Tranquillo
     Protrusion-Retraction Dynamics of an Annular Lamellipodial Seam . 73

I.10 *Vladimir A. Teplov, Yuri M. Romanovsky, Dmitri A. Pavlov and Wolfgang Alt*
Auto-oscillatory Processes and Feedback Mechanisms in *Physarum* Plasmodium Motility 83

I.11 *Volker Lendowski and Alex Mogilner*
Origin of Actin-induced Locomotion of *Listeria* 93

I.12 *Gül Civelekoglu and Edith Geigant*
Models for the Formation of Oriented F-actin Structures in the Cytoskeleton 101

Discussion & Open Problems 111

Chapter II
Dynamics of Cell Interaction with the Environment
Coordinator: Robert T. Tranquillo

Introduction 115

II.1 *Andrzej Grebecki*
Cell-Substratum Interactions of *Amoeba proteus*: Old and New Open Questions 117

II.2 *Micah Dembo, Tim Oliver, Akira Ishihara and Ken Jacobson*
Imaging Traction Stresses 123

II.3 *Michael G. Vicker*
Chemotaxis and Chemokinesis of *Dictyostelium* Amoebae: Different Accumulation Mechanisms Induced by Temporal Signals and Spatial Gradients of Cyclic AMP 133

II.4 *Robert T. Tranquillo and Wolfgang Alt*
Receptor-mediated Models for Leukocyte Chemotaxis 141

II.5 *Richard B. Dickinson*
A Model for Cell Migration by Contact Guidance 149

II.6 (Box) *Richard B. Dickinson*
Derivation of a Cell Migration Transport Equation from an Underlying Random Walk Model 157

II.7 *Robert T. Tranquillo and Victor H. Barocas*
A Continuum Model for the Role of Fibroblast Contact Guidance in Wound Contraction 159

II.8 *Galina Solyanik*
Wound Healing and Tumour Growth – Relations and Differences – .. 165

Discussion & Open Problems 167

Chapter III
Dynamics of Cell-Cell Interactions
– Collective Motion and Aggregation –
Coordinator: Philip K. Maini

Introduction 171

III.1 *Alex Mogilner, Andreas Deutsch and Julian Cook*
Models for Spatio-angular Self-organization in Cell Biology 173

III.2 *Angela Stevens and Frank Schweitzer*
Aggregation Induced by Diffusing and Nondiffusing Media 183

III.3 *John C. Dallon, Hans G. Othmer, Catelijne Van Oss, Alexandre Panfilov, Paulien Hogeweg, Thomas Höfer and Philip K. Maini*
Models of *Dictyostelium discoideum* Aggregation 193

III.4 *Nicholas J. Savill and Paulien Hogeweg*
A Cellular Automata Approach to the Modelling
of Cell-Cell Interactions 203

Discussion & Open Problems 209

Chapter IV
Dynamics within Tissues – Morphogenesis and Plant Movement –
Coordinators: Sharon R. Lubkin and Lev V. Beloussov

Introduction 213

IV.1 *Lev V. Beloussov, Jürgen Bereiter-Hahn and Paul B. Green*
Morphogenetic Dynamics in Tissues: Expectations of
Developmental and Cell Biologists 215

IV.2 *Lev V. Beloussov*
Mechanical Stresses in Animal Development: Patterns and
Morphogenetical Role 221

IV.3 *Sharon R. Lubkin*
Mechanisms for Branching Morphogenesis of the Lung 229

IV.4 *Zygmunt Hejnowicz and Andreas Sievers*
Tissue Stresses in Plant Organs: Their Origin and
Importance for Movements 235

IV.5 *Paul B. Green*
Self-Organization and the Formation of Patterns in Plants 243

IV.6 (Box) *Steven C. Rennich and Paul B. Green*
The Mathematics of Plate Bending 251

IV.7 *Alexander A. Stein, Mechthild Rutz and Hanna Zieschang*
Mechanical Forces and Signal Transduction in Growth and
Bending of Plant Roots 255

IV.8 *Jerzy Nakielski*
Growth Field and Cell Displacement within the Root Apex 267

IV.9 *Nicolas Rivier, Benoit Dubertret and Gudrun Schliecker*
The Stationary State of Epithelial Tissues 275

Discussion & Open Problems 283

References 285

Group Picture 316

Addresses 319

Index 323

General Introduction

Wolfgang Alt, Andreas Deutsch, Graham Dunn

> *Wenn in der Physik ganz neue "Mathematiken" notwendig waren, die den Scharfsinn der mathematischen Physiker bis zur Erschöpfung anspannten, so kommt es einem eigentlich recht unwahrscheinlich vor, daß bei der Behandlung der kompliziertesten ganzheitlichen Gebilde der Natur – der Organismen – die bloße Anwendung elementarer Denkweisen der Physik und physikalischen Chemie genügen sollte ... Nur das innige Zusammenwirken des Biologen, des theoretischen Physikers, des Mathematikers und Logikers wird das Problem der "Mathematisierung" der Biologie lösen können.*
>
> Ludwig von Bertalanffy (1932)

What is Life?

What characterizes *life* better than its capacity for *self-organization*? Self-organization, in its original biological meaning, might be defined as the formation, maintenance and reproduction of "entities" that autonomously and continuously shape their structure in relation to and in interaction with a varying environment (habitat). Consequently, biology is characterized by the wide and rich spectrum of such entities on very different spatial and informational scales, reaching from prions and viruses, over cells and organisms, to social entities such as swarms or herds.

Biological *cells* are generally considered to be the most fundamental of these autonomous entities of life, occuring both as "independent" unicellular organisms and as "compartments" (Lat. *cella*: con'cealed' storing or living room) within multicellular plant or animal *tissues* (Lat. *textum, textura*: connected arrangement of constituent parts such as fibres, rods, or cells). We should recall that the concept of *cellular tissue*, which in the narrower sense means a connected array of cells, is also used in a wider sense to include any texture or connected environment (such as extracellular matrix, medium or substratum) in which cells are situated and carry out their activities of sending or receiving signals, exerting or experiencing forces and, ultimately, performing various kinds of motions. Motion, the most dynamical aspect of life, is also the most obvious expression of cellular self-organization and the main topic of this book.

Dynamics of cell and tissue motion

When we chose the keyword *dynamics* (Grk. *dynamis*: force, ability, influence) to lead the book's title, we had in mind not only to express and promote the general importance of force transduction and stress fields for any kind of motion within cells and tissues but also to emphasize that the contents deal largely with kinetic processes and interactions rather than with static structures, and that most theories and models presented herein rely on dynamical systems in the broader mathematical sense.

It is the dichotomy of cells, existing and moving either as autonomous units or as components of a "social" complex, that forms the central subject of the book and leads to a whole series of questions ranging from "What are the forces, actions and reactions characterizing the motion of single cells?" to "What are the conditions and principles of tissue formation, maintenance and deformation?" In multicellular animals, the aggregative motion of cells is an essential precondition for the formation of many tissues, and a paradigmatic "system" is provided by the life cycle of *Dictyostelium discoideum*. These cellular slime moulds exist as free-living, crawling amoebae (Grk. *amoibé*: change, alteration) until starvation induces the aggregative phase when the amoebae migrate in spirals or streams to collect in a central mound of cells which eventually develops into a *slug* capable of autonomous migration. The formation and locomotion of the slug thus results from the differentiation and coordinated activity of its constituent cells and may be considered a prototype of the self-organizing tissue and organ development that occurs during embryogenesis. The study of cellular slime moulds might therefore clarify the link between the locomotion of free cells and the generally slower developmental motions and shape changes of organogenesis, at least in animals. In contrast, the mechanisms of organogenesis in plants can be very different and rather related to the growth of immobile cells inducing slow tissue deformation. However, it was Darwin (1875) who early demonstrated that fast motion on the level of organs or organisms is not restricted to the animal realm – his study on carnivorous plants gives a beautiful demonstration of rapid plant tissue movement.

Overview on themes and methods

The series of Chapters in the book follow the conceptual framework of starting with the lowest level of organization, the "roots" of intracellular dynamics and molecular mechanisms, and climbing through the "thicket" of cell-matrix or cell-cell interactions until finally reaching the more global "panorama" of whole organs or organisms. These are the titles of the four Chapters:

- Chapter I: Motile Dynamics at the Cellular Level
  – Cytoplasmic Motion and Cell Shape –

- Chapter II: Dynamics of Cell Interaction with the Environment

- Chapter III: Dynamics of Cell-Cell Interactions
  – Collective Motion and Aggregation –

- Chapter IV: Dynamics within Tissues
  – Morphogenesis and Plant Movement –

The structure of each Chapter is to start with an **Introduction** that briefly reviews the background of the field and the current state of the art, poses the main problems under investigation and provides an overview of the subsequent contributions. Some of these are conceived as **Boxes**, which briefly present self-contained methodological or conceptual topics, whereas others may include such Boxes. Each Chapter ends with a **Discussion** section and a survey of outstanding topics to be investigated as **Open Problems**. The **References** are collected in a cumulative list at the end of the book, followed by a **List of Addresses** and an **Index** of representative keywords.

The *locomotion of cells* is the principal theme of the book, and Chapters I and II cover various theoretical tools, observational techniques and models that are designed to elucidate the functional interactions of motor components, within the cell as well as in the surrounding extracellular matrix or medium. New methods of adaptive cell image processing (see I.1 & 6) and model-supported analysis of cell shape and protrusions (I.2-5) as well as cell traction (II.2) are used to reveal more detailed information about the spatio-temporal characteristics of "amoeboid motility".

The decision to confine our considerations to amoeboid or "crawling" locomotion is not arbitrary but has its basis in the ubiquitous nature of the amoeboid motor, distributed in the cytoplasm and mainly consisting of a filamentous meshwork (in particular F-actin and myosin) that can be dynamically assembled and disassembled as when needed. In contrast, other cellular motors such as flagella or cilia have more specific functions that depend on their given molecular and geometrical architecture, so that their modelling is restricted to these particular cases. The modelling of actin-mediated motility is much further reaching and concerns, for example, the universal question of how pseudopodial protrusion activity around the cell periphery, which can occur with remarkable regularities in time and space (see I.1 & 2), is coordinated within the cell, maybe by diffusing chemical messengers (I.3), by mechanical stress or hydrostatic pressure (I.9), or by suitable combinations (I.10). Also, the molecular regulations and micromechanical effects of cytoskeletal filaments, such as their crosslinking and alignment (I.11 & 12) or their functional role in *cell adhesion* (II.1 & 7) are still under investigation and could be different in, for example, amoebae and tissue cells. Even so, the dynamical principles revealed for the actomyosin system are likely to be of general importance and also applicable to other phenomena of cell and tissue motion, such as *cytokinesis* (I.7 & 8), *wound healing* (II.7) and the *morphogenesis* of animal and plant tissues (see Chapter IV).

In many of these cases contractility, in concert with further mechanical properties of the cytoskeleton and the cell cortex, creates *stress fields* that are able to transmit forces across the cells and eventually, via adhesion sites and the extracellular matrix, also to the environment and to other cells. Of particular interest for modelling in this context is the ability of a tissue cell not only to follow extracellular structure or stress lines, the phenomenon of *contact guidance* (II.5), but also to produce such anisotropies in its environment by polarity-dependent cell forces. This amplifying feedback is typical for motile responses and occurs, for example, with the adaptive chemotactic response of cells migrating in temporal or spatial concentration gradients (II.3 & 4). In general, such detailed models for the motile response of individual cells to their environment, including mutual effects between cells, are necessary for any substantial derivation of transport or diffusion equations (see Box II.6, and III.2) and of interactive cellular automata (III.1 & 4) that describe *coordinated migration* in cell populations and the related phenomenon of *cell aggregation*, the two themes of Chapter III. Various discrete or continuous simulation techniques are applied in order to reproduce, for example, the (hitherto only experimentally well-known) aggregation patterns in *Dictyostelium* (III.3).

The biomechanical framework, however, changes considerably if cell-induced forces, pressures and stresses are explicitly taken into account for mathematical modelling, as they already have been on the intracellular level in Chapter I, see above, and on the level of cell-substratum interaction with surfaces (II.2) and fibrous gels (II.7). Pursuing this theme, the theories and models developed in Chapter IV are particularly devoted to the important theme of *global stress fields* (see IV.1) that are maintained either by direct cell-cell contacts, as they frequently occur during embryogenesis for example, or by indirect force transduction via an extracellular fibrous tissue, such as the dermis or basement membrane in animal epithelia, or the network of cell walls in plant epithelia. With the help of classical continuum mechanics adapted to the specific structures or geometries, and complemented by functional relations for physiological or genetic control, it can be shown how a global coordination of local cell traction or (turgor) pressure and cell growth or division rate leads to changes in state and shape of the whole organ/tissue, for example, in gastrulation (IV.2) or bifurcation of embryonic tissues (IV.3), in bending of plant stems and roots (IV.4 & 7) or in buckling and budding of primordia at the apical growth cone of sprouts (IV.5). Finally, by considering also the dynamics of cell division and their dependence on stress fields and neighbouring cells, changes in the geometrical or topological configuration of cellular tissues are described, such as in root apices (IV.8) or, in the particular case of stationary states, in general epithelia (IV.9).

Outlook

One central problem connecting most of the topics in the book is the investigation of principles and mechanisms that are responsible for creating, maintaining or changing the geometrical *shape* of a single cell or a cellular tissue. There are striking analogies in the phenomena of shape formation on different organizational levels and on different spatial or temporal scales, for example when comparing

- the indenting furrow in cell division (I.7 & 8) and gastrulation (IV.2) which in both cases proceeds locally (and autonomously) by inward bending of the *cortex* being an active contractile layer of polymer fibres or an interactive sheet of growing and deforming cells, respectively;

- the oscillatory dynamics of plasmodial drops or veins in *Physarum* (I.10) and the early aggregations in myxobacteria or *Dictyostelium* (Chap. III) which all show streaming and pulsations like a viscous and contractile (multiphase) fluid; either the cytoplasm with its multiple consistency (endo/ectoplasm, cytosol/gel) of fibres being embedded in an aqueous phase, or the slug-like mound of cells gliding or crawling in self-produced slime, respectively;

- the shape changes and locomotion of the giant unicellular *Amoeba proteus* (II.1) and of the multicellular *Dictyostelium* slug (III.4) both repeatedly lifting up their leading tips from the substratum and re-attaching them from time to time.

Thus the question arises whether mathematical models with their abstracting potential are able to reveal common principles or universal mechanisms of the underlying dynamical processes. There are some indications that this is so, since the contributions and discussions in the book tend to use a similar selection of analytical tools and universal modelling methods, whether dealing with polymers and fibres, or with migrating and growing cells, so that formulated dependencies or relationships are more easily comparable in the different cases. Mostly we find the application of continuity and force balance equations describing densities, concentrations, vector and tensor fields and the dynamics of free boundaries; also random processes and stochastic simulations have been applied in a closely parallel manner to describing individual behavior and interactions of both molecules and cells. Furthermore, the universal *physical principles* of classical mechanics and hydromechanics are applied within models for cell and tissue motion that rely on stresses, and they can be compared to competing models that rely on the mediation by diffusing chemical messengers. On a more abstract level, even purely *geometrical or topological* arguments are used, such as for describing the development of densely packed plant or animal tissues, where cell divisions play an important role.

A further important problem that can be traced throughout most of the book is the property of *polarity* or *directionality* of both kinds of "entities": polymers on the level of intracellular cytoskeleton or extracellular matrix, and cells on the tissue level. Suitable mathematical models can help here, too, in order to reveal similarities and differences between phenomena on widely different spatial scales. Not only have refined stochastic models and simulation algorithms been used to reproduce the observed phenomena of cooperative alignment, mutual orientation and the formation of corresponding domains, but also continuum models from the physics of nematic fluids have been consulted so that, in an old tradition of comparing wonderful biological phenomena with wonderful physical examples of self-organization, the concept of "living nematic liquid crystals" was introduced (Gruler & De Boisfleury-Chevance 1994).

Challenges for the future include understanding the mathematical and physical principles involved in processes that superficially appear to be related, such as *cancer invasion* and *wound healing*, which on a qualitative level have striking similarities (II.8). Modelling of the kind described here might help to reveal the relations underlying such processes and also, more importantly, to reveal what subtle differences in cell communication and response might account for the "antisocial" behaviour of the malignant cell "rabble" compared to the coordinated and "purposeful" efforts of the healing cells. The concept of *contact inhibition of locomotion* was formulated many years ago, and verified experimentally, to account for the cessation of cell movement following wound closure whereas the "uncontrolled" infiltration of malignant cells into normal cell populations was attributed to a unilateral failure of this response on the part of the malignant cells, which was also confirmed by observing cells colliding in culture. Two outstanding problems associated with this response are ideal candidates for investigation by the methods described here. Firstly, how do the differing patterns of cell population behaviour depend on differences in the contact interactions between individual cells? And, secondly, can the models of contractile dynamics within single cells be extended to simulate the rapid localized paralysis and subsequent contraction that sometimes occur when two normal cells collide?

Further topics for future research include the evolutionary aspects of motion and aggregation/compartmentalization. Viewing different aggregation strategies from the perspective of *natural selection* might lead to interesting models and simulations. Research in this direction should also help to elucidate the relations between the gene and the behavioural (phenotypic) level. The contributions throughout the book stress the importance of the environment which both *shapes and is shaped* so that there seems to be no principal distinction between external and internal motion; instead of resting in some stationary equilibrium state, more frequently cells operate in a "dynamical" state by continuously responding to their actively or passively changing environment.

Finally, we hope that the research efforts in the biological areas described here may also have fruitful implications for further mathematical developments, in particular for bridging the gaps between individual-based descriptions and macroscopic equations, or for the derivation and analysis of new prototypes of nonlinear problems such as those with moving boundaries.

Chapter I

Motile Dynamics at the Cellular Level
– Cytoplasmic Motion and Cell Shape –

Coordinators: Graham Dunn and Wolfgang Alt

> *Seit der Schrift Max Schultzes bildet das Protoplasma den Mittelpunkt des biologischen Naturbildes, da es bereits für sich allein Lebenserscheinungen zeigt und das Leben bedingt. ... Die Stoffteilchen und materiellen Kräfte treten im Protoplasma unter besonderen Umständen zusammen, um Gefüge und Verrichtungen der lebenden Substanz anzunehmen, wobei ein neues gesetzmäßiges Verhalten Platz greift, welches das allgemeine physiko-chemisch überlagert.*
>
> Johannes Reinke (1922)

Introduction

Cell locomotion by crawling over a substratum, as opposed to propulsion by cilia or flagella, is widespread among protozoans and dominates metazoan cell migration. The phenomenon can be investigated in single cells, such as solitary amoebae or isolated tissue cells in culture, by posing questions at three organisational levels:

(A) What are the ultimate transducers of chemical energy into mechanical energy – the molecular motors – and how are these distributed, interlinked with structural elements and controlled at the molecular level?

(B) Which internal forces and movements determine the shape and spreading of a cell and its protrusion and retraction of processes?

(C) Which mechanisms enable a cell to translocate over the substratum and direct and control its locomotion?

Of course, the properties at all "higher" levels are direct consequences of the properties at the elementary level A and so it would seem that, if we could fully understand the molecular level, we should be able to predict the properties at the next and all subsequent levels. There are two flaws in this approach. Firstly, understanding the molecular level requires more than identifying specific components and their interactions; we need also to understand their functional dynamics. This is by no means easy since it now appears that single functions are often distributed over several parallel pathways of molecular interactions, and single interactions may partake in several functions. Secondly, it is becoming increasingly clear that predicting how these dynamics give rise to the properties that emerge at the next level also is not as easy as once was thought, because the emergent properties of even simple non-linear systems can be very complex and difficult to predict.

In this Chapter, and in most of the book, we advocate that a better approach is to attack the problem of cell movement on a wide front using the combined tools of the molecular biologist, biophysicist and biomathematician. This involves observing properties and seeking explanations at all levels simultaneously. In this Chapter we will focus on the properties of level B, but any satisfactory explanation of cell motility requires an understanding of the relationship between levels. Thus an understanding of how internal forces, movements and shape changes are generated by molecular interactions constitutes an explanation of cell motility at the level of the A-B relationship. An equally valid explanation of cell motility, however, lies at the B-C relationship level in which the mechanisms that propel and direct the cell are described in terms of internal movements, forces and mechanical properties. It is possible, of course, to seek either explanation independently of the other.

A necessary initial stage is to describe the behaviour and to search for patterns within each level. The first three papers (I.1-3) are concerned with observing dynamic patterns that can arise at level B and with identifying possible relations to adjacent levels. In (I.1), Winklbauer, Selchow, Boller & Bereiter-Hahn propose that differences between mesodermal cells and keratocytes of *Xenopus* in their pattern of lamellar protrusion are responsible for a difference in the persistence of migratory behaviour at level C. The second and third papers (I.2 and I.3) both deal with quantifying protrusive behaviour as a function of time and angular position in relation to the cell centre. Hinz & Brosteanu (I.2) have analysed the ruffling and protrusive activity of normal and oncogene-transfected human keratinocytes whereas Vicker & Xiang (I.3) have quantified the extension of pseudopods in *Dictyostelium*. Despite the wide biological differences between these cell types, strikingly similar patterns of dynamic behaviour emerge that are periodic in both space and time. Here is the obvious analogy to the spatio-temporal patterns that can arise by self-organisation in an excitable medium, as exemplified by the Beloussov-Zhabotinsky reaction. Both papers discuss the probable role of the actin meshwork as a major component of the excitable medium but they differ in their speculation about the roles of actin-associated proteins or G-actin to F-actin transitions. Theoretical methods for predicting how patterns at level B can arise from molecular properties at level A, some of which are pioneered later in this Chapter, will be required to decide the issue of whether these aspects do differ fundamentally between different cell types.

The molecular/genetic approach is a highly productive way of discovering hard facts at level A and much can be done with well established methodologies. At levels B and C, however, the facts become increasingly "fuzzy" and probabilistic in nature, and the methods required to discover them are often less developed and less standardized. The next three contributions (I.4-6) consider the theory of gathering objective, quantitative data at level B. The problem of analysing the spatio-temporal dynamics of the cell periphery is handled by Brosteanu, Plath & Vicker (I.4) by using the auto- and cross-correlation functions to examine the changes of closed shapes such as cell outlines. Dunn, Weber & Zicha (I.5) approach the problem of shape changes from a different aspect by describing various methods for measuring the rates and dispositions of protrusion and retraction and discussing how these antagonistic activities are related to the translocation of the cell at level C. Algorithms for extracting and classifying the three-dimensional shape and structure of a cell from two-dimensional image slices are presented by Inkielman & Doroszewski (I.6).

The most direct approach towards understanding how the properties of one level give rise to the properties that emerge at the next level is to manipulate the properties at the lower level and observe the resulting changes to the higher level. While this technique often identifies a causal relationship between levels, it does not usually yield an explanation of how it arises. For this we need a model and the remainder of the Chapter is devoted to exploring these causal relations by physico-mathematical modelling. There is a danger in mathematical modelling that almost any process can be modelled by juggling a sufficient number of parameters – but this gives us little confidence of the model's basis in reality. Some of the following papers illustrate the safeguards against this that can be built into the models.

The first two of these models concern cytokinesis which is possibly the simplest and most predictable aspect of cell motility. He & Dembo (I.7) analyse the first cleavage of an echinoderm zygote which is remarkably regular and reproducible. Using the reactive interpenetrating flow formalism, they assign values to eight biomechanical parameters of the actin/myosin meshwork and predict the outcome using classical field theory. This model is a tour de force of biological modelling. Not only does it predict a highly realistic picture of many aspects of the dynamics of cleavage, but the values of six of the eight parameters are fixed by biophysical measurements made at the molecular level and so the scope for parameter juggling is severely restricted. A second test of a good model is a variant of Occam's razor: that the model should use a minimal number of parameters in relation to the degrees of freedom of the system that it seeks to explain. In He & Dembo's model the parameters are honed to the bare minimum and the contractility of the meshwork, in particular, is represented by a simple scalar value with no consideration given to the highly anisotropic arrangement of filaments in the contractile ring. That the model works is a vindication of this bold simplifying assumption. In the second model of cytokinesis, Svetina, Božič & Žekš (I.8) concentrate on the properties of the cell membrane and its underlying actin/myosin cortex and show that considerations of the bending energy of this complex can lead to shape changes characteristic of cytokinesis. This model illustrates another desirable property of a good model: it makes the testable prediction that cytokinesis involves at least two stages with differing energy requirements.

The following paper by Alt & Tranquillo (I.9) returns to the problem of explaining the periodic, spatio-temporal patterns that can arise in the lamellar region of a cell spread like a fried egg. Here the model is a one-dimensional system of partial differential equations and the parametrisation, as with He & Dembo's model, is again at the level of the filamentous and fluid phases of the actin-based meshwork. This model exemplifies the need for sophisticated mathematical modelling since the dynamic properties and patterns that emerge at level B are wholly unexpected in any predictions based on common sense. It also illustrates another important purpose of modelling: to demonstrate that complex dynamics can arise from simple systems.

The remaining three papers deal with models of sub-cellular aspects of actin-based dynamics. Teplov, Romanovsky, Pavlov & Alt (I.10) model the so-called shuttle streaming in the plasmodial strands of *Physarum*. An important difference that emerges between modelling diffusion/reaction chemical systems and modelling cytoplasmic motility is that mechanical coupling, either hydrodynamic or viscoelastic coupling or both, as well as chemical feedback may be essential components of the contractile dynamics of cytoplasm. Mechanical stresses and pressures can have rapid and long-range influences which may lead to dynamics that are not possible in purely biochemical systems and this paper presents evidence that mechanical signalling needs to be taken into account in explaining auto-oscillatory contractions.

Lendowski & Moligner (I.11) turn their attention to the parasitic bacterium *Listeria* which propels itself through its host cell's cytoplasm by organising the host actin into a continuously polymerizing and depolymerizing tail by a process which may be closely related to cellular protrusion. They show that a rectified diffusion or thermal "ratchet" model can be considerably improved by incorporating a term for the elasticity of the actin filaments or, alternatively, by explaining it with an analogous term for swelling of the network – again demonstrating that we should not always neglect mechanical considerations.

Finally, Civelekoglu & Geigant (I.12) further descend the scale of organisation to model a dynamic process at the basis of cell motility: the self-assembly of actin structures. In contrast to the previous two papers, here they show that the molecular interactions between actin filaments and actin-binding proteins can account for the formation of various microfilament structures while ignoring their mechanical properties and external mechanical forces. This paper demonstrates that there is much to learn about the molecular level, beyond merely identifying molecules, interactions and biochemical pathways, before we can understand how it gives rise to the fascinating range of dynamic properties that emerge at level B.

I.1

Embryonic Mesoderm Cells and Larval Keratocytes from *Xenopus*: Structure and Motility of Single Cells

Rudolf Winklbauer and Andreas Selchow (Köln)
Beate Boller and Jürgen Bereiter-Hahn (Frankfurt)

1 Introduction

The shape of a cell reflects an equilibrium of forces and cytoplasmic fluxes (Bereiter-Hahn et al. 1995). The forces are acting along structural elements as are adhesion sites and cytoskeletal fibrils; these fibrils may interact with each other, with motor molecules and with integral membrane proteins. Crawling of cells can be regarded as a shape generating process distant from equilibrium. The differences between the non-equilibrating forces provide the motive force for crawling. If this description is true, then the same events driving locomotion should be responsible for shape generation and *vice versa*.

Single epithelial or fibroblast-like cells in culture are characterized by a relatively thick part, the cell body, and a thin layer of peripheral cytoplasm, the lamella. The latter is devoid of larger organelles, i.e. mitochondria. These parts are the descriptors of cell shape. On the basis of the above statement that crawling can be comprehended as a continuous process of shape generation apart from the equilibrium of forces, an intimate dependence should exist between the size and the shape of the cell body and the formation of the lamella. This view of cell shape and locomotion seems to be contradicted by the motile behaviour of lamellae separated from the cell body experimentally (Euteneuer & Schliwa 1984). Some recent hypotheses on cell crawling also emphasize the lamella as the only part of a cell involved in locomotion (Small 1994a,b). Comparison of shape generation and lamella development, in different but closely related types of cells, with the organisation of their cell bodies should allow some further insights into cell body/lamella relationships.

Embryonic mesoderm cells, and also keratocytes from tadpoles of *Xenopus*, develop lamellae which are distinct from the cell body. In the embryonic cells, the cell body is tightly filled with yolk platelets which are no longer found in tadpole keratocytes. For some experiments a third type of *Xenopus* cells is included, XTH-2 cells, a line of endothelial origin (Schlage et al. 1981). These cells also develop a central dome shaped cell body surrounded by a very thin layer of cytoplasm which, however, contains mitochondria and rough endoplasmic reticulum and therefore does not correspond to the lamellae in the two other cell types. At the very periphery also XTH-2 cells form some small lamellae.

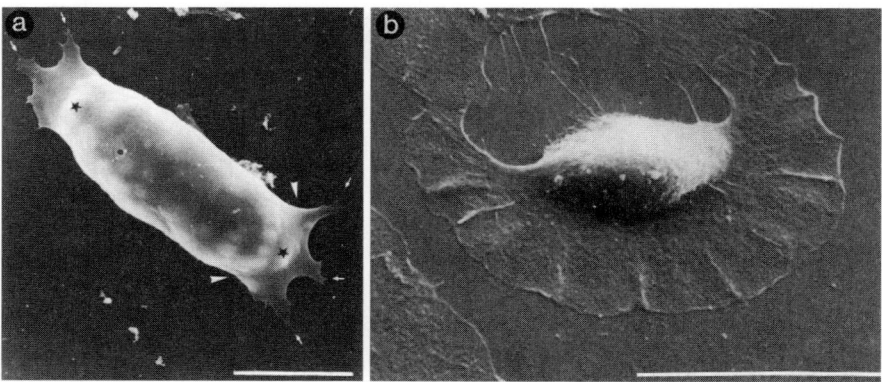

Figure 1: Scanning electron micrographs of an isolated *Xenopus* mesoderm cell (a) and a *Xenopus* tadpole epidermal keratocyte (b). Asterisks: yolk platelets; arrows: filiform processes of lamellae (e.g. defined by arrowheads). Bars = 25 μm.

2 Results and discussion

2.1 External morphology

From their morphology, mesoderm cells and keratocytes appear characteristically distinct, mainly due to the different shape and arrangement of cytoplasmic protrusions which extend in both cases from an elongated, but rather bulbous cell body (Fig. 1). Mesoderm cells contain a number of large yolk platelets which increase their volume about two-fold. In both cell types, the cell body is not flattened against the substrate. Like other fast moving cells, keratocytes or mesoderm cells do not form focal contacts (Selchow & Winklbauer, in preparation, Bereiter-Hahn et al. 1981). The relatively weak substrate adhesion may contribute to the bulbous shape of the respective cell bodies. The most interesting difference between the two cell types is the geometry of their protrusions. Mesoderm cells possess a variable number of small lamellae (Winklbauer & Selchow 1992, Selchow & Winklbauer, in preparation). Most often, two processes form at opposite ends of an elongate cell, giving it a bipolar morphology (Fig. 1a). Keratocytes typically have a single lamella, which is very large, however, and extends on one side of the elongated cell body over its whole length (Fig. 1b). Thus, direction of movement is perpendicular

1.1 Embryonic Mesoderm Cells and Larval Keratocytes

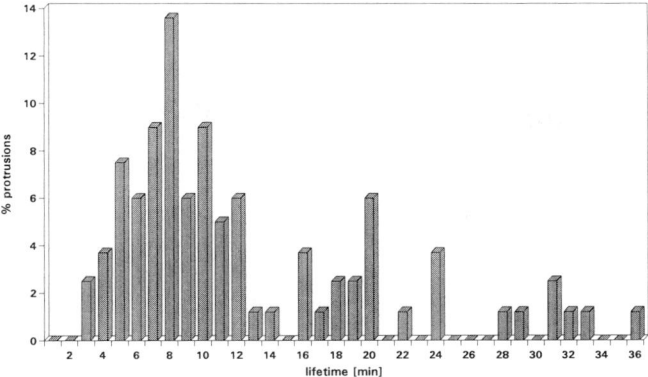

Figure 2: Frequency distribution of life times of mesoderm cell lamellae on a fibronectin coated substrate (n = 89).

to the long axis of the cell body in keratocytes, and parallel to it in mesoderm cells. In keratocytes, the area of the lamella is up to about twice that of the cell body, whereas in mesoderm cells, it is less than a quarter of it (e.g. Fig. 1). This means that the relative size of the lamellae differs by an order of magnitude between the two cell types.

2.2 Protrusive activity

The large canoe-shaped lamellae of keratocytes are usually maintained for hours without being retracted, or dividing. Consequently, the movement of these cells is smooth and persistent. Velocities of up to 30 $\mu m/\min$ can be attained (Bereiter-Hahn et al. 1981). In contrast, protrusions of mesoderm cells are short-lived (Fig. 2). Their mean life time is only 9-10 min, although life times may range from 3 min to half an hour. Consequently, the frequency distribution of life times is strongly skewed. It has a major peak at 8 min and tapers towards longer times (Fig. 2). The number of protrusions per mesoderm cell is not only altered by the retraction of established processes or the extension of new ones, but also by the frequent splitting of lamellae (Winklbauer & Selchow 1992). This rapid turn-over of lamellae is essential for translocation. The protrusions of mesoderm cells usually antagonize each other by pulling in opposite directions. Substantial movement occurs only when the distribution of lamellae changes by extension, retraction, or splitting. Movement ceases as soon as a new equilibrium configuration of protrusions is attained (Winklbauer & Selchow 1992). This discontinuous, non-persistent mode of translocation is in sharp contrast to the steady gliding of keratocytes. Mesoderm cells can reach peak velocities of 25 $\mu m/\min$ during bursts of translocation thus being in the same range as those of keratocytes. Average speed, however, is only about 3 $\mu m/\min$ (Winklbauer et al. 1992). Thus, differences in the stability and distribution of protrusions result in a strikingly different long-term migratory behavior. Therefore, we tried to elucidate what determines the characteristic size and persistence of the lamella.

2.3 Cytoskeletal architecture of the cell body

A submembraneous, cortical layer of actin and myosin is most prominent in keratocytes and XTH-2 cells as well, and less developed in the mesoderm cells (Figs. 3c,d). Towards the interior of the cell this layer is supported by an extensive network of microtubules. Both in embryonic mesoderm cells and in keratocytes these form a basket enclosing the whole cell body, and a few microtubules only emanate into the lamella (Figs. 3b, 4a-c). In XTH-2 cells, however, the microtubules do not form a basket-like structure; rather they are following the medium facing cell surface from the centrioles situated above the nucleus and extend into the periphery of a tapering cytoplasmic layer. In addition a bottom layer of microtubules is present (Figs. 4d-f).

Microtubule and actomyosin fibrils enclose the cytoplasmic organelles including the nucleus, mitochondria, yolk granules and other granular components. Intermingled into these organelles, intermediate filaments are found in keratocytes (cytokeratin) and XTH-2 cells (vimentin-type intermediate filaments). In mesoderm cells cytokeratin filaments have also been identified by immunofluorescence (using pan-cytokeratin antibody LU5, Böhringer, Mannheim F.R.G), which participate in the formation of the peripheral web and are missing in the more central parts of the cell body (Fig. 3a) (Selchow & Winklbauer, in preparation).

2.4 Probing the role of microtubules in shape generation

The interactions between the cortical shell components in the cell body can be studied by colcemid treatment which destroys the microtubules close to the cortex – some other microtubules may survive this treatment but they are never found in the cortex region. In XTH-2 cells colcemid induces spreading of the cell body towards the periphery, its circumference increases. This has been interpreted as a result of loss of anchoring the actin cortex to the microtubules extending toward the periphery (Bereiter-Hahn *et al.* 1995). In addition, diminished fluxes of cytoplasmic material from the periphery towards the cell centre may be responsible for the shape changes. If, on the other hand, keratocytes are treated with colcemid, the opposite phenomenon is observed: The projection area of the cell body shrinks by about 15% and the lamella becomes extended by approximately 20% (Fig. 5). This behaviour can be interpreted on the basis of the hydraulic pressure hypothesis used for the explanation of cell shape and locomotion (Bereiter-Hahn *et al.* 1981, Strohmeier & Bereiter-Hahn 1987, Bereiter-Hahn & Lüers 1994). By contraction the actomyosin-fibrillar web in the cortex exerts pressure on the cytoplasm extending the lamella. The microtubular basket counteracts this pressure, forming an abutment for the cortical actomyosin fibrils. If this basket is destroyed,

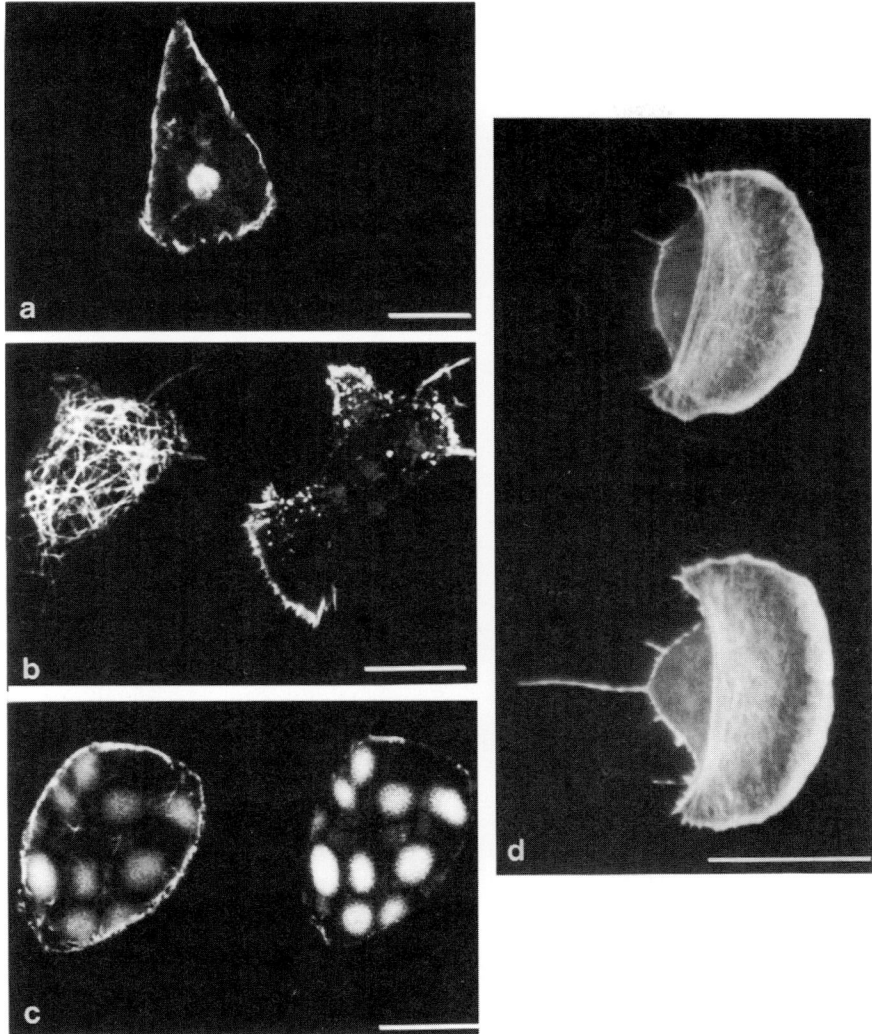

Figure 3: Optical sections made by confocal laser scanning microscopy of isolated *Xenopus* mesoderm cells (a-c) and two keratocytes (d) with immunofluorescence demonstration of pan-keratin (a), of microtubules (b,c) and TRITC-phalloidin staining for F-actin (b,c,d). a: Distribution of pan-keratin in a mesoderm cell. Staining is restricted to a cortical cytoplasmic layer. b, c: Double staining of the same mesoderm cells for microtubules (left part of the images) and F-actin (right part of the images) close to the substratum (b) and further apart in the cell body region (c). Microtubules are restricted to the cortical cytoplasm in the cell body (c) and to its basal part (b). d: F-actin distribution in two keratocytes. F-actin is concentrated in the cortex of the cell body and in the periphery of the lamella. Bars: 25 μm.

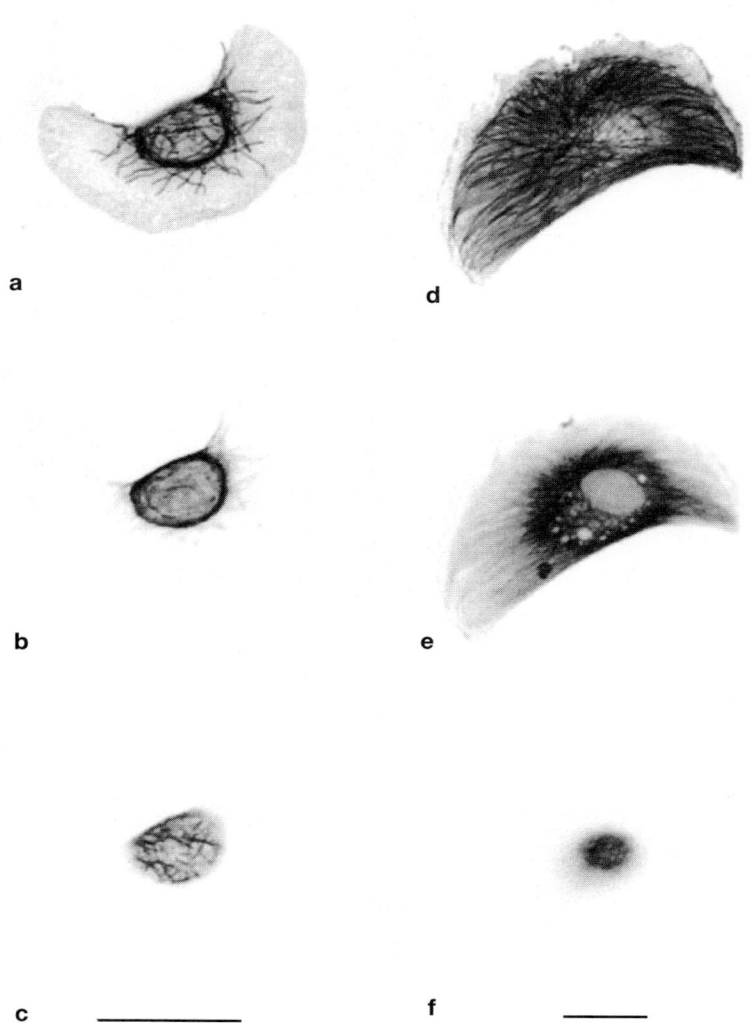

Figure 4: Microtubule distribution in a keratocyte (a-c) and an XTH-2 (endothelial) cell (d-f) as revealed by indirect immunofluorescence (first antibody: YL1/2) and imaging with a confocal laser scan microscope (Leica TCS). (a) and (d) are optical sections close to the substratum, (c) and (f) are sections close to the upper surface of the cell body (distant from the substratum), (b) and (e) traverse approximately through the middle of the cells. The basket like arrangement of microtubules in the keratocyte is obvious in the series (a)-(c). In the XTH-2 cell most of the microtubules are located close to the adhering cell surface (d), and along the upper cell surface they are evenly distributed from the cell center towards the periphery (e), (f). Bars: 25 μm.

the resistance against the contractile force of the cortex is lowered, and the actomyosin cortex squeezes cytoplasm out of the cell body into the lamella which extends thereby (Fig. 5).

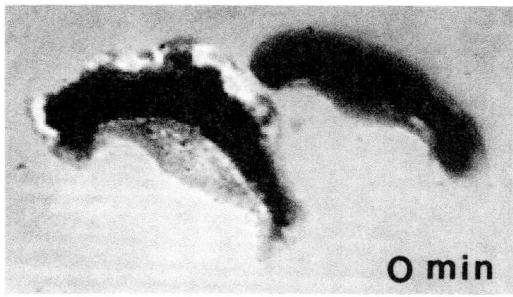

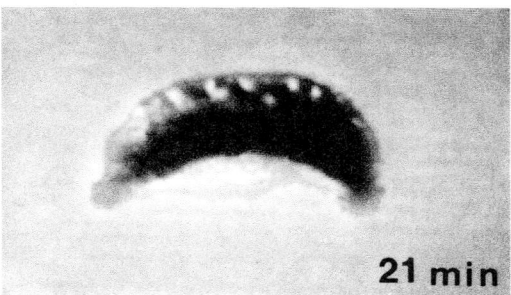

Figure 5: Reaction of single moving keratocytes to colcemid (5 μM). The cell 21 min after colcemid addition is the same as the left one at the moment of addition of the drug (0 min). RIC-images, the area of the cell body can easily be distinguished from the lamella zone. The cell body area diminished under the influence of colcemid by 15%.

2.5 Model for the cell body/lamella relationship

The differences in size and stability of lamellae in mesoderm cells and keratocytes can be related to the organisation of the cell body region. The colcemid experiments described above revealed that

- the morphological (and dynamic) reactions following destruction of the microtubules depend on the arrangement of the microtubules in the cell body ("basket-arrangement" versus extension into the periphery)
- the microtubular basket counteracts the contraction of the cortical actomyosin layer and thus limits the size of the lamella.

In the case of the embryonic mesoderm cells, the large amount of yolk granules counteracts the contraction of the cell cortex and thus limits the relative amount of cytoplasm which can be squeezed out of the cell body to form lamellae. As a consequence the lamellae are unstable, the extending forces and fluxes may easily be overcome by retractive forces developed in the lamella / cell body transition region (Bereiter-Hahn & Lüers 1994), and very small disturbances from the substratum induce local retractions. This prevents continuous movement in one direction requiring the development of a stable lamella. In keratocytes, the contraction of the actomyosin cortex may be limited by the peripheral microtubules and after they disappear the nucleus and surrounding organelles limit the minimum size of the cell body, thus a large portion of cytoplasm may be found in the lamella which becomes stabilized by its size.

Acknowledgements Support by the Deutsche Forschungsgemeinschaft to J.B.H. (Be 423/16), R.W. and A.S. and by the Gesellschaft der Freunde und Förderer der Johann-Wolfgang-Goethe-Universität Frankfurt am Main (to J.B.H.) is greatfully acknowledged.

I.2

Periodicity in Shape Changes of Human Epidermal Keratinocytes

Boris Hinz (Bonn)
Oana Brosteanu (Leipzig)

Analysing cell motion and cell shape dynamics is of major importance for basic research and for medical application. We examined normal human epidermal keratinocytes (nHEK) and cells transfected with viral oncogenes (trHEK) (Barbosa & Schlegel 1989) to reveal typical patterns in spontaneous cell behaviour using new methods of image processing and statistical autocorrelation analysis. Actin-driven migration is necessary for normal cells to cover wounds or for carcinoma cells to invade foreign tissues. Thus, revealing the basic mechanisms of cell motion may finally lead to medical prevention of metastasis or acceleration of wound healing processes.

Under culture conditions where formation of cell-cell contacts is prevented, keratinocytes grow as isolated cells. A single keratinocyte spread on a surface consists of a cell body (Fig. 1A) containing the nucleus and cell organelles, and being surrounded by flat organelle-free lamellae. At the edges of lamellae the cell constantly protrudes small filopodia and lamellipodia which subsequently retract, lift up from the substrate and form centripetally moving ruffles (see Winklbauer et al. I.1 this volume).

1 One-dimensional analysis of single lamellae dynamics

The protrusive activity of single lamellae was examined using one-dimensional analysis of phase contrast image sequences of HEK along section lines transversal to the cell edge extending into the nucleus (Fig. 1A). The luminance profile of structures along the section line was stored at regular intervals (Fig. 1B) and plotted in three-dimensional diagrams where dark structures are represented as valleys (Fig. 1C). Drawing level lines of luminance values only below a certain threshold results in two-dimensional topographic plots (Fig.1D). In these plots ruffles are characterized by low luminance values and constant movement starting at the cell edge (continous line in Fig. 1D), which is detected by a new algorithm (Alt et al. 1995). This allows the combined analysis of cell edge and ruffle movement.

After fast protrusion of a new lamellipodium to a length of 0.8 to 1.7 μm, retraction of the cell edge follows resulting in the separation of a ruffle which is

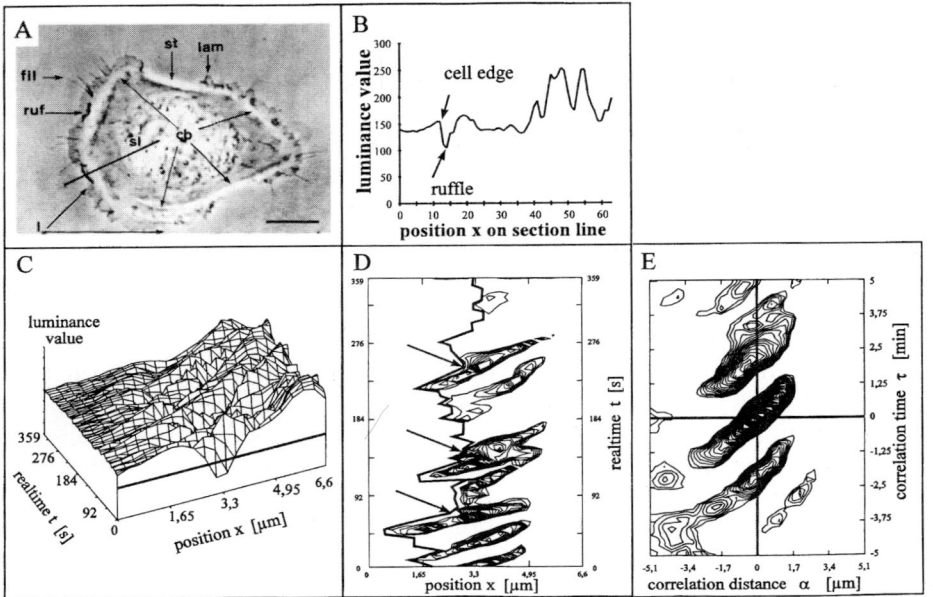

Figure 1: From digitized keratinocyte images to autocorrelation analysis of ruffle movement. (A): Phase contrast image of a typical human keratinocyte; cb: cell body, fil: filopodia, l: lamella, lam: lamellipodia, ruf: ruffles, st: stretched cell border. (B): Luminance profile of structures along a section line (sl) recorded in regular intervals and plotted over time (C). (D): Level lines of luminance values plotted below a threshold (dark line in C). (E): Autocorrelation analysis of long term observations of ruffle dynamics.

constantly moving over the cell surface towards the cell body. At the separation point of ruffle and cell edge the retraction movement stagnates for a short time (arrows in Fig. 1D) and a new lamellipodium forms at the same location on the cell periphery. Ruffle velocity decreases with increasing distance from the cell edge and the structure finally disappears at the cell body. Even in short sequences a strong periodicity in ruffle appearence and movement is obvious. To obtain quantitative results, long term sequences of up to 4 h were analysed by autocorrelation analysis (Alt et al. 1995). Autocorrelation was related to a reference line, and the level lines above the expectation value (Fig. 1E) were plotted. Maximum correlation between two points is represented by the center of topographical line arrangements and stands for high probability of a second ruffle appearing at distance α and correlation time τ over the whole observation period. Free areas represent none or negative correlation to the reference point. The topographical line arrangement around the reference point (line $\alpha = 0$) represents the characteristic time and space behaviour of all observed ruffles in one sequence and is therefore typical for this cell region. Ruffles of all observed cells (15 nHEK and 15 trHEK) start with a slight acceleration, are moving most of their course with constant velocity, and finally slow down near the cell body. A new ruffle usually occurs at the same time when the former is getting slower.

Measuring ruffle velocity at different temperatures (Alt et al. 1995) for normal and transfected keratinocytes revealed differences between the two cell types. trHEK protrude and retract lamellipodia, transforming into ruffles, with significantly higher velocities than nHEK. Consistent with this observation, ruffles of trHEK also appear in higher frequencies (0.43 min^{-1} on average) compared to nHEK (0.24 min^{-1} on average).

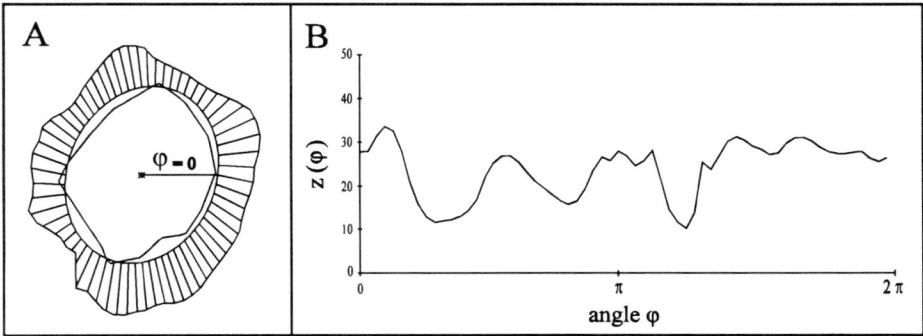

Figure 2: Determination and presentation of normal cell outline extensions. (A): Final result of automatic cell outline detection with a weighed ellipse adapted to the digitized cell body (inner line). (B): Transformation of cell outline data to normal coordinates ($z(\varphi)$) at 180 discrete angular positions.

2 Two-dimensional analysis of patterns in lamellar protrusion and retraction

The main question in analysing the cell periphery was to reveal if there are certain periodic patterns in lamellar activity around the whole cell and if spontaneous motility of keratinocytes is non-random. Differences between the spontaneous behaviour of normal and transfected cells were shown. Phase contrast video sequences of up to ten hours duration were digitized in 24 sec intervals. The cell outlines of one sequence were then detected automatically by a special image processing program (Werner Heiße, Dept. of Theoretical Biology, University of Bonn) and further processed to diminish the amount of data (Alt et al. 1995).

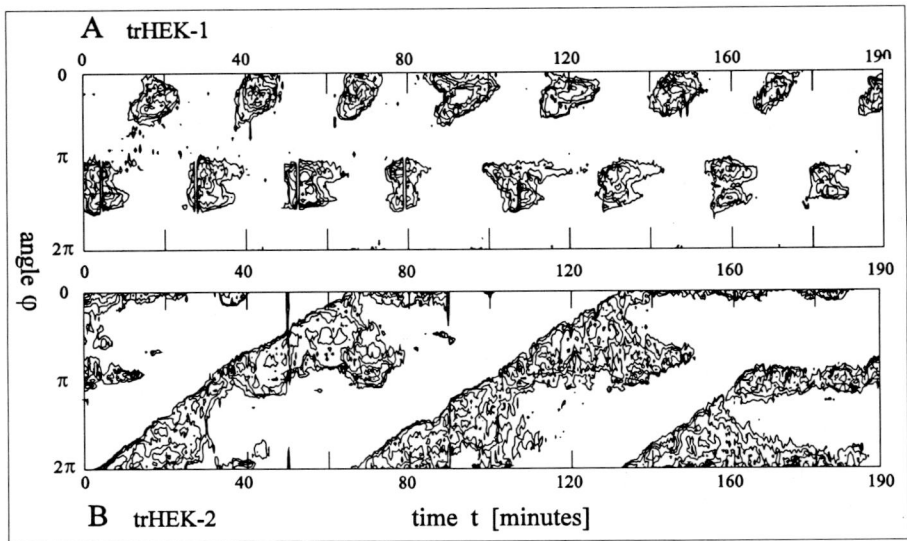

Figure 3: Topographic plot of parametrized cell outlines of keratinocytes over angle φ and time t. (A): Pulsating lamellipodial extension and retraction pattern of a trHEK. (B): trHEK with one lamella rotating counterclockwise around the cell.

To allow quantitative analysis, cell outlines were parametrized in respect to the weighed momental ellipse (Fig. 2) and presented in topographic plots (Fig. 3) analogous to Fig. 1. Angular Fourier and autocorrelation analysis was applied to detect periodic patterns in cell motility (Fig. 4). These methods are described by Brosteanu et al. (I.4 this volume) in detail.

All 30 observed cells formed lamellae over a long period in at least two regions of the cell periphery which were held mainly constant (Fig. 3A: lamellae at $\frac{\pi}{2}$ and $3\frac{\pi}{2}$). Lamella formation in the case of two predominant regions on the cell outline always occured in an antipodal manner. Retraction of one lamella correlated with extension of a new lamella on the other side. In the case of three or four regions of lamella formation, the largest possible distance between these regions was kept ($\frac{2}{3}\pi$ or $\frac{1}{2}\pi$) and lamellae appeared in clockwise or counterclockwise sequences around the cell. Transfected keratinocytes differed from normal keratinocytes (not shown) mainly in having more predominant regions of lamella formation and showing a higher frequency of lamella formation and retraction. Beside this typical behaviour of HEK, two special cases among the observed trHEK demonstrated an impressive regular motility pattern. In the first case (Fig. 3A, Figs. 4A-C) the cell pulsates with very high frequency between two stages of lamellar extension (15 min/extension). The path described by the first two Fourier coefficients of the cell

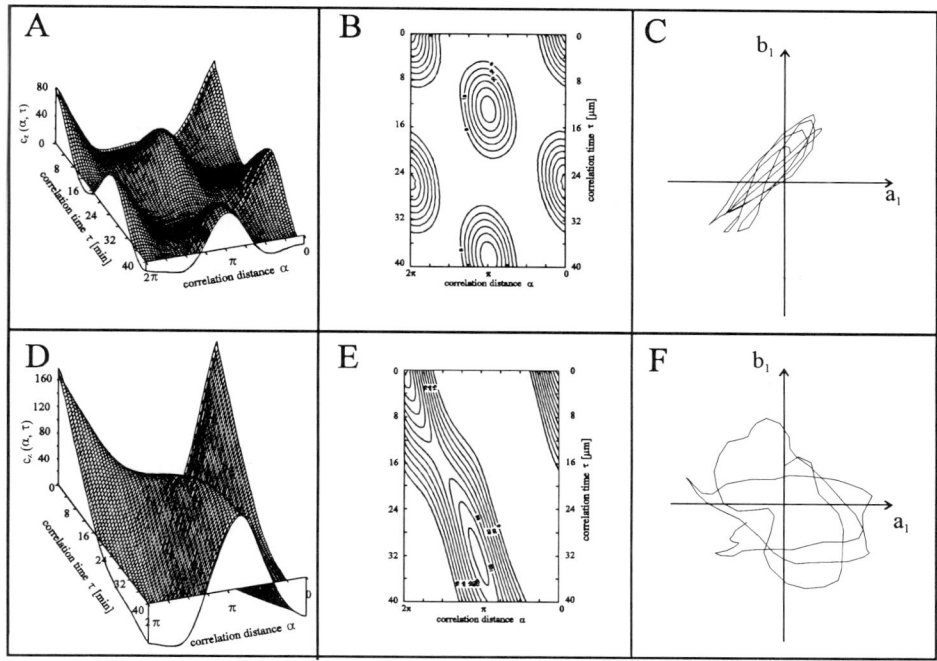

Figure 4: Autocorrelation and angular Fourier analysis of lamellar dynamics. (A), (D): Autocorrelation function $c_z(\alpha, \tau)$ for the corresponding data in Figs. 3A,B plotted in three-dimensional diagrams over correlation time τ and correlation distance α. (B), (E): Topographic plots of correlation values above zero. (C), (F): Corresponding paths described by the first two Fourier coefficients a_1 and b_1 of the cell shape during the observation period.

shape (Fig. 4C) reveals this very regular transition between the two stages, while the autocorrelation function (Fig. 4A) quantifies the mean width ($\frac{3}{8}$ cell circumference), duration (10 min) and periodicity (25 min) of the alternating antipodal lamellae. Also, a change in location of about 1/16 cell circumference per 10 min becomes visible (Fig. 4B). In the second case (Figs. 3B, 4D-F), there is no obvious region of preferred lamella formation, but one single lamella rotating counterclockwise around the cell periphery with a velocity of about one turn per hour. The rotation is reflected in the path of the first two Fourier coefficients (Fig. 4F) which describes counterclockwise cycles. The autocorrelation function (Fig. 4D) quantifies not only the mean drift, but also a cyclic increase and decrease of the width and outward extension of the lamella with a period of 32 min. The last is due to the "rear end" of the lamella, which broadens in two predominant locations (π and 2π) and persists even when the rotation stopped after 160 min (Fig. 3B).

3 Conclusion

Centripetal movement of actin containing ruffles with a velocity of 3.66 μm/min on average is similar to observations of a general backward actin flow with rates of about 6-12 μm/min in the periphery of different cell types (Symons & Mitchison 1991, Forscher et al. 1992, Cao et al. 1993). Even amoeboid cells such as *Chaos carolinensis* exhibit similar actin flow patterns (Grebecki 1990) but with much higher speed and frequency which is consistent with a generally higher migration rate of these large cells. Constant velocity of ruffles and their periodic behaviour suggest a steady intracellular retraction force, pulling the lamellar F-actin towards the cell body. Only if this force is counterbalanced by outward pressure and protrusion of new lamellipodia, can lamellae be maintained. This leads to a steady cycle of actin-assembly at the lamellar tip, actin-flow towards the cell body and disassembly of the actin network at the base of the lamella. Transduction of this steady retraction force to a substrate could lead to cell migration if the actin system is connected to the extracellular matrix by transmembrane linker proteins like integrins. In fact, migrating keratinocytes give the impression of the cell body pulling itself on a rope (the actin system) towards a pole (cell matrix adhesion sites); see also Dembo et al. II.2 this volume.

Our results in analysing cell shape changes clearly demonstrate that the spontaneous behaviour of HEK is in no case chaotic or random but follows regular patterns. Even the observed cancerogenic cells maintain a periodic behaviour which differs mainly by higher frequency. We propose that this periodicity is only due to cytomechanical properties of the cell including (a) contractility of the actin network presumably caused by association with myosin II, (b) intracellular hydrostatic pressure, (c) actin concentration differences with high concentration at the cell body and a more loose network in lamellae, (d) local weakening of the actin cortex by fragmentation and depolymerization, and (e) stabilization of the network by crosslinking proteins or polymerization processes. This point of view is further supported by mathematical models which reproduce the observed motility patterns using the parameters mentioned above, working with the simplifying assumption of a homogeneous distribution of actin-binding proteins (gelsolin, filamin, etc.) in sufficient concentration but excluding regulatory substances as Ca^{++}, profilin or thymosins (Alt et al. 1995). Every feature of the actin system in these models solely depends on the F-actin concentration distribution, and a hydrostatic pressure drop between cell body and tip is regulated by concentration and velocity of the network. In a short résumé we suggest lateral coordination of lamellar extensions as being based on the interaction of two physical properties of the dynamical cortical network: (1) the local regulation of hydrostatic pressure at the plasma membrane induced by contractile flow of the actin network and (2) local stabilization and weakening of the F-actin system at the lamellar tip by assembly/disassembly and visco-contractile tensions (compare Alt & Tranquillo I.9 this volume).

I.3

Self-organized F-actin Autowaves Govern Pseudopodium Projection and the Non-random Locomotion of *Dictyostelium* Amoebae

Michael G. Vicker and Wei Xiang (Bremen)

> *A movement can only develop, naturally, from another movement.*
>
> *(Leibniz)*

1 Introduction

The crawling locomotion of eukaryotic cells has been the subject of intensive investigations aimed at understanding and eventually regulating critical life processes, including various morphogenetic movements, wound healing, inflammatory reactions and metastasis. Nevertheless, at every organizational level of the cell, unanswered questions about locomotion remain concerning its molecular mechanism, how cells stop and start, how oriented movements such as chemotaxis are controlled by the cell and by extracellular signals, and on its relationship to phagocytosis and to DNA synthesis.

In accord with the way most other fundamental cellular functions are viewed today, cell locomotion is consistently treated as a stochastic, chemical concentration-dependent process (reviewed in Killich *et al.* 1993). However, no general theory of cell locomotion has emerged based on this view and recent efforts to measure and compare cell movement (e.g. between normal and malignant cells) have, despite their increased automation and objectivity, been inhibited by a serious, inherent difficulty arising from it. Thus, the validity of even seemingly simple parameters, such as cell speed, locomotory persistence, etc., is doubtful, because of the artificiality of any mean behaviour if cell motion is random (Abercrombie & Heaysman 1953). Furthermore, little attention has been paid to the significance of the fast rates of pseudopodia appearance and extension compared to the slower cell centroid movement rate. Both rates may vary during locomotion or may differ between cell types and depend to some degree on the data acquisition rate. Without a theory of cell locomotion, any particular time scale for data acquisition is essentially subjective. Yet, acknowledgedly little of the information obtainable from the images of locomoting cells is used in such analyses (Noble 1990).

Explanations of the mechanism of cell locomotion have also been sought at the molecular level, and various models have been advanced, based either on the proposed functions of bundles or orthogonal networks of actin filaments or on actomyosin sol-gel transformations (Schleicher & Noegel 1992, Condeelis 1993, Zigmond 1993, Peskin et al. 1993, Stossel 1994). Most current discussion involves the postulated locomotory role of reversible actin filament (F-actin) assembly induced by interactions with phosphoinositides and calcium (Janmey 1994) and some of the myriad actin binding proteins (ABP) including gelsolin, thymosin and profilin. However, although until recently considered the chief modulators of F-actin assembly dynamics, the roles of some *Dictyostelium* ABPs are puzzling. Thus, many transcript-negative, single and multiple mutants of these ABPs, including those for ponticulin (Hitt et al. 1994), severin (André et al. 1989), coronin (De Hostos et al. 1991), myosin I, myosin II, α-actinin, 120 kDa ABP and gelation factor (Wessels et al. 1988, Witke et al. 1992, Peterson & Titus 1994), extend pseudopodia and locomote, though in many cases abnormaly. In other cases, ABPs like talin are correctly positioned, but temporally out of phase with stimulated F-actin assembly (Kreitmeier et al. 1995). Evidence of F-actin turnover at the leading cell margin is substantial (Wang 1985, Theriot & Mitchison 1991, Symons & Mitchison 1991, Redmond & Zigmond 1993).

We review here our recent work looking at cell locomotion and its actin basis (Killich et al. 1993, 1994, Vicker et al. 1996). The locomotion of *Dictyostelium discoideum*, an amoeba frequently serving as a locomotory and developmental model (Schleicher & Noegel 1992), was examined using a synergetic approach in which as much information as possible from the spatio-temporal, 2-D cell image was exploited. The dynamics of the intracellular distribution patterns of F-actin was also examined after a single impulse of the natural *Dictyostelium* cell chemoattractant cyclic AMP (cAMP). The actin response is one of the earliest to a cAMP pulse (McRobbie & Newell 1984, Condeelis et al. 1988, Hall et al. 1989, Europe-Finner et al. 1989): the F-actin concentration doubles above pretreatment levels and the F-actin relocates to a subcortical shell within 6 s, then rapidly disassembles and finally enters a short phase of F-actin assembly oscillations beginning about 30 s after cAMP addition. Neutrophils behave similarly (Omann et al. 1989). Pseudopodia extension is synchronous with the F-actin state and chemotaxis has been shown to specifically require this reaction (see Vicker 1994). These results indicate that cell locomotion is not random and that pseudopodium extension is an expression of activation waves of F-actin assembly, which may also provide the cell with a fundamental oscillator.

2 Methods and materials

Axenic *Dictyostelium discoideum* AX-2 cells were grown in HL-5 medium in shaken suspension (Sussman 1987), washed thrice in potassium phosphate buffered salts solution (PPBSS, pH 6.5). Cells were incubated on glass or PPBSS-agar for either

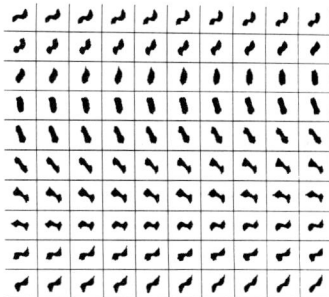

Figure 1: *Dictyostelium* cell shape during locomotion on glass. Video images were digitized each 3 s from left to right and transformed into black-and-white representations. The turning behaviour of this cell is generated primarily by the superposition of traveling waves at its periphery. The frame width equals 16.25 μm.

< 6 h at 21°C or 16 h at 6°C. Effects of pulsatile cAMP excretion were avoided by using cells before development to the aggregation stage and, for locomotory analysis, by placing only 1-5 cells in 2 μl PPBSS in a 50 μm space between a glass slide and a coverslip. Filming commenced 15 min later at 21°C. For studies of intracellular actin, 5000 cells in 50 μl PPBSS were placed on acid washed glass coverslips and incubated at 21°C for 15 min until motile. Cells were then incubated in 1.0 μM cAMP (manufactured by Sigma, Diesenhof) and fixed at intervals in 3.7% formaldehyde for 15 min before specific staining for F-actin with 0.8 μM phalloidin-TRITC (Sigma) for 30 min (Howard & Meyer 1984). Preparations were mounted in 80% glycerol. All solutions were made in PPBSS.

Live cell video images were acquired by a 20x, 0.25 N.A., F-LD phase contrast objective, a green filter and minimal illumination. The images were processed using FG-100-AT-1024 video digitizer (Imaging Technology; Woburn, Mass., USA) as previously described (Killich *et al.* 1993). The cell centroid was determined at 3 s intervals after transformation from raw, grey scale to binary (black-and-white) images. For circular (radial) mapping, the cell's 2-D contour was ascertained as cell centroid-to-edge radii starting at a defined angle and the original polar coordinates were transformed to Cartesian ones. Rare reentry problems were treated by subtracting the innermost cell-free space to obtain one absolute radius at the relevant angle. This produced little or no distortion of the results. Cells extending one pseudopodium produce two radial maxima about 180° apart, because the centroid becomes positioned inbetween.

Phase contrast and fluorescent images of stained cells were recorded on Kodak T-MAX 400 film for 15 or 30 seconds, respectively, using a Zeiss 100x objective, N.A. 1.3, with Zeiss fluorescence filter combination BP 546, FT 580, LP 590 was

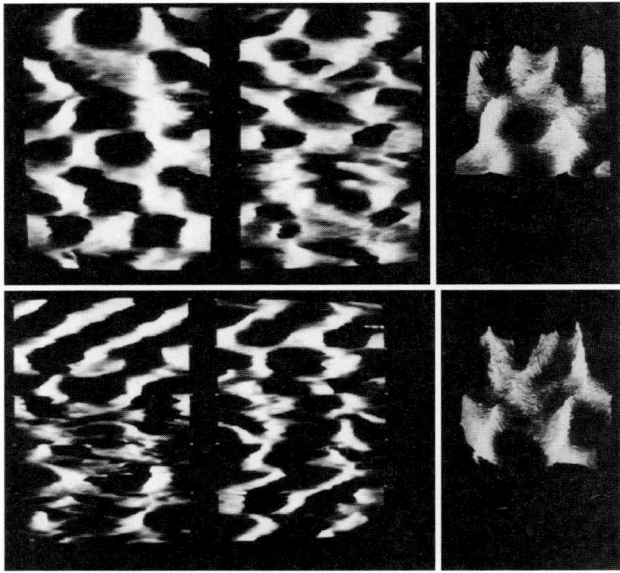

Figure 2: Circular maps of two *Dictyostelium* cells imaged in 2-D each 3 s during locomotion on glass. Panel upper left: Stationary waves at the cell periphery are perhaps most conspicuous during the first 15 min. The horizontal axis indicates 2π radians circumcellular and the vertical axis indicates time (s) beginning at the lower right and continuing to the figure half on the right. Each small, lateral tick marks 300 s. The grey scale from black to white indicates increasing cell centroid to margin radii. Panel lower left: Travelling waves erupt at 15 min and again at 40 min. Panels far right: Stationary and travelling waves (upper and lower images, respectively) are depicted as 3-D landscapes with the z, x and y axes representing the cell radius, 2π radians, and time = 300 s, respectively (in part from Killich *et al.* 1993).

used with a 50 W Hg lamp. Inner and outer diameters of ring structures staining for F-actin were measured by optical micrometry.

3 Results and discussion

The *Dictyostelium* cell surface, like that of neutrophils, appears to project pseudopodia in all directions incoherently (Fig. 1), looking somewhat like a sack containing a few playful kittens, at least under direct observation. However, cell motion is not random. Time series of 2-D circular maps constructed from the periphery of locomoting cells show clear classical standing and travelling wave interference (*schlieren*-like) patterns (Fig. 2). Coherent wave patterns may persist for seconds or minutes and then degenerate into featurelessness as cells round-up (to produce the horizontal stripes in Fig. 2) or into phases of complex patterns. Abercrombie *et al.* (1970a) demonstrated the rhythmic features of the advancing fibroblast margin and neutrophil wave motion was reported by De Bruyn (1946) 50 years ago.

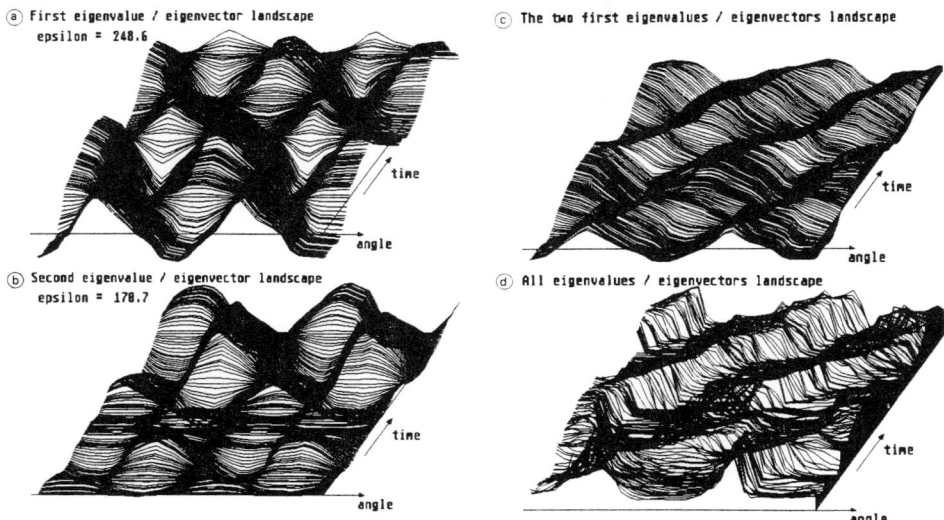

Figure 3: Karhunen-Loéve decomposition: reconstruction of the two primary modi (c) and all 36 modi composing the shape of the cell in Fig. 2 (lower left panel (d): travelling waves) between 20-30 min of locomotion. The abscissa represents the circumcellular angle while rotating 2π radians around the cell centroid (from Killich *et al.* 1994).

Simulation of the cell shapes observed required a closed, basal circular or elliptical line, expressed as a function of the mode number, and parameters of the rate and angle of wave rotation, its centrifugal amplitude and time. The superposition of two such functions sufficed to depict cell form very closely. Cell and simulated shapes express modes analogous to those generated mechanically in bounded oscillatory systems. Further analysis of these patterns employed the Karhunen-Loéve decomposition (Bestehorn *et al.* 1989, Killich *et al.* 1994 and I.4 this volume). A simple travelling wave pattern, like that in Fig. 2, was generated by 2-3 oscillatory wave modi (Fig. 3). The strength of both these primary modi and the less significant modi (i.e., in determining the momentary cell form) was time dependent. Thus, all cells also evinced periods of more complex locomotion with 6 or more determining modi (Killich *et al.* 1994).

The mechanism of wave (i.e. pseudopodia) generation was examined by recording the distribution of F-actin in cAMP-stimulated and unstimulated cells. A cAMP pulse induced cells to round-up, retract their pseudopodia, extinguish all

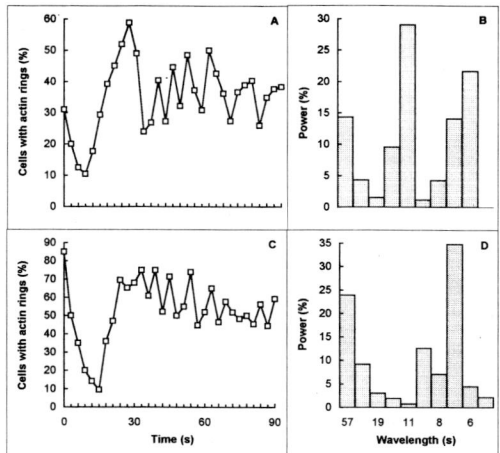

Figure 4: The proportion of cells containing F-actin rings each 3 s following an applied impulse of 1 μM cAMP to aggregative-stage *Dictyostelium* cell populations. (A) 1766 NC-4 cells (33-86/interval) and (C) 721 AX-2 cells (20-32/interval) were fixed during locomotion on glass and stained for F-actin. Fourier analysis applied to the NC-4 (B) and AX-2 (D) data 30 s after cAMP addition shows that the F-actin ring oscillation frequency in the cell strains is similar (from Vicker *et al.* 1996).

existing F-actin structures while forming newly assembled F-actin into a subcortical sphere (essentially phase-resetting the oscillator) and then relax as new F-actin structures reappeared. To investigate these dynamics, we measured the inner and outer diameters of one type of F-actin structure, rings, which are relatively easy to define (Figs. 4 & 5). Both NC-4 and AX-2 cell strains react to a cAMP impulse initially by a transient disappearance of almost all rings and then an oscillation in F-actin ring generation and size. Fourier analysis indicates that the oscillation frequency is similar in the two cell strains.

Logistic regression of ring diameter vs. rim width indicated that rings grow from very small dimensions and that the rim broadens proportionally in a typical sigmoidal fashion (Vicker *et al.* 1996). Markov and Fourier analyses of the data by Wosniok (1987) indicated substantial ring growth and decay rates and estimates of ring front propagation of 3.1-17.5 μm/min in both cell strains. Liklihood values derived from various time dependent and independent Markov models to the data indicated good fits using models with similar ring front speed ranges (Vicker *et al.* 1996). These speed estimates are similar to those recorded for locomoting *Dictyostelium* cell centroids, e.g. (Vicker 1994).

F-actin rings have been imaged in stimulated fibroblasts (Blume-Jensen *et al.* 1991, Hedberg *et al.* 1993, Li *et al.* 1993) and in *Dictyostelium* (Condeelis *et al.* 1988, De Hostos *et al.* 1991, Peterson & Titus 1994), but their significance has not previously been interpreted. Aside from rings, we detected several other F-actin structures by 2-D and 3-D microscopy in *Dictyostelium*, including target patterns, ring fragments, degenerated rings, single spirals and paired, contrarotating spirals (Fig. 5) and apparent remnants of ring collisions, showing the mutual, local annihilation of the structures. These forms were usually associated with the plasma membrane and were not related to either contractile vacuoles – which rarely stain for F-actin (Zhu & Clark 1992) – or to the "comet tail" F-actin assembly found to propel some microorganisms through the cytoplasm (Theriot *et al.* 1992, Kocks *et al.* 1993) or on the surface (Forscher *et al.* 1992) of eukaryotic cells, because comet tailing requires a nucleator of F-actin assembly on the microorganism; see also Lendowski & Mogilner 1.11 this volume. No such nucleator is evident in *Dictyostelium* actin waves.

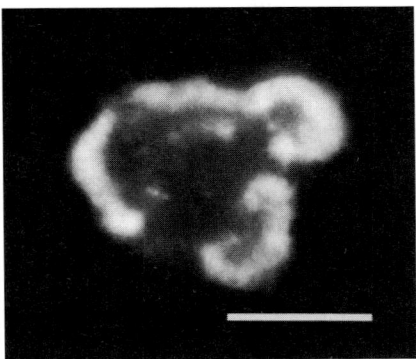

Figure 5: A single *Dictyostelium* AX-2 cell, fixed while locomoting on glass and then stained for F-actin with phalloidin-TRITC. Two autowave fronts each forming a contrarotating double spiral are visible at the leading margin of the cell on the right. The scale bar indicates 10 μm.

The F-actin patterns in *Dictyostelium* show striking morphological resemblence to reaction-diffusion waves in chemical systems, e.g. the Belousov-Zhabotinski (BZ) reaction (Winfree 1973, Welsh *et al.* 1983, Linde & Engel 1991), and each actin pattern compares to one of the stages in wave form evolution, e.g. ring breakage and double spiral development. Presently, the only known physical origin of these structures is by a reaction-diffusion process. The apparent dynamics of actin structures in *Dictyostelium* amoebae supports our suggestion that they are 3-D activation fronts of reversible F-actin assembly propagating as toroidal scroll waves (Winfree & Strogatz 1984). Waves commonly appear as remnants undergoing extinction or in patterns produced by topological sections, due to 2-D microscopy or to physical limitation within the cytoplasm. Rings represent sections through spherical propagation fronts, because of the physical unlikelihood that molecules might organize themselves into ring objects by any other means. Actin waves might also generate the lamellipodium-like ("ruffled membrane") rings, which have been described diversely as "phagocytic food cups" (De Chastellier & Ryter 1977, Ryter & Brachet 1978) or F-actin and coronin-containing "crowns" (De Hostos et al. 1991). Lamellipodia and their retrograde "ruffles" (Abercrombie et al. 1970b) probably represent the distally and proximally moving rims of actin wavefronts, respectively. Findings that actin only assembles at the extreme, leading margin of lamellipodia (Wang 1985, Theriot & Mitchison 1991) might be due to the loss of smaller "Triton X-100-soluble" actin filaments during fixation.

Significantly, tubulin, the other main intracellular filament-forming proteinnucleotide, shows reaction-diffusion wave behaviour *in vitro* (Mandelkow et al 1989) and Ca^{2+} autowaves have been vizualized in activated frog eggs (Lechleiter & Clapham 1992) and cardiocytes (Lipp & Niggli 1993). However, calcium seems to have little to do with F-actin waves, because both are temporally and physiologically uncorrelated after cell stimulation (Omann et al. 1989).

The self-organized formation and disintegration of F-actin oligomers, averaging 200 nm in *Dictyostelium* (Podolski & Steck 1990), but possibly shorter, is likely to facilitate rapid wave propagation. Therefore, cytoplasm may be considered an unstable, excitable medium within which wave passage depends on 3 actin states (Dufort & Lumsden 1993a,b) alternating in spatio-temporal order: 1) regenerated, unstable G-actin-ATP capable of polymerization, 2) a steep front of F-actin formation and filament age-dependent ATP-hydrolysis and 3) depolymerization and release of ADP-actin from the filament ends to produce G-actin-ADP (Korn et al. 1987), which rapidly exchanges ADP for ATP to regenerate to state 1. These 3 states correspond to conditions characterizing self-organized, chemical autowaves (Winfree 1974; Winfree & Strogatz 1984). A filamentous wavefront ought to possess relatively high tensile properties, compared to the BZ reaction front, but like the BZ reaction it should, as a reaction-diffusion wave, generate no mechanical force by itself. However, the plasma membrane does not appear to block actin waves, because rings often extend just as far into the cell as their distal half projects out of it while forming a pseudopodium. Pseudopodium extension may then be further driven by cytoplasmic streaming elicited in the vicinity of the wave. Such streaming, induced by reactant concentration differences, characterizes the BZ reaction (Miike et al. 1989; Matthiessen & Müller 1995). Thus, we suggest that this new dynamical form of actin governs pseudopodium extension, locomotion and cell shape in *Dictyostelium* and, possibly, in other crawling cells: a point for further investigation. In addition, the actin oscillator might provide the cell with a timing principle involved in the regulation of intercellular signal receptor activity, intercellular frequency coordination and intracellular rhythms, such as the cAMP oscillator (Gerisch & Hess 1974) and circadian rhythms.

I.4

Mathematical Analysis of Cell Shape

Oana Brosteanu (Leipzig)
Peter J. Plath and Michael G. Vicker (Bremen)

Cell motility involves translocation of the cell's centroid as well as changes or distortions in the cell's shape. Clues about the mechanism of cell movement may be obtained from information about its shape changes in time. The changes occur in multiple dimensions and can be highly periodic, however they may elude superficial observation. The techniques outlined in this contribution might help to reveal otherwise undetectable periodic shape changes.

The methods described here are appropriate for the analysis of spatio-temporal changes of closed shapes, such as cell outlines. They require an angular parametrization $z(\varphi, t)$ of the outline (for every angle φ and time t) refering to a suitably chosen point or area lying inside the cell. The methods are particularly suitable for qualitative and quantitative description of the lamellar dynamics of single motile cells with a distinct cell body and a flat periphery around it (e.g. human epidermal keratinocytes) but also for more amoeboid cells.

1 Representation of the outline

The parametrization of the outline depends on the particular problem to be analysed. In order to describe single motile amoeboid cells, the following parametrizations are suitable:

For an unbiased analysis of the spatio-temporal dynamics of the cell periphery, the representation of the cell outline has to be chosen independently of cell translocation. A parametrization $z(\varphi)$ in polar coordinates around the centroid of the cell seems a good choice. The location of the centroid is given by the two first order physical moments (m_{10}, m_{01}) of the cell:

$$m_{lk} := \frac{\int_Z x^l y^k f(x,y) dx dy}{\int_Z f(x,y) dx dy} \qquad 0 \leq l, k \tag{1}$$

where Z is the cell area and $f(x,y)$ the mass distribution within Z. For cells with distinct cell body and flat lamellae, the choice of two different constant weights for cell body and flat periphery, reflecting the mass difference between them (e.g. $f(x,y) = 1$ for the cell body, $f(x,y) = 1/20$ for the lamellae) reduces the bias on the location of the centroid due to widely extended lamellae. In case the cell body is indistinct, $f(x,y) \equiv 1$ may be chosen. If the cell has a flat lamella, but data on shape and location of the cell body are not easily available, its outline may be approximated by using information from the peripheral cell outline and the location of the cell body at the beginning of the time sequence. The cell body may be regarded as an inert cell region that changes shape and position only if its outline comes into conflict with the peripheral cell outline because of global shape changes or cell translocation, for details see (Brosteanu 1994).

If the cell has an elongated shape, the bias due to deviation from a circular shape may be reduced by representing the cell outline $z(\varphi)$ for every angle φ as distance in the normal direction from the weighed momental ellipse of the cell; see Hinz & Brosteanu, I.2 this volume, Fig. 2. The weighed momental ellipse is defined by the centralized (around the centroid) second order physical moments, μ_{20}, μ_{02} and μ_{11},

$$E_Z := \left\{ (x,y)^T \mid \mu_{02}x^2 - 2\mu_{11}xy + \mu_{20}y^2 = 4\mu_{20}\mu_{02} - 4\mu_{11}^2 \right\}. \qquad (2)$$

This weighed momental ellipse has the same moments of second order as Z and approximates the shape and orientation of the cell body (cf. Alt et al. 1995).
A topographic line plot of the three-dimensional data set of $z(\varphi,t)$, with protrusions represented as elevated regions, gives a preliminary insight into the lamellar dynamics; see Hinz & Brosteanu (I.2 this volume, Fig. 3).

2 Angular Fourier analysis

The angular Fourier analysis of the outline

$$z(\varphi,t) = \frac{1}{2}a_0(t) + \sum_{k=1}^{\infty} a_k(t)\sin(k\varphi) + b_k(t)\cos(k\varphi) \qquad (3)$$

$$= \frac{1}{2}a_0(t) + \sum_{k=1}^{\infty} c_k(t)\cos(k\varphi - \psi_k(t)) \qquad (4)$$

provides a decomposition of the shape in symmetrical, k-modal components given by $\sin(k\varphi)$ and $\cos(k\varphi)$. The time course of the coefficients a_1 and b_1 (and perhaps a_2 and b_2) reflects the main shape changes. A plot of the path of a_k versus b_k reveals characteristic features such as rotating waves and standing pulsating waves (see Hinz & Brosteanu, I.2 this volume, Figs. 4C & F). In the case of rotating waves, Eqn. 4 with decomposition into shifted cosine functions may be used to describe the angular drift of the wave (by the shift coefficient ψ_k), the coefficient $c_k(t)$ quantifies the contribution of the k-modal component to the global cell shape at time t. Thus, $c_1(t) \gg c_k(t)$ ($k \neq 1$) indicates situations with a single outstanding lamella, whereas $c_2(t) \gg c_k(t)$ ($k \neq 2$) shows coexistence of two antipodal lamellae.

3 The Karhunen-Loève expansion

One particularly interesting way to extract the slaving modes from long sets of data is by reducing them by the Karhunen-Loève expansion (Fuchs et al. 1988, Friedrich & Uhl 1992, Krischer et al. 1993, Killich et al. 1994). For this purpose, one averages the distances $z(\varphi,t)$ from the origin (e.g. the cell centroid) to points on the two-dimensional boundary or margin (φ,z) in the angular direction φ for each of a series of "snapshot" images of the cell, taken at frequent intervals of time. The actual distances $z(\varphi, t)$ may then be decomposed into the temporal averages $D(\varphi)$ and their actual deviations $d(\varphi, t)$, which in fact represent the shape of the cell's two-dimensional periphery:

$$z(\varphi, t) = D(\varphi) + d(\varphi, t). \tag{5}$$

The set of distances $d(\varphi_i, t_m), i = 1, 2, \ldots, N$, forms a N-dimensional vector $\mathbf{d}_\varphi(t)$, where N is the total number of points on the cell's margin, measured at times $t_m, m = 1, 2 \ldots, M$. This vector $\mathbf{d}_\varphi(t)$ represents the temporal development of the cell's margin in the direction φ. The temporal developments of the margin in any two different directions φ_i and φ_j are explicitly correlated. The strength of this correlation is the scalar product of the vectors

$$\mathbf{d}_i(t) = (d(\varphi_i, t_1), \ldots, d(\varphi_i, t_M)) \text{ and } \mathbf{d}_j(t) = (d(\varphi_j, t_1), \ldots, d(\varphi_j, t_M))$$

divided by M:

$$r_{ij} = \frac{1}{M} \cdot \mathbf{d}_i(t) \mathbf{d}_j(t) = \frac{1}{M} \cdot \sum_{m=1}^{M} d(\varphi_i, t_m) \cdot d(\varphi_j, t_m); \quad i, j = 1, 2, \ldots, N. \tag{6}$$

This allows one to obtain a so-called correlation matrix $R = (r_{ij})$. N eigenvalues $\varepsilon_k, k = 1, 2, \ldots, N$, belong to this matrix as well as N corresponding eigenvectors

$$\mathbf{s}_k(\varphi) = (s_k(\varphi_1), \ldots, s_k(\varphi_N)). \tag{7}$$

This set of N eigenvectors $\mathbf{s}_k(\varphi)$ represents an optimal set of discrete functions, which embody a description of temporal changes in the cell's shape.

A measure of the contribution of each eigenvector $\mathbf{s}_k(\varphi)$ to the actual shape of the cell's margin is the scalar product $c_k(t)$ of the eigenvector $\mathbf{s}_k(\varphi)$ and the vector

$$\mathbf{d}_m(\varphi) = (d(\varphi_1, t_m), \ldots, d(\varphi_N, t_m))$$

which is constructed from all deviations $d(\varphi_i, t_m), i = 1, 2, \ldots, N$, at time t_m. If $c_k(t)$ is close to zero, the eigenvectors $\mathbf{s}_k(\varphi)$ and $\mathbf{d}_m(\varphi)$ are almost perpendicular. This circumstance indicates that the contribution of $\mathbf{s}_k(\varphi)$ to the actual shape of the cell margin is negligible, because the two-dimensional shape of the cell is the sum of the eigenvectors $\mathbf{s}_k(\varphi)$ weighted by $c_k(t)$. Otherwise, if the vectors $\mathbf{s}_k(\varphi)$ and $\mathbf{d}_m(\varphi)$ are nearly parallel, the $c_k(t)$ values will differ strongly from zero and it will be necessary to take into account the contribution of $\mathbf{s}_k(\varphi)$ to cell shape (from P. J. Plath in (Killich et al. 1994), see also (Vicker & Xiang, I.3 this volume)).

4 Angular-temporal autocorrelation

The correlation patterns of the outline $z(\varphi, t)$ are described by the autocorrelation function:

$$c(\alpha, \tau) := E_{\varphi t} \left[(z(\varphi, t) - E_{\varphi t}[z(\varphi, t)]) \cdot (z(\varphi + \alpha, t + \tau) - E_{\varphi t}[z(\varphi, t)]) \right] \tag{8}$$

where $E_{\varphi t}[\cdot]$ is the expectation value over φ and t. $c(\alpha, \tau)$ provides a measure for correlation of the outline in angular distance α ($0 \leq \alpha < 2\pi$) and temporal distance $\tau \geq 0$. $c(0, 0)$ is the variance of the outline extensions $z(\varphi, t)$. High values of the instantaneous angular correlation $c(\alpha, 0)$ characterize the width of protrusions, whilst high values of the temporal correlation at the same angular position, $c(0, \tau)$, characterize the duration of protrusions. Local maxima or elevated areas of $c(\alpha, \tau)$ at $\alpha > 0$ and $\tau > 0$ show typical spatio-temporal periods between protrusions. The autocorrelation analysis reveals typical features of cell shape changes over time. Periodic or regular patterns as alternating standing waves or travelling waves are detected and quantified; see Hinz & Brosteanu (I.2 this volume, Fig. 4).

5   A comparison of the methods

Angular Fourier and Karhunen-Loève analyses are both based on the decomposition of cell shape using either sine and cosine functions or the shape eigenvectors, respectively. Both methods yield the temporal dynamics and relative contributions of each of the modal components which sum up to determine the shape of the cell periphery, and are suitable for qualitative comparison between different cells and between different "snapshots" of a cell. The Karhunen-Loève expansion, in its dependence on the eigenvectors of the individual cell, is of particular usefulness in the identification, quantification and comparison of the shape-relevant protrusion in case of cells with irregular shapes, where Fourier analysis shows no dominating component. On the other hand, Fourier analysis has the advantage that the k-modal components involved in the decomposition do not depend on the individual cell, and so Fourier analysis allows for quantitative comparison of the dominating modal components between cells. Autocorrelation analysis gathers information on the angular-temporal dynamics of the shape and enables not only detection of periodic patterns, but also quantification of the mean periodicity of shape changes.

I.5

Protrusion, Retraction and the Efficiency of Cell Locomotion

Graham Dunn*(London)
Igor Weber (Martinsried)
Daniel Zicha (London)

1 Introduction

Protrusion and retraction are two conspicuous aspects of the phenomenon of amoeboid locomotion or of the crawling motility of tissue cells. At the most general level, protrusion is the gain of territory that the cell makes during a short time interval and retraction is the loss of territory during the same interval. This territory may be considered to be the three-dimensional space occupied by the cell or, particularly if the cell is crawling over a planar substratum, it may be considered to be the two-dimensional projection onto the substratum of this occupied space or it may even be considered as the area of the substratum to which the cell is actually attached.

Different forms of microscopy and image analysis are required to observe the regions of protrusion and retraction in each instance and it is only with recent advances in light microscopy and in digital image processing that it has become feasible to gather large amounts of data on the protrusion and retraction behaviour of cultured cells. In particular, high-speed, optical-sectioning differential interference contrast (DIC) microscopy has enabled changes in the three-dimensional space occupied by a moving cell to be reconstructed (Murray *et al.* 1992); digitally recorded interference microscopy with automatic phase-shifting (DRIMAPS) has allowed the dry mass of protrusion and retraction regions to be estimated and has greatly improved the accuracy of measuring protrusion and retraction areas in thinly spread fibroblasts (Dunn & Zicha 1995); and reflection interference contrast microscopy (RICM) has enabled the protrusion and retraction of cellular regions adherent to the substratum to be followed in moving *Dictyostelium* cells (Weber *et al.* 1995).

Our aim in this paper is to promote the measurement of protrusion and retraction as an essential tool in the investigation of the dynamics of cell motility and for assessing the effects of molecular/genetic manipulation of the cell's motile

---

*Author for correspondence concerning this article

machinery. We will first describe various measures of protrusion and retraction and consider the effect of sampling the data at different frequencies. Next we will show how the translocation of a cell can be resolved into two components depending on the separate contributions of protrusion and retraction. One advantage of studying protrusion and retraction as separate processes, as opposed to analysing the translocation of cells directly, is that it offers an insight into the roles of the different molecular processes that are involved in cell motility. Thus protrusion is associated with adhesion to the substratum and presumably with assembly of cytoskeletal components whereas retraction is more likely to be associated with the antagonistic processes, deadhesion and disassembly, as well as with the contraction of specific regions of the assembled cytoskeleton.

Examining the relationship between protrusion and retraction can give an insight into more global molecular processes, such as the transport of materials – possibly in a disassembled state – from regions of retraction to regions of protrusion. We will show how the speed of cell translocation can be resolved into a component that depends on the magnitudes of protrusion and retraction and another that depends on their polarity or relative disposition. This can give an indication of whether any change in cell speed is due to a change in the volume of materials transported and/or to a change in the effectiveness with which they are transported from the rear to the front of the cell. A related question is whether a cell's locomotion is efficient. Protrusion and retraction must consume a great deal of the cell's metabolic energy and, if their chief function is to enable the cell to translocate, then it is obviously more efficient for the cell to achieve a higher translocation for a given amount of protrusion and retraction. Our approach to this question is to examine how effectively a cell explores the substratum without apparently wasting effort by repeatedly exploring the same region.

Finally we will consider the analysis of the dynamic relationship between protrusion and retraction and the insights that this provides into the control of the cell's locomotory machinery. Evidence emerged some years ago that protrusion and retraction can be closely interlinked during a specific behavioural event; Chen (1979) and Dunn (1980) demonstrated that a wave of increased protrusive activity follows shortly after retraction of the tail of a chick heart fibroblast. Chen (1981) later showed that the retraction phase of this "retraction-induced spreading" is partially due to an active contraction and Brown & Dunn (1989) showed that very rapid internal flows of cellular material can occur during the spreading phase. More recently, Dunn & Zicha (1995) and Weber et al. (1995) have investigated the cross-correlation structure of protrusion and retraction at different lag times and have shown that they are generally interlinked during the locomotion of chick fibroblasts and *Dictyostelium* cells respectively. Weber's article reveals that the cross-correlogram is a particularly sensitive device for detecting subtle differences in motility exhibited by mutants lacking combinations of actin-crosslinking proteins. Here, we will show how the cross-correlation between the two processes, at different time lags, can reveal control mechanisms and possible cause/effect relationships in the dynamics of the cell's locomotory machinery.

Since this paper is chiefly concerned with the functional roles of protrusion and retraction in the crawling locomotion of cells, which requires that the cells be attached to a solid substratum, we can ignore any aspects of protrusion and retraction that do not lead to the gain or loss of substratum territory. In the case of cells crawling on a planar surface, this enables us to simplify the measurement of protrusion and retraction by considering only the two-dimensional geometry of the cell's projection onto the substratum or of its regions of attachment to the substratum. It must not be forgotten, however, that natural substrata are rarely planar and that protrusion and retraction, in the most general usage of these terms, have functional roles in several cellular activities, such as endocytosis, in addition to spreading and translocation.

2 Measures of protrusion and retraction

We will use nomenclature of the set theory to denote regions of the substratum. If region S_1 is the set of all points on the substratum covered by a spread cell at time t_1 then the region $\bar{S}_1$ is the set of all points on the substratum not covered by that particular cell at time t_1. If time t_2 is later than time t_1, we can define the retraction, R, protrusion, P, and common, C, regions over the time interval $<t_1, t_2>$ as the following intersections of sets:

$$R = S_1 \cap \bar{S}_2; \quad P = \bar{S}_1 \cap S_2; \quad C = S_1 \cap S_2.$$

Thus R is the region that lies within S_1 but not in S_2; P lies within S_2 but not in S_1 and C lies within both S_1 and S_2. Note that the regions R, P, and C are non-overlapping but each region need not be contiguous and any, but not all, of the three regions may be the empty set. If the cell has displaced so far during $<t_1, t_2>$ that C is the empty set then $R = S_1$ and $P = S_2$. It is generally best to choose short time intervals such that C is never the empty set when sampling protrusion and retraction. Measurements of these various regions can be based on the moment series (Dunn & Brown 1987) and we will first consider two functions based on the zero and first order geometrical moments: the scalar function $a[X]$ gives the area of region X and the 2-D vector function $\mathbf{c}[X]$ gives the position of the areal (geometrical) centroid of region X. We will use the following notation to denote the results of applying these two functions to the various regions (bold characters denote vectors):

scalars: $\quad s_1 = a[S_1], \quad s_2 = a[S_2], \quad p = a[P], \quad r = a[R], \quad c = a[C],$
vectors: $\quad \mathbf{s_1} = \mathbf{c}[S_1], \quad \mathbf{s_2} = \mathbf{c}[S_2], \quad \mathbf{p} = \mathbf{c}[P], \quad \mathbf{r} = \mathbf{c}[R], \quad \mathbf{c} = \mathbf{c}[C].$

The most commonly used measures of protrusion and retraction are defined as:

$$\text{protrusion area} = p,$$
$$\text{retraction area} = r$$

(Dunn 1980, Dunn & Brown 1987, Soll et al. 1988, Dunn & Zicha 1995, Weber et al. 1995). Vector measures which incorporate information on the disposition of protrusion and retraction, as well as their magnitude, have been defined as:

$$\text{protrusion vector } \boldsymbol{P} = p(\boldsymbol{p} - \boldsymbol{s_1}),$$
$$\text{retraction vector } \boldsymbol{R} = r(\boldsymbol{s_2} - \boldsymbol{r})$$

(Weber et al. 1995). These measures are illustrated in Fig. 1.

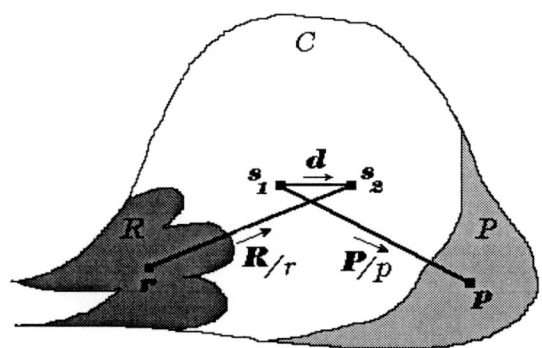

Figure 1: The retraction region, R, is shown in dark grey and the protrusion region, P, in light grey. The centroids of these regions are shown as $\boldsymbol{r}$ and $\boldsymbol{p}$ respectively and the centroids of the two consecutive cell outlines as $\boldsymbol{s_1}$ and $\boldsymbol{s_2}$. The vector drawn from $\boldsymbol{r}$ to $\boldsymbol{s_2}$ is the retraction vector, $\boldsymbol{R}$, divided by the area of retraction, r, whereas the vector drawn from $\boldsymbol{s_1}$ to $\boldsymbol{p}$ is the protrusion vector, $\boldsymbol{P}$, divided by the area of protrusion, p.

Weber et al. (1995) point out that these four measures are equally applicable in the case where S_1 and S_2 represent the regions of substratum to which the cell is attached at times t_1 and t_2 respectively as determined by RICM. In this case neither S_1 nor S_2 need be contiguous and each may be broken into many small regions of close or focal adhesion. Since regions of attachment can change much more rapidly than the cell outline, C is often the empty set even for short time intervals. In this case r and p merely represent the successive areas of attachment and so the vectors $\boldsymbol{P}$ and $\boldsymbol{R}$ are often more useful for quantifying changes in the regions of attachment.

It is essential when comparing protrusion and retraction data from different experiments to use the same time interval or sampling frequency. This is because the area of protrusion (or retraction) over a double sampling interval, $< t_i, t_{i+2} >$, is not simply the sum of the areas of protrusion (or retraction) over the two intervals $< t_i, t_{i+1} >$ and $< t_{i+1}, t_{i+2} >$. In fact each consists of this sum minus the areas of the two regions of alternating occupancy:

$$\overline{S_i \cap S_{i+1}} \cap \overline{S_{i+2}} \quad \text{and} \quad S_i \cap \overline{S_{i+1}} \cap S_{i+2}.$$

Thus the area of protrusion (or retraction) over a double sampling interval is generally smaller and never larger than the sum of the areas of protrusion (or retraction) over the two included intervals. This argument becomes more complicated for higher multiples of the basic sampling interval but we have found that the function:

$$F[\tau] = \alpha \tau \exp[-\tau/\beta] + \gamma,$$

where τ is the length of the multiple sampling interval, models the *relative area* of protrusion (*protrusion area* expressed as a fraction of total cell area) very well in the case of chick fibroblasts. Fig. 2 shows the effect of sampling frequency on the mean *relative area* of protrusion, p/s, over a continuous run of approximately 4h from a chick fibroblast, with the function $F[\tau]$ fitted by the least squares method.

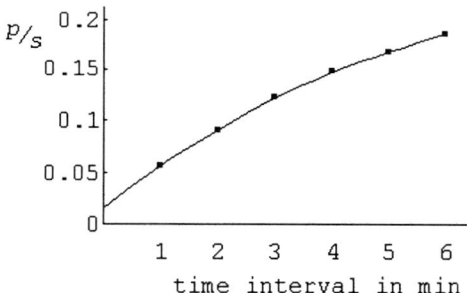

Figure 2: The mean *relative area* of protrusion, p/s, plotted against the sampling time interval over a continuous run of approximately 4h from a chick fibroblast, with the function $F[\tau]$ fitted by the least squares method.

The fit yields values for the three parameters of $\alpha = 0.044$ min^{-1}, $\beta = 13.6$ min and $\gamma = 0.016$. The results for the mean *relative area* of retraction are very similar ($\alpha = 0.043$ min^{-1}, $\beta = 12.8$ min, $\gamma = 0.015$). We interpret the parameter α as the basal rate of relative protrusion (or retraction) per unit time, β is a characteristic time and γ is a constant level of additional noise which is independent of the length of the sampling interval. The value of the noise term is about 1.5% of the total cell area and this may be due to image noise in the pixels that have low grey values at the cell's edge or it may be due to biological noise caused by high frequency oscillations of the cell's margin. We favour the latter explanation for most of the noise since the image noise can be estimated by measuring the *relative areas* of protrusion and retraction using fixed cells which in the DRIMAPS system, gives values that are much lower than the noise level for living cells. This method of sampling protrusion and retraction over multiple sampling intervals may eventually prove to be a useful way of analysing their dynamics but we include it here simply to emphasise the need for a standardised sampling interval.

3 Relation of protrusion and retraction to cell motility

Protrusion and retraction over the time interval $< t_1, t_2 >$ together determine the cell's change in area, change in shape and the displacement of the cell's areal centroid during the time interval. The changes resulting from a given protrusion and retraction are quite easy to calculate if measures of cell spreading, shape and translocation are based on the geometrical moment series as advocated by Dunn & Brown (1987). Raw moments have a simple addititive property such that any moment of a region that is the union of two non-overlapping regions is the sum of the corresponding moments of the two regions. In the case of the zero-order moment, which is simply the area of a region, this rule is obvious and gives:

$$s_1 = c + r \quad \text{and} \quad s_2 = c + p.$$

Thus the change in spread area of the cell during the interval $< t_1, t_2 >$, which is $s_2 - s_1$, is simply expressed in terms of protrusion and retraction as $p - r$. If the cell is to maintain a constant spread area then p must be exactly matched to r at all times. Fig. 3 shows that chick fibroblasts treated with low doses of colcemid can achieve a remarkable match. The traces have been smoothed with a 10-min moving average to remove the high frequency noise discussed in the previous section. As the cell grows, the spread area increases smoothly despite considerable fluctuations in the values of p and r. This suggests that spread area may be actively controlled and we will discuss the analysis of this control mechanism later.

Another special case is provided by the first order moments and, in the notation used here, the two first-order moments of the region P, for example, are the two components of the vector $\boldsymbol{pp}$ and so by applying the addition rule we get:

$$s_1\boldsymbol{s_1} = \boldsymbol{cc} + \boldsymbol{rr} \quad \text{and} \quad s_2\boldsymbol{s_2} = \boldsymbol{cc} + \boldsymbol{pp}.$$

This enables us to rewrite the protrusion and retraction vectors, $\boldsymbol{P}$ and $\boldsymbol{R}$, as:

$$\boldsymbol{P} = \boldsymbol{pp} - p\boldsymbol{s_1} = s_2\boldsymbol{s_2} - \boldsymbol{cc} - p\boldsymbol{s_1},$$

$$\boldsymbol{R} = r\boldsymbol{s_2} - \boldsymbol{rr} = r\boldsymbol{s_2} - s_1\boldsymbol{s_1} + \boldsymbol{cc}.$$

By summing $\boldsymbol{P}$ and $\boldsymbol{R}$ and substituting for $\boldsymbol{s_1}$ and $\boldsymbol{s_2}$ we get:

$$\boldsymbol{P} + \boldsymbol{R} = (s_2 + r)\boldsymbol{s_2} - (p + s_1)\boldsymbol{s_1} = (p + c + r)(\boldsymbol{s_2} - \boldsymbol{s_1}).$$

I.5 Protrusion, Retraction and the Efficiency of Cell Locomotion

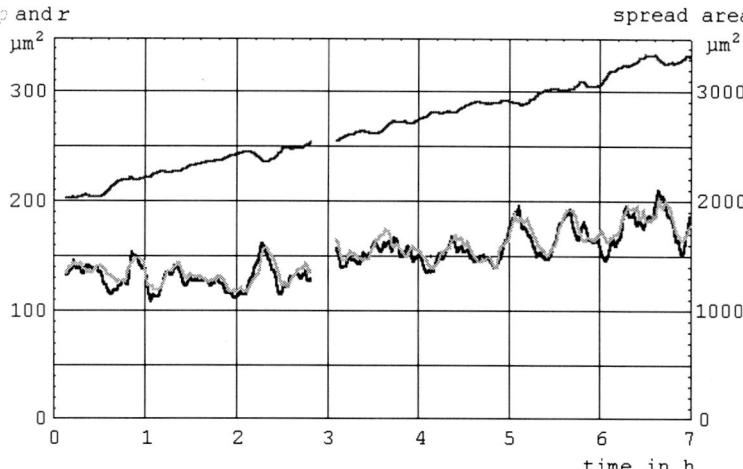

Figure 3: The upper trace is the spread area of a secondary chick heart fibroblast treated with 5×10^{-7} g/ml colcemid. The two lower traces are the corresponding *protrusion area* (light grey) and *retraction area* (black) sampled over 1-min intervals. The traces are smoothed by a 10-min moving average to reduce noise.

Since the vector $s_2 - s_1$ is the displacement, d, of the cell's geometrical centroid during the time interval $< t_1, t_2 >$, as shown in Fig. 1, this equation relates the locomotion of the cell to the magnitude and position of its protrusion and retraction activities: $d = (P + R)/(p + c + r)$. Therefore the displacement of the cell's geometrical centroid is the vector sum of P and R divided by the area of the total region covered by the cell during the time interval $< t_1, t_2 >$.

This equation enables us to resolve the cell's locomotion into two components that probably result from different molecular mechanisms. Fig. 4 shows the displacement track (dark grey) of a *Dictyostelium* cell resolved into a protrusion track (light grey) and a retraction track (black) obtained by joining the vectors $P/(p + c + r)$ for the protrusion track and by joining the vectors $R/(p + c + r)$ for the retraction track.

The tracks start at the bottom of the figure and it can be seen that, for this cell, the *protrusion vectors* contribute more to the overall displacement than the *retraction vectors*. A magnified region shows the displacement track resolved into alternating protrusion and retraction vectors for each 5 seconds step. Two intervals of equal duration are marked by the three asterisks and these show a clear difference in protrusion/retraction dynamics. In particular, the zigzag portion of the track indicates that the protrusion and retraction vectors are not always coaligned and may act independently in determining the cell's direction of travel.

Figure 4: Left: The displacement track (dark grey) of a *Dictyostelium* cell resolved into a protrusion track (light grey) obtained by joining the vectors $\boldsymbol{P}/(p+c+r)$ and a retraction track (black) obtained by joining the vectors $\boldsymbol{R}/(p+c+r)$. Right: A magnified region of the displacement track shows it resolved into alternating protrusion and retraction vectors for each 5 seconds step.

4 The polarity of protrusion and retraction

We have shown that the cell's displacement is determined by the magnitudes of protrusion and retraction and by their relative disposition. The aim of this section is to separate these two aspects of locomotion so that they may be studied individually. Using the methods of the last section, the change in first order moments during the interval $<t_1, t_2>$ is given by:

$$s_2\boldsymbol{s_2} - s_1\boldsymbol{s_1} = p\boldsymbol{p} - r\boldsymbol{r}.$$

In the case when p is closely matched to r, which is not uncommon for real cells, we have $p \approx r \approx a$ and $s_1 \approx s_2 \approx s$ and so we can approximate the last equation as:

$$s(\boldsymbol{s_2} - \boldsymbol{s_1}) \approx a(\boldsymbol{p} - \boldsymbol{r})$$

which gives:

$$\boldsymbol{d} \approx (\boldsymbol{p} - \boldsymbol{r})a/s.$$

This shows that the direction of the displacement vector $\boldsymbol{d}$ is approximately the same as the direction of the vector $\boldsymbol{p} - \boldsymbol{r}$ and so we can ignore their directions and deal only with their magnitudes:

$$|\boldsymbol{d}| \approx |\boldsymbol{p} - \boldsymbol{r}|a/s.$$

Therefore the magnitude of displacement is the product of two factors: the *relative area* of protrusion and retraction, a/s, and the separation of their centroids, $|\boldsymbol{p} - \boldsymbol{r}|$

which we will call the *polarity* of protrusion and retraction. Determining the separate contributions of these two factors to changes in the speed of cell translocation could lead to an important insight into the mechanism of speed variation under different experimental conditions. Fig. 5 shows this decomposition into factors applied to spontaneous fluctuations in displacement for the cell which was also used as an example in Fig. 3.

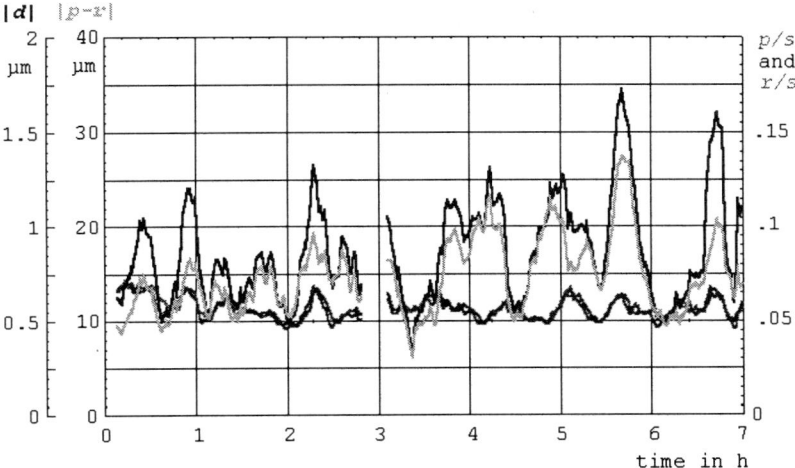

Figure 5: The spontaneous fluctuations in displacement (black) for the cell which was also used as an example in Fig. 3. The two almost coincident traces shown in dark grey are the *relative areas* of protrusion and retraction whereas the *polarity* is shown in light grey. The traces are smoothed by a 10-min moving average to reduce noise.

It can be seen that the fluctuations in displacement (black) are matched much more closely by the fluctuations in *polarity* (light grey) than by the fluctuations in the *relative areas* of protrusion and retraction (dark grey). The *relative areas* remain fairly constant at about 5% of the cell's total area and the cell varies its speed chiefly by varying the *polarity* of protrusion and retraction activity. A tentative interpretation is that this cell changes its speed not by changing the amount of disassembled materials that it transports from sites of retraction to sites of protrusion but by changing the distance over which it transports them.

This method of decomposing speed into two factors is strictly valid only during periods when the areas of protrusion and retraction are equal, but it is a useful approximation even when this condition is violated. As a test of validity, Fig. 6 shows the equation expressed graphically for a real cell where a is taken as the mean of r and p and s is taken as the mean of s_1 and s_2.

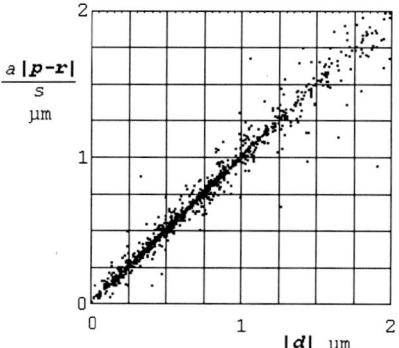

Figure 6: A scatter plot of $a|\boldsymbol{p} - \boldsymbol{r}|/s$ against $|\boldsymbol{d}|$ for a chick heart fibroblast where a is taken as the mean of r and p, and s is taken as the mean of s_1 and s_2. Each point represents the measures taken over a 1-min time interval.

5 The efficiency of cell locomotion

Polarity is a measure of the effectiveness of protrusion and retraction. It describes how far a cell translocates for a given area of protrusion and retraction relative to the total spread area. *Polarity* does not have a definite maximal value, however, since the maximal separation distance of $\boldsymbol{p}$ and $\boldsymbol{r}$ is determined by the length of the cell along its axis of travel and this could theoretically increase without limit. In this respect, cells such as fish keratinocytes, which have a shape that is characteristically foreshortened along the axis of travel, might be considered to be ineffective translocators even though they glide smoothly over the substratum. Increasing elongation in the axis of travel, however, tends to reduce the width of the cell and thus to reduce the area of substratum explored for a given displacement of the cell. Since the area of substratum explored may be an important aspect of the function of cell locomotion, we also need a measure of the effectiveness of protrusion and retraction that takes this into account. An obvious approach is to compare the total area of substratum explored by the cell, i.e. the area of the cell's track, with the cumulative areas of protrusion and/or retraction that the cell used to explore the track. This measure would take a maximum value of 1 if the cell had protruded only once over each section of its track and so we will consider it to be a measure of the *efficiency* of cell locomotion.

The efficiency of cell locomotion is best visualised by constructing an efficiency track as shown for a Walker carcinosarcoma cell treated with colchicine in Fig. 7. Here the grey scale represents the total number of protrusions over each point on the substratum. It was constructed by assigning a *protrusion frequency* value to each point equal to the number of regions, P_i, to which the point belongs where the P_i are the protrusion regions taken from consecutive time intervals

$< t_i, t_{i+1} >; (i = 1, 2, 3, \ldots)$. Since the cell no longer occupies the track, the number of retractions from each point must equal the number of protrusions over the point and so it makes no difference whether the efficiency track is constructed by considering protrusions or retractions. The *efficiency* of locomotion, referred to a specific region of the efficiency track, can now be more precisely defined as the reciprocal of the mean *protrusion frequency* for all points within the region. It is a cumbersome measure, however, since it refers to a specific region rather than a specific time and it is not always clear how the regions should be sampled, particularly since regions where tracks intersect (or where a track intersects itself) should be avoided. In any case it is usually more useful to make qualitative comparisons of the efficiency tracks from different experimental treatments or different cell types. Pseudocolour coding of the protrusion frequencies can sometimes show more detail than grey-scale coding.

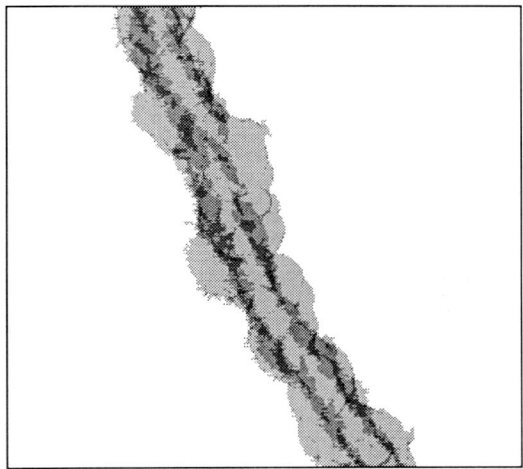

Figure 7: The efficiency track of a Walker carcinosarcoma cell treated with colchicine. The grey scale represents the total number of protrusions over each point on the substratum.

6 Dynamic relationship between protrusion and retraction

The mutual interdependency of two simultaneous time-series can be expressed using the cross-correlation function (Diggle 1990). Fig. 8A shows the cross-correlogram of the 1-min relative protrusion areas (p/s) and the 1-min relative retraction areas (r/s) of our example fibroblast at different lag times over the full 7-h period.

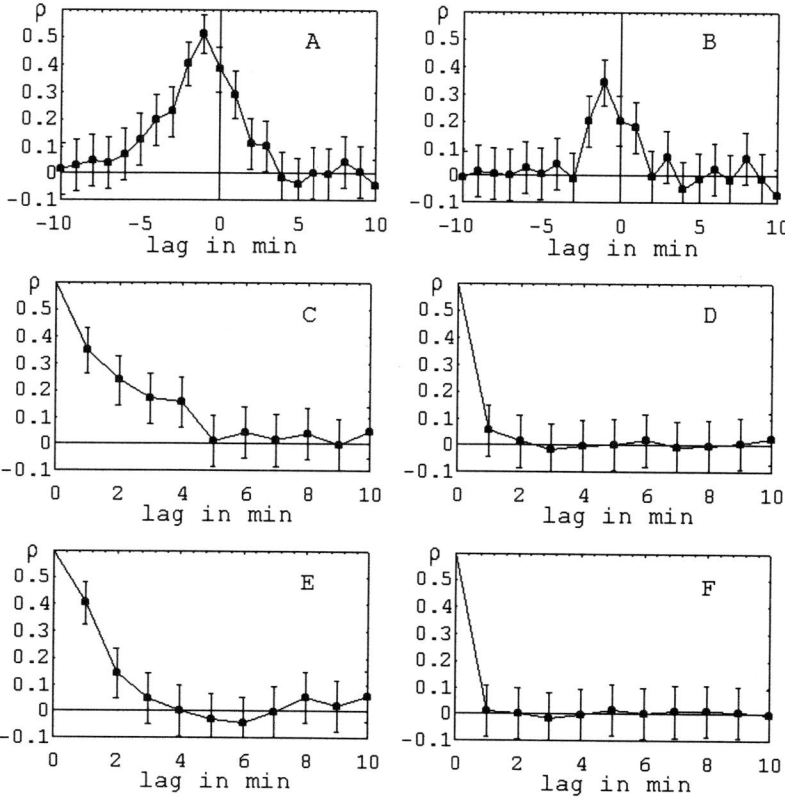

Figure 8: A-B: Cross-correlogram of the 1-min relative protrusion areas and the 1-min relative retraction areas before (A) and after (B) pre-whitening. C-D: Autocorrelogram of the 1-min relative protrusion areas before (C) and after (D) pre-whitening. E-F: Autocorrelogram of the 1-min relative retraction areas before (E) and after (F) pre-whitening. See text for more details.

The peak value of the correlation coefficient in Fig. 8A is around $\rho = 0.5$, but the smoothed data of Fig. 5 suggest that a much higher value might have been expected. The discrepancy is due to the noise that we discussed in Section 2 and in fact the smoothed values used for Fig. 5 show a peak cross-correlation approaching $\rho = 0.9$. However, smoothing introduces significant autocorrelations into the two series which can give rise to a misleading cross-correlogram. Even in the absence of smoothing, any trends or autocorrelations in the two series can give rise to spurious cross-correlations and should be removed before calculating the cross-correlogram (Diggle 1990). Although the protrusion and retraction areas both show a pronounced tendency to increase as the cell grows (Fig. 3), taking relative areas removes these trends (Fig. 5) without recourse to arbitrary trend-removal procedures. Each series shows significant autocorrelations (Figs. 8C & E), however, and these were removed by fitting an autoregressive model of order 11 (AR[11]) to each using the Yule-Walker method (a routine is provided in the *Mathematica* package *Time Series Pack*, see Diggle 1990). The residuals from this procedure form what are known as pre-whitened series and Figs. 8D & F shows that significant autocorrelations are absent from the two pre-whitened series.

The end product of this procedure is the pre-whitened cross-correlogram of Fig. 8B. This shows that protrusion and retraction have a significant interdependency over a period of four different lag times (the error bars are 95% confidence limits for the correlation coefficient). The most striking aspect of the correlogram is that the correlations are not symmetrical about a time lag of zero and in fact the peak occurs at a lag of -1 min which means that retraction areas show their greatest interdependency with protrusion areas of 1 min later. In fact this lag can be detected in Fig. 3 where it is apparent that the fluctuations in retraction tend to lead slightly the fluctuations in protrusion. An obvious interpretation, but by no means the only one, is that the fluctuations in retraction cause the fluctuations in protrusion. This would suggest that the area of spreading of this cell is controlled by a mechanism which changes the rate of protrusion in order to compensate rapidly for any fluctuations in the rate of retraction.

7 Discussion

The simple procedure of measuring the projected areas and geometric centroid positions of protrusion and retraction can lead to a much greater insight into the dynamics of cell locomotion than measures of motility based solely on the translocation of the cell. We have shown two methods for "resolving" the translocation into components that are presumably dependent on distinct mechanisms. In the first method, the separation of a cell track into a protrusion track and a retraction track indicates that protrusion and retraction may behave independently in determining the cell's direction of travel. In the second method, the separation of speed of locomotion into a component that depends on the quantity of protrusion and retraction (a/s) and another that depends on their relative disposition (*polarity*) reveals that, in our example at least, changes in cell speed are largely due to changes in *polarity*. This decomposition reveals that *polarity* is a measure of how effective a given relative rate of protrusion and retraction is in producing cell translocation. The *efficiency tracks*, on the other hand, show how effectively each region of the substratum is explored in relation to the local rate of protrusion or retraction. Although closely related, *polarity* and *efficiency tracks* are not fully interdependent and different experimental treatments may show differing patterns of affecting them.

A full analysis of the characteristics of protrusion and retraction also entails investigating their interdependency. In the case of our example chick fibroblast, the pre-whitened cross-correlogram of protrusion and retraction appears straightforward to interpret and suggests a causal dependency of protrusion on retraction. In other cell types and experimental treatments that we have studied, however, the cross-correlogram often appears more complex and does not suggest a simple mechanistic interpretation. Nevertheless, there is a significant interdependency between protrusion and retraction in all the cases that we have studied, covering a wide range of cell types, and there can be little doubt that this is a manifestation of processes that are central to the mechanism of cell spreading and locomotion. Even in the absence of a mechanistic interpretation, the cross-correlogram appears to be a sensitive assay for subtle changes to the locomotion of cells caused by experimental treatments. Cells that show no detectable differences in their speed or persistence of translocation often show consistent and conspicuous differences in their cross-correlograms of protrusion and retraction.

In summary, we envisage that the analysis of protrusion and retraction will have two main roles in cell behaviour research: as an assay system and as an insight into the underlying dynamics of cell motility. In our experience, even a full analysis of the characteristic pattern of cell translocation is, at best, an insensitive assay of experimental treatments and quite serious disruptions of the cell's locomotory machinery, such as depolymerisation of the majority of cytoplasmic microtubules or deletions of a combination of actin binding proteins, may lead to no detectable changes. In these cases, an analysis of the protrusion and retraction characteristics provides a further opportunity for detecting changes in the motile characteristics of the cell. Interpreting measures of protrusion and retraction in terms of cellular mechanisms will be much more difficult but observations such as the intimate interdependency of protrusion and retraction provide clues that cannot be ignored in formulating hypotheses about the functioning of the cell's locomotory machinery.

Acknowledgement We thank Professor Hans-Uli Keller for supplying the colcemid-treated carcinosarcoma cells.

I.6

Microscopic Image Classification Based on Descriptor Analysis

Marcin Inkielman and Jan Doroszewski (Warszawa)

1 Information contained in microscopic images

Computer acquisition, storage and processing of microscopic images makes easier and more precise an objective comparison of various cell features in different experimental conditions. In most analysis systems (Soll & Wessels 1988, Diaspro *et al.* 1990), the image acquisition is performed with the use of a video camera joined to the frame grabber and computer. The images are collected in form of pixel tables. In most applications the useful information constitutes only a small part of the information the image contains as whole; therefore it is essential to elaborate a method of the computer image analysis which would make it possible to distinguish and to process especially significant cell features while neglecting the unimportant ones.

The direct form of cell image acquired with a computer-based microscopic system reflects the distribution of luminosity (and/or, in some cases, of colour), but does not embrace the original spatial relations of the objects under examination. The fact that the image pixels are interrelated by an artificial structure of the pixel table has two major inconvenient consequences: on the one hand, a huge computer memory is required, because all parts of images, even unimportant such as background, have to be stored with high resolution to avoid loss of significant information and, on the other hand, it is difficult to compare the images from the view point of specific features of the objects, e.g., shape. Therefore, it is necessary to transform the primary, directly acquired image into a standard form of image description closely interrelated with chosen aspects of original images. If the class of examined objects is sufficiently narrow it is often possible to define multiple image aspects which are not important in a given case and to discard them in the process of image description. This approach enables one to develop methods which considerably decrease computer memory required to store images and makes image comparison faster and easier. The method of a two- and three-dimensional cell image analysis presented in this paper is based on this concept. It has been applied in the study concerning the adhesion of L1210 cells (lymphocytic leukemia, mouse) as examined with the use of a fluorescent microscope.

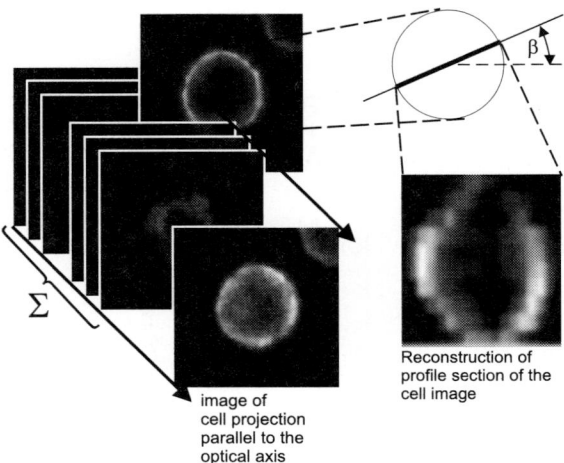

Figure 1: Organisation of three-dimensional cell structure, reconstruction of the cell profile section and its projection parallel to the optical axis.

2   Applied acquisition and pre-processing methods

All images were observed with a fluorescence microscope (NIKON Diaphot 300) using a big optical magnification ($\times 100$), registered with the use of a video camera (HAMAMATSU model C2400-08), digitized by a simple frame grabber (PE-VON model PV120) and stored in computer memory in form of collections of 3D data. Every data collection represents a single isolated L1210 cell and is composed of $\Sigma (=$ eight to sixteen) 2D image slices (collected in $2\mu m$ steps). Each slice has the same resolution and dimension (Fig. 1). Before the main analysis, several pre-processing procedures have to be applied in order to enhance images quality (Combettes 1995) and to correct brightness intensity distribution distorted by the out-of-focus planes. In the presented example, to enhance image quality, the following methods were used:
a) histogram correction (elimination of non-linearity of brightness/signal conversion of the camera and the frame grabber),
b) image averaging, low-pass filtering (gaussian filter) and non-matching pixels removal algorithm based on a median filter (reduction of the noise to signal factor),
c) background distortion reduction
d) simplified deconvolution method (elimination of out-of-focus information – this method is considerably faster than the exact method described by Carrington and Fogarty (1987)) and
e) final contrast/brightness correction (expansion of image brightness histogram).
Fig. 2 presents successive phases of a 3D image restoration procedure.

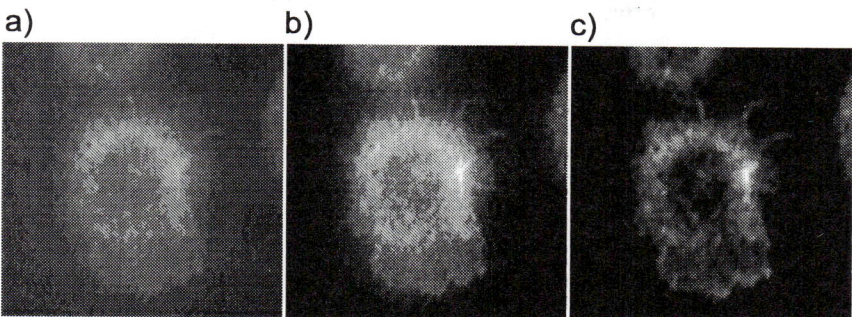

Figure 2: Successive phases of image enhancement, a) original image, b) after grey scale linearization, c) final image after pseudo deconvolution.

3 Decomposition of a two-dimensional cell image

After image quality enhancement, the cells studied with a fluorescence microscope can be easily distinguished from the image background (Fig. 3). The first step of developed 3D cell structure analysis is the cell shape detection for each 2D image slice as illustrated in Fig. 3. To detect the cell edge a simple algorithm examining brightness and gradient distribution is used. In order to better illustrate successive phases of descriptor extraction from a 2D structure artificial images are used as phantoms (Fig. 4). An extracted contour defines the interior and exterior of a cell. It is stored in the form of a sorted list of points (nodes) belonging to the edge and separated with a constant distance. The exterior is ignored in the further analysis and the interior is divided in $N \times K$ parts (where $N = 64$ is the number of nodes and $K = 10$ is the number of sub-contours). Form and position of each part are strictly determinated by the cell contour (as illustrated on Fig. 4 for the phantom "FC"). Then the image of the cell is decomposed into three sets of data:

1. immediate shape descriptors,
2. function describing contour complexity,
3. 2D function describing cell surface complexity.

In the present example the first group is represented by the surface area S, mean radius R_0, coordinates of shape centre (x_0, y_0) and the angle of its elongation main axis α. These parameters are calculated using standard object moments' analysis as described in (Kulikowski 1985, Dunn & Brown 1987, Doroszewski et al. 1993). The function describing the contour complexity $R(n)$ is defined as:

$$R(n) = \sqrt{\frac{\pi}{S}} \cdot \sqrt{(x(n) - x_0)^2 + (y(n) - y_0)^2}$$

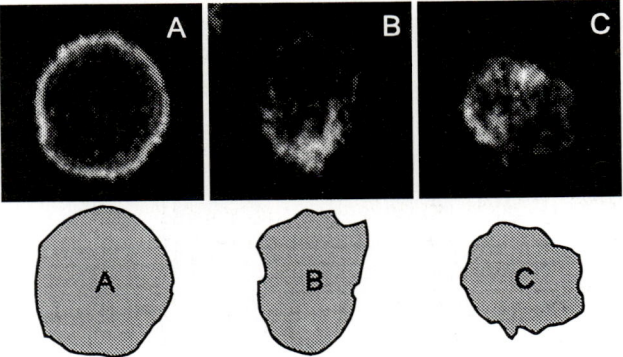

Figure 3: An example of cell images observed with a fluorescence microscope after image quality enhancement and automatically extracted cell contours.

where $x(n)$, $y(n)$ are coordinates of node n ($n \in [0, N]$), N is the number of nodes of the contour and S is its surface area calculated using the moments' method. As the shape is more similar to a circle, the $R(n)$ function is constant and close to 1 for each node n.

The surface brightness complexity is described by a table $S(n, k)$. Element (n, k) of this table contains information about the mean brightness of the neighbourhood of node n on the sub-contour k. Using parameters S, a, (x, y), function $R(n)$, and the table $S(n, k)$ it is possible to reconstruct the original cell image in a complete way, i.e. without any loss of information. The obtained dataset may be treated as a new form of the cell image. The main advantage of this approach is that images are stored in a normalized form and may be easily compared. Several standard statistical and analytical methods may be used for this comparison; methods based on neural computation, however, are especially well suited for this purpose (Kerr & Bartlett 1995).

4 Quantitative evaluation of cell aspects

Parameters S, α, (x_0, y_0), function $R(n)$, and the table $S(n, k)$ completely describe the 2D cell image. Some important cell aspects may easily be defined using $R(n)$ and $S(n, k)$. The function $R(n)$ makes it possible to define parameters characterizing deviation from the circular shape, which may be defined as follows:

$$SCR = \frac{\sqrt{\frac{\sum_{0 \leq n < N} N \cdot R^2(n) - \left(\sum_{0 \leq n < N} R(n)\right)^2}{(N-1)N}}}{\frac{1}{N} \sum_{0 \leq n < N} R(n)}$$

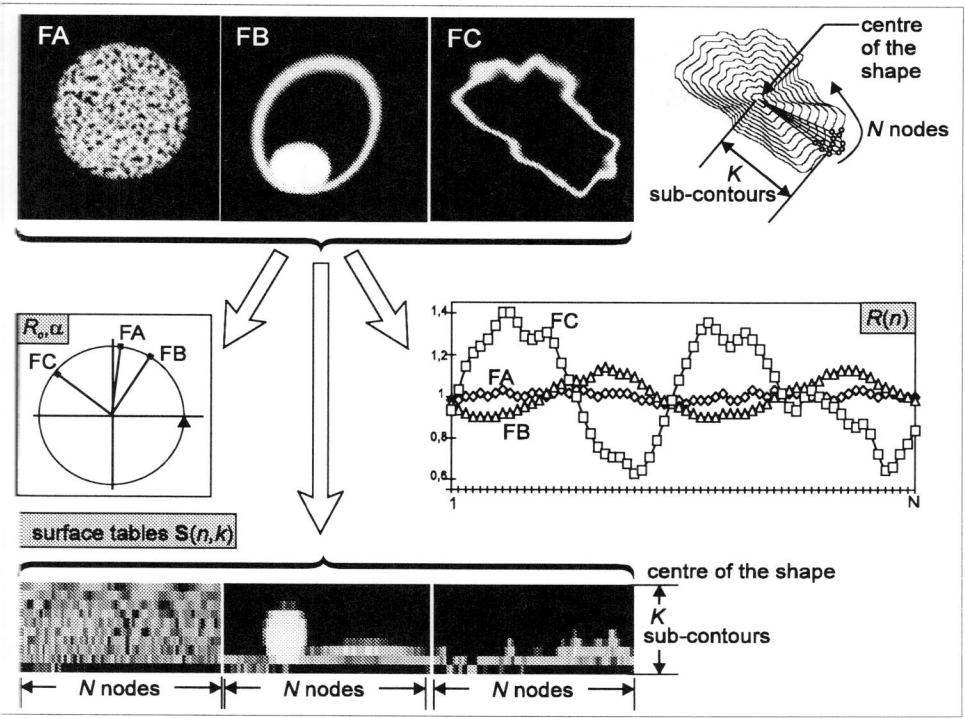

Figure 4: Illustration of image decomposition for three phantom cells (see text for details).

$$CR(i) = \frac{\sum_{i\leq k<\frac{N}{2}} |h_k[R(n)]|}{\sum_{0\leq k<\frac{N}{2}} |h_k[R(n)]|},$$

where $h_k[R(n)]$ is the k-th Fourier harmonic of the function $R(n)$. $CR(1)$ and SCR have a similar value for most contours and describe the basic irregularity of the shape whereas $CR\,(i>1)$ reflects subtle irregularities of the contour. For estimation of the cell interior brightness distribution, the function $I(k)$ and the coefficient F may be used. They are defined as follows:

$$I(k) = \frac{\sum_{0\leq n<K} S(n,k)}{N}, \qquad F = 2 \cdot \frac{\sum_{0\leq k<K/2} I(k)}{\sum_{0\leq k<K} I(k)}.$$

If the value of F is close to 1, the cell image is "filled" and has a bright edge. If F is close to 0 then the cell interior is dark; F greater than 1 corresponds to the concentration of brightness near the centre of the cell.

The following functions $D(k)$, $CS(i)$ and parameter SCS describe the complexity of cell shape in a similar manner as SCR and $CR(i)$ describe the irregu-

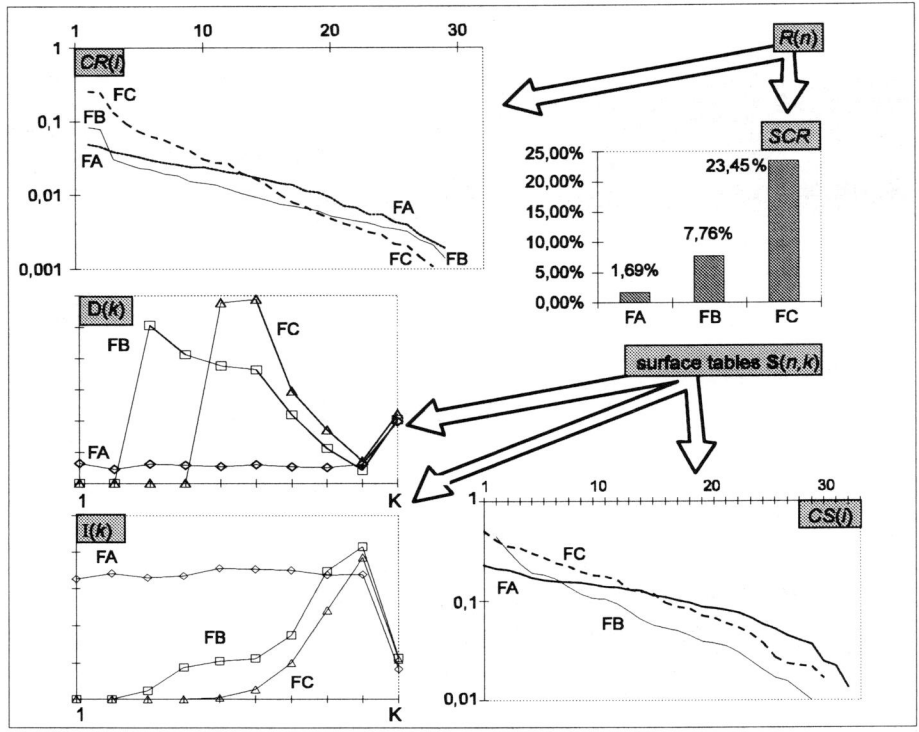

Figure 5: Quantitative representation of chosen aspects of phantom cells from Fig. 4.

larity of the contour:

$$D(k) = \sqrt{\frac{\sum_{0\leq n<N} NS^2(n,k) - (\sum_{0\leq n<N} S(n,k))^2}{(N-1)N}}$$

$$SCS = \frac{1}{K}\sum_{0\leq k<K} D(k)$$

$$CS(i) = \frac{\sum_{i\leq k<N}|h_k[R(n)]|}{\sum_{0\leq k<N}|h_k[R(n)]|}.$$

A graphic representation of the above defined coefficients and functions is shown in Fig. 5 for three phantoms from Fig. 4. The table of Fig. 6 illustrates the quantitative comparison between phantom and original cell images.

I.6 Microscopic Image Classification

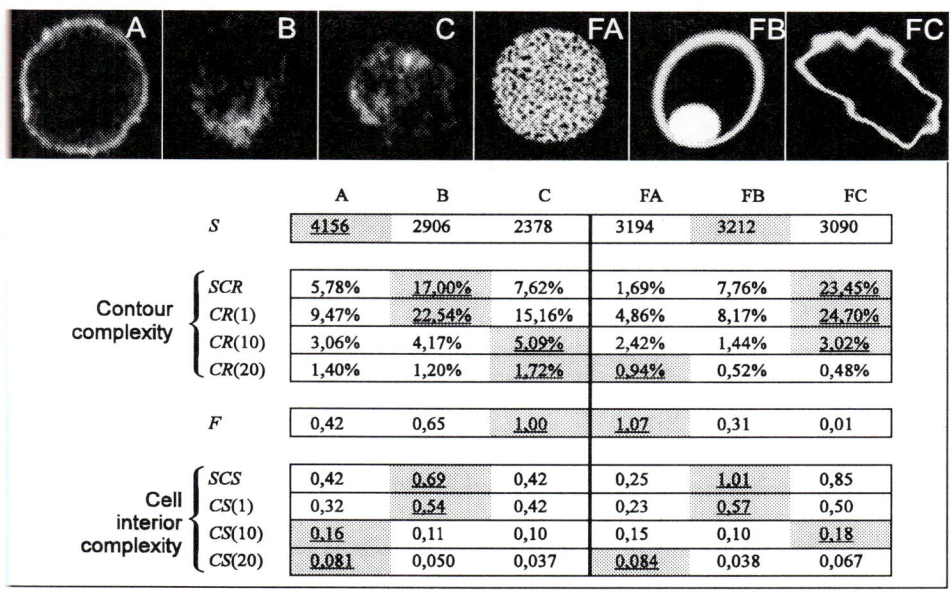

Figure 6: Quantitative comparison of chosen cell aspects proceeded for three phantom and three original cell images.

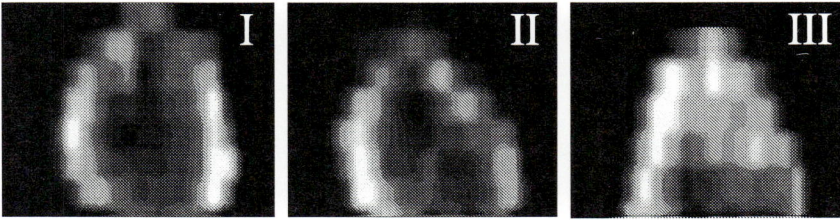

Figure 7: Reconstruction of profile section of three L1210 cells adhering the surface with different forces (cell "I" the less adherent, cell "III" the most adherent). Each horizontal line represents 2 mm.

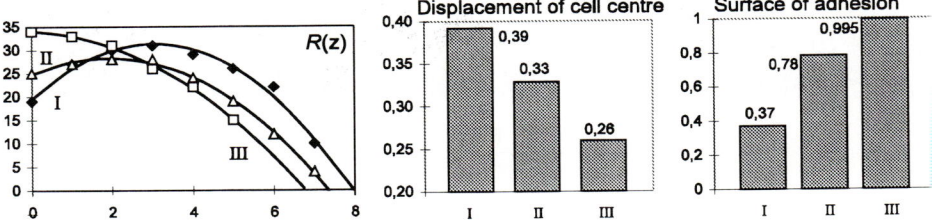

Figure 8: Example of quantitative comparison of adhesion level for L1210 cells from Fig. 7.

5 Analysis of the three-dimensional cell structure

The above described method of cell aspects quantification was developed primarily in order to enable an analysis of the influence of adhesion on the three-dimensional structure of L1210 cells. Fig. 7 presents reconstruction of a profile-section of three cells. Cells are adhering to the surface with different strength.

The parameters defined in previous sections make a graphic representation of several cell aspects possible as the function of distance from surface of adhesion, e.g., the first graph of Fig. 8 illustrates the relation between the mean radius of consecutives cell slices and their distance from the surface of adhesion for each cell of Fig. 7. These relations are approximated by a second order polynomial function. Using such approximating functions it is for example possible to estimate the displacement of the centre of mass influenced by the adhesion and to calculate a normalized area of adhesion for each cell.

Acknowledgement This work was financed by Grant CMKP-S-26/95.

I.7
A Dynamical Model of Cell Division

Xiaoyi He and Micah Dembo* (Boston)

1 Introduction

The first division of the fertilized or parthenogenically activated echinoderm egg is remarkable for its regularity and reproducibility. In the past these attributes have encouraged a variety of experimental studies of echinoderm cytokinesis so that at present a substantial amount of information has accumulated; see the reviews by Rappaport (1986) and Mabuchi (1986). From this work it has been clear for some time that the initiation of cytokinesis involves a very simple and unidirectional flow of information. Speaking schematically, one may say that the initial orders specifying the time and place of cleavage are supplied by the mitotic apparatus and particularly the asters. These instructions are then acted upon by an intelligent mechanical engine constituted in the cell cortex by contractile cytoskeletal filaments.

In the initial situation, (i.e. well before the putative "signal" for cytokinesis), the actin of the cytoskeleton has a rather uniform or diffuse staining pattern but with some minor concentration in a thin cortical layer (Wang & Taylor 1979, Cline et al. 1983, Kitanishi-Yumura & Fukui 1989, Fukui 1990). This initial situation is apparently quite stable in the face of substantial mechanical or chemical perturbations. Even direct exposure to a maximal astral signal by micromanipulation fails to produce any prompt effects of a dramatic sort. Indeed it requires several minutes of continuous exposure to the astral signal before noticeable furrowing begins. During this "latent" or initial phase of cytokinesis there has been reported to be a mechanical "stiffening" of the cell cortex. There is also a slight flattening of the egg at the equator so that it is no longer a perfect sphere (Rappaport 1961, Rappaport & Ebstein 1965, Hiramoto 1978, Rappaport 1986). On the ultrastructural level, these outward changes coincide with a progressive condensation and remodeling of the diffuse cytoskeleton to produce a zone of more concentrated actin/myosin in the form of a belt at the cell equator (Schroeder 1968, 1972). To avoid confusion we should emphasize that all available data indicate that the definition and organization of the axis and equator of cleavage is purely a function of the mitotic apparatus. These features are not intrinsic to the actin cytoskeleton. Thus if the mitotic apparatus is physically translated or rotated, then the cytoskeleton will

---

*Author for correspondence concerning this contribution

respond by assembling a contractile ring that has been moved in a corresponding fashion (Rappaport & Ebstein 1965).

Corresponding to the subtle physical changes detectable during the latent phase of cytokinesis, the engine responsible for ultimately carrying out cytokinesis is undergoing a progressive functional maturation during which it gradually gains more and more freedom for autonomous action. This functional maturation process is apparently complete by the time overt furrowing or negative curvature of the egg surface occurs (Rappaport & Ebstein 1965). Thus once the contractile ring and associated cytoskeletal structures are fully assembled or "mature", there is no longer an absolute need for any further astral signalling. Thereafter, the actin/myosin cytoskeleton acts as an independent mechanical automaton fully programmed and energized to pinch the egg into two daughter cells by a sphincter-like action (Hiramoto 1965, 1978, Rappaport 1967, 1969). If the astral signal is physically interrupted before complete maturation, then division will be aborted and the cytoskeleton will "relax" and return to its initial uniform state. If the signal is restored after such an interrupted division then the cytoskeleton can respond by assembling a new ring. Despite its important actions, the chemical or physical medium of the ephemeral astral message has never been elucidated (Rappaport 1986).

2 Theory

To develop some precise and quantitative analysis that can encompass both the mechanics and control of cytokinesis, the first requirement is a sufficently general language. For this purpose we will use the so-called reactive interpenetrating flow, or RIF, formalism (Dembo et al. 1984, Dembo & Harlow 1986, Dembo 1994). In this approach we regard the cytoplasm as a composite of two interpenetrating media, an aqueous solvent and a network of actin/myosin filaments (i.e. the cytoskeleton). The solvent is assumed to be a passive incompressible fluid; its motion governed only by pressure gradients and by friction with the network filaments in accord with Darcy's law. The network phase is modeled as an isotropic non-Newtonian fluid with viscosity that is allowed to vary temporally and spatially due to changes in the degree of interfilament cross-linking (i.e. gel-sol transitions). Motion of the network is driven by passive interfilament repulsions, (sometimes called "solvation" or "osmotic" forces) and also by active contractile forces. In addition to this the network mass is assumed to be in continuous chemical flux as a result of polymerization/depolymerization processes. The transport equations governing the time evolution of the cytoskeleton and aqueous solvent in a sea urchin egg as a result of these many processes are summarized in Table I. This Table also gives boundary conditions and the three constitutive laws that provide all the important elements of biological control for our model of cytokinesis. These are the dynamical rules for the network polymerization rate J, the network viscosity M, and the network contractility Ψ.

Although we refer to Ψ as the "contractility", this quantity should actually be understood as an effective pressure that sums the stress resultants from the active and passive forces tending to aggregate network filaments, together with the various repulsive forces that tend to keep these same filaments apart. The ability to express such complex many-body interactions by a single scalar field is essentially a consequence of the assumption of network isotropy (otherwise the network contractility would become a tensor related to the anisotropy tensor). This view of contractility as a simple scalar is in stark contrast to some of the usual dogmas concerning cytokinesis. One of the latter holds that the functionality of the contractile ring requires and depends on the organization of highly anisotropic and specialized arrangements of filaments, analogous to the arrangement of skeletal muscle, e.g. see (Svetina et al. I.8 this Chapter). Since our model of the cytoskeleton entirely neglects filament anisotropy, the reader can judge for himself (cf. Sect. 3), whether or not this commonly observed architectural feature is truly fundamental to the physiological dynamics of the cytoskeleton.

In the constitutive law for contractility given in Table I, the contributions of the passive interfilament repulsions are quantified by a simple constant, s. The active part of the network stress is assumed to be proportional to another constant, c, to the network density, θ_n, and also to the concentration of a chemical messenger, m. According to this formulation the contractility will be negative (i.e. dominated by the swelling term), whenever the regulatory signal and/or the cytoskeletal density is sufficiently small. Only if both θ_n and m are large, will the cytoskeleton actually become truly "contractile".

Without some means of encoding spatial and temporal information, and communicating this information to the distributed components of the cytoskeleton, it is impossible for a theoretical model to account for either the cortico-medullary differentiation of the premitotic egg or for the subsequent specialization of the cytoskeleton with respect to the mitotic axis. In the equations of Table I the involvement of the chemical messenger, m, in the contractile part of the constitutive law for Ψ serves as the sole vehicle for providing this functionality. In our proposed mechanism, the messenger m should not be confused with the famous "astral" signal of Rappaport and coworkers. Rather, m is actually a "second message" that is modulated or induced by the primary astral message. In our conception m is constitutively produced by all segments of the plasma membrane. It then diffuses into the cell interior where it is rapidly degraded (the necessary diffusion-reaction equation is given in Table I). Because the second message influences Ψ, the main dynamical effect of these processes is to increase myosin activity in a thin zone immediately underneath the plasma membrane. This secondarily causes a cortical accumulation of the network and an effective surface tension which keeps the egg round. In the absence of the first message (i.e. in the absence of the true signal from the asters), there is nothing intrinsic to distinguish one boundary segment from another and the cortex is stable and uniform. To achieve the triggering of cytokinesis the signal from the mitotic apparatus has merely to superimpose a smooth angular perturbation on the background rate of second messenger production so that there is a small percentage increase at the equator (see discussion of boundary conditions in the legend of Table I for more details).

$0 = \nabla \cdot [\mathbf{V}_n \theta_n + \mathbf{V}_s \theta_s]$	incompressibility
$0 = \partial_t \theta_n + \nabla \cdot [\mathbf{V}_n \theta_n] - J$	network continuity
$0 = \kappa^{-1} \theta_n \theta_s [\mathbf{V}_n - \mathbf{V}_s] - \theta_s \nabla P$	Darcy's law
$0 = \nabla \cdot [\theta_n M (\nabla \mathbf{V}_n + \mathbf{V}_n \nabla)] - \nabla P + \nabla \theta_n \Psi$	momentum balance
$0 = \delta_c^2 \nabla \cdot \nabla m - m$	messenger diffusion-reaction
$J = (\theta_{eq} - \theta_n)/\tau_n$	network polymerization rate
$M = \mu_n \exp[\theta_n/\theta_{eq}]$	network viscosity
$\Psi = -s + c\theta_n m$	network contractility

Table I: The reactive interpenetrating flow model. The indicated equations govern the physics of five dependent variables of the RIF model ($\theta_n, \mathbf{V}_n, \mathbf{V}_s, P$ and m) inside a region Ω. The volume fractions of the network and solvent phases are denoted θ_n and $\theta_s = 1 - \theta_n$, respectively; $\mathbf{V}_n$ and $\mathbf{V}_s$ are network and solvent velocity vectors, P is the pore pressure, m is the nondimensional concentration of a messenger that controls myosin activity. The boundary of Ω can generally be represented as the solution of an equation $\Gamma(\mathbf{x}, t) = 0$. If $\mathbf{n}$ is the outward unit normal then the evolution of the boundary is obtained from the kinematic condition; $\partial_t \Gamma = -|\nabla \Gamma| \mathbf{V}_n \cdot \mathbf{n}$. The other necessary boundary conditions are the usual "free surface" stress condition $P\mathbf{n} = \theta_n M (\nabla \mathbf{V}_n + \mathbf{V}_n \nabla) \cdot \mathbf{n}$, a condition enforcing zero solvent flux across the plasma membrane, $\mathbf{V}_s \cdot \mathbf{n} = \mathbf{V}_n \cdot \mathbf{n}$, and finally a Dirichlet constraint on the messenger; $m = 1 + \varepsilon(t) \sin^2(\phi)$. The term $\varepsilon(t) \sin^2(\phi)$ in the boundary concentration results from the perturbation in messenger production at the plasma membrane caused by the asters. The angle ϕ is the latitude of a point on the boundary (0 at the poles and $\pi/2$ at the equator). The triggering due to the asters is said to be turned "off" if $\varepsilon(t) = 0$ and is "on" with amplitude 5 when $\varepsilon(t) = const = 0.05$.

The inclusion of both swelling and contractile contributions within the constitutive law for Ψ is physically plausible and also endows our model with the characteristics needed to explain the initial stability of the cytoskeleton against random perturbations while also accounting for its propensity to undergo a critical transition after a sufficiently sustained period of astral stimulation. In short, we propose that the mechanism leading to irreversible condensation of the contractile ring arises from a simple type of mathematical bifurcation, originally pointed out and analysed some time ago (Dembo et al. 1984). According to this analysis, if swelling everywhere dominates contraction ($\Psi < 0$) then small perturbations in network density tend to dissipate. In contrast, if contraction dominates ($\Psi > 0$).

I.7 A Dynamical Model of Cell Division

then any small accumulations or nonuniformities of the network act as foci that attract even more network etc. As a result, if Ψ starts out as being everywhere negative and then becomes positive in a belt surrounding the cellular equator, there will ensue an autocatalytic or snowballing collapse of the network towards this location and the contractile ring will be the consequence.

If acting alone, cytoskeletal contraction and swelling will eventually produce complete phase separation or else a redistribution of the network phase into a configuration in which stresses are statically balanced. In either event there will then be an irreversible cessation of motility. To prevent this outcome it is necessary to invoke a simultaneous chemical reaction of network assembly and disassembly. If such a reaction occurs then it is possible for the actin cytoskeleton to achieve states of true continuous motility. The simplest working cycle of such a continuous cytoskeletal engine is:

contraction $\rightarrow$ disassembly $\rightarrow$ diffusion/convection $\rightarrow$ reassembly $\rightarrow$ contraction ....

The occurrence of the necessary chemical turnover of the cytoskeleton is by now very well established in most motile cells (Cao & Wang 1990, Zigmond 1993). In our calculations we will take the rate law for network formation and breakdown to be the simplest one consistent with the needed dynamical effect (i.e. linear mass action uniform in space with no extraneous control or influence by diffusible messages). As a result of this constitutive law (see Table I), the total mass of cytoskeleton in the egg will be constant but there will be a steady tendency to redistribute network away from places where it is concentrated in favor of places where it is depleted.

Dynamical models of the cytoskeleton frequently fail because the cohesive forces prove to be insufficient to prevent the actin network from being torn apart as a result of its own internal stresses (Dembo *et al.* 1986). In the current calculations the dynamical solution to the problem of such cohesive failure is achieved by programming the actin cross-linking proteins of the cytoskeleton so as to ensure the needed consistency in places of high contractile stress, while also allowing for the needed low resistance to flow in other places. This programming is achieved in a simple and physically plausible way by including a gelation factor, $\exp[\theta_n/\theta_{eq}]$, as part of the constitutive law for cytoskeletal viscosity (see Table I). Such density dependent gelation also explains why the cortical cytoskeleton is always observed to stop aggregating when the interfilament spacing in the ring reaches a value of ~ 200Å (Schroeder 1968, 1972). In other words, gelation or crosslinking acts to prevent the density of the filaments in the ring from increasing past a definite upper limit.

As formulated in Table I, the equations governing cytokinesis according to the RIF model involve a total of eight physical constants. All of these can be easily obtained from existing experimental measurements as summarized in Table II. Several of the values in Table II are worth special mention since they give some important quantitative feeling for the cytoskeletal physics. First the average volume fraction of the cytoskeleton is very low ($\theta_{eq} \sim$ half of one percent). This means that the network has a lot of room for condensation and that it is rather easy for water to percolate through the cytoskeletal pores. Despite being very diffuse, the cytoskeleton is still very cohesive. This can be seen from the fact that the effective viscosity of the cytoplasm (i.e. the product $\theta_{eq}\mu_n$) turns out to be on the order of 300 poise, (i.e.$\sim 10^5$ times greater than the viscosity of water!). Finally the characteristic chemical lifetime of the cytoskeleton ($\tau \approx 50$ sec) is short compared to the characteristic duration of cytokinesis (700 sec). Thus on average a filament of the cytoskeleton undergoes many cycles of assembly/disassembly during the course of cell division.

Symbol	Value	units	remarks
R	5.0×10^{-3}	cm	1
θ_{eq}	5.0×10^{-3}	–	2
τ_n	5.0×10^{1}	s	3
μ_n	6.0×10^{4}	$poise$	4
δ_c	5.0×10^{-4}	cm	5
κ	6.4×10^{-12}	$cm^2/poise$	6
s	1.0×10^{6}	dyn/cm^2	7
c	5.6×10^{7}	dyn/cm^2	7

Table II: Parameter values. **1)** The radius of the sea urchin egg (Hiramoto 1969b). **2)** The characteristic volume fraction of the cytoskeleton (Pollard 1981). **3)** Typical time scale for a cycle of assembly – disassembly of the cytoskeleton (Zigmond 1993). **4)** Cytoskeletal viscosity based on the magnetic particle method. The value given is obtained by reanalysis of the raw data (Hiramoto 1969a), assuming $\theta = \theta_{eq}$ and using slip boundary conditions between the particle and the cytoskeleton. **5)** The typical distance diffused by the 2nd messenger during its lifetime should be on the order of the thickness of the cell cortex. Thus, based on images generated by anti-actin immunofluorescence staining (Kitanishi-Yumura & Fukui 1989, Fukui 1990) we use a value of 10% R. **6)** The permeability constant of the network is calculated according to the hydrodynamic formula of Happel & Brenner (1991) for creeping flow of a solvent through a periodic array of long cylindrical filaments; $\kappa = \pi a^2 (\ln(\pi/\theta_{eq}) - 1)/8\mu_s$. To obtain κ using this equation we let the viscosity of the solvent phase of the cytoplasm be $\mu_s = 0.03$ poise (Wang et al. 1982), and the radius of an actin filament be $a = 3 \times 10^{-7}$ cm (Schroeder 1972). **7)** With all other parameters fixed, c and s can be obtained by a shooting method. The necessary conditions are that the maximum density of cytoskeleton in the initial resting cortex should be just a few percent below the critical point for stability and secondarily that the rate of contraction near the half way point of division should be in agreement with the kinematic data of Hiramoto (1968).

3 Results

The main advantage of the RIF model over more qualitative descriptions is that it is possible to numerically solve the equations of Table I and compare the detailed results with experiments. We use a Galerkin finite element method with adaptive mesh for this purpose (Dembo 1994). Fig. 1 shows the typical result for the equatorial and polar radii of the egg and also the instantaneous constriction force being exerted by the contractile ring. The initial condition of this computation is generated by allowing cytoskeleton of the egg to relax until a stable steady state is reached. Then at $t = 0$, astral triggering of 5% is initiated and thereafter maintained constant. Parameter values in this and all other computations are the ones given in Table I. The data shown in Fig. 1 are from an event recorded by Hiramoto (1963).

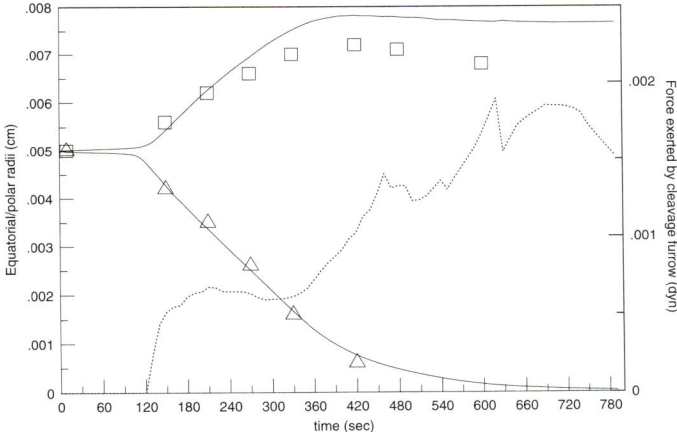

Figure 1: Time course of cytokinesis when astral triggering is turned on with amplitude of 5% at $t = 0$ and is thereafter maintained constant. Curves show the values of the equatorial and polar radii as a function of time in a calculation based on the equations of Table I and using the parameter values of Table II. Also shown are the computed values of the total contraction force being generated by the cleavage furrow at various times (dashed curve). Data points correspond to measurements of equatorial ($\triangle$) and polar ($\square$) radii by Hiramoto (1963). The characteristic kinematic stages of cytokinesis are the "latent" phase lasting from 0 to 120 sec, the "cleavage" phase lasting from 120 until 420 sec and the "final" phase lasting from 420 sec until complete cleavage.

Following the initialization of astral signalling our model predicts a period of about 120 sec during which very little seems to be happening. Obviously there is a correspondence between this behavior of the model and the mysterious latent phase of cell division mentioned previously. Furthermore, the predicted duration of the latent period in our calculation is in excellent agreement with the durations observed experimentally. Although the computed changes of the polar and equatorial radii during the latent phase are small, they are still highly significant. In fact these correspond closely with many experimental reports that eggs "flatten' to a barely detectable degree just before overt cleavage begins (Hiramoto 1956, 1963, 1968, Rappaport & Ebstein 1965). At the end of the latent period there is an abrupt onset of positive contractile force at the equator (see dashed curve in Fig. 1). Thereafter the polar and equatorial radii enter a period of rapid change that lasts about 300 sec. As can be seen from Fig. 1, the rate and extent of the changes in the furrow radius are in excellent agreement with Hiramoto's data. The model does however seem to systematically overestimate the polar radius of the cell by about 7%. This discrepancy could arise if the spindle axis of the cell were orientated at a slightly oblique angle with respect to the observer.

The phase of rapid cleavage reduces the equatoral radius of the egg to the point that only a narrow bridge connects the incipient daughter cells, (the radius of this bridge is about 10% of t the initial cell radius). At this point ($t \sim 420$), the RIF model predicts a marked slowing in the rate of cleavage. The ensuing interval, which we call the final phase, lasts about five minutes and consists of a painful exponential constriction of the intracellular bridge. Simultaneously there is a barely perceptible but significant retraction of the polar dimension of the incipient daughter cells. This small polar retraction reflects a more noticable large scale geometrical rearrangment in which the daughter cells are seen to "nestle' closer to each other. For echinoderm eggs it is difficult to observe the equatorial radius during the final phase of cytokinesis because the bridge is deeply hidden in the cleft between the two daughter cells. Nevertheless several authors have provided qualitative confirmation of the persistent existence of an intercellular bridge long after the bulk of cytokinesis has been completed, (Schroeder 1972, Fukui & Inoué 1991). The retraction of the polar dimension predicted by the model is clearly confirmed by Hiramoto's data. There is good agreement concerning the extent of the retraction, (3%) and also with respect to the time at which the maximum polar radius occurs ($\sim$410 sec).

In the calculation of Fig. 1 the integrated contractile force exerted by the ring maintains a relatively steady level during the middle phase of rapid constriction. The force during the final slow phase of cleavage peaks at a value that is almost triple the force during the middle phase. The general magnitude of the dynamic contraction forces predicted in Fig. 1 can be checked against measurements of the force exerted by the contractile ring under stall or isometric conditions. The latter forces have been measured in the sea urchin using calibrated microneedles (Rappaport 1967, Hiramoto 1978). Considering the many parameter values that go into the estimate, it is surprising to report that the maximum force of Fig. 1 is exactly the value expected from the experimental studies (i.e. 1/2 the force deflecting the needle).

Fig. 2 summarizes the results of a series of computations designed to reproduce the crucial experiments of Rappaport and Hiramoto on the interruption of astral signalling. In this regard Fig. 1 can be regarded as the control in which the perturbation caused by the asters is simply initiated and then maintained. In Fig. 2

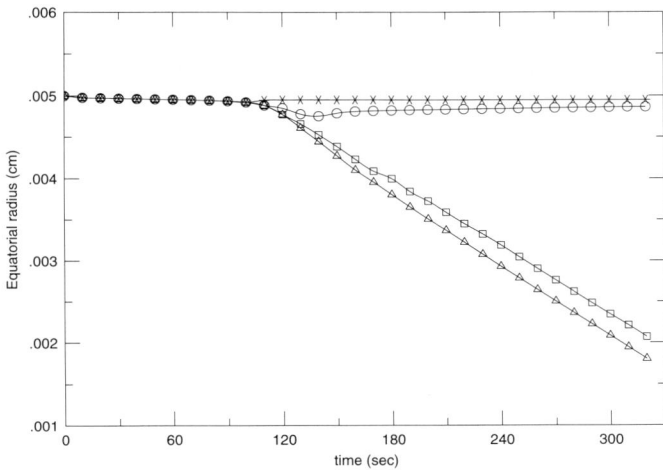

Figure 2: Effect of interrupting the astral signal on the progress of cytokinesis. Starting with the standard initial condition calculations were started by turning on the astral signal with 5% amplitude. Triggering was then either maintained constant as a control, ($\triangle$), or else turned off after various intervals. If triggering is turned off at 100 sec, ($\times$), then the equatorial radius stopped decreasing and eventually returned to its starting value. If the signal was turned off after 110 sec, ($\bigcirc$), then, although furrowing activity continued for a short while, the cleavage was abortive and eventually the cell returned to its starting condition. On the other hand if triggering is maintained for 120 sec before being turned off, ($\square$), then the equatorial radius continued to decrease and the cytokinesis eventually reached completion. The final stages of cleavage under these circumstances are of abnormal geometry and they are also slightly delayed compared to a control in which triggering is maintained constant.

we confirm that if the signal is turned off at a sufficiently early stage then the egg fails to undergo cleavage and returns to its initial state. On the other hand, if the perturbation has been "on" for a sufficient period, then cleavage will continue in an almost normal fashion even after the signal is cut off.

The duration of signalling required to program the cytoskeleton for successful cytokinesis in the calculation was 120 sec. These results confirm essential agreement between the predictions of the RIF model and the aster manipulation experiments of Rappaport (1961) and Rappaport & Ebstein (1965). The latter show that if the aster stimulus is physically interrupted after less than 90 sec then no detectable furrow or contractile ring appear. Furthermore, if the signal is interrupted after (180 sec) then the cytoskeleton behaves as if it were completely "committed" or "programmed" to division and cleavage proceeds in an almost normal fashion. This basic result is also shown by experiments of Hiramoto (1956) in which the

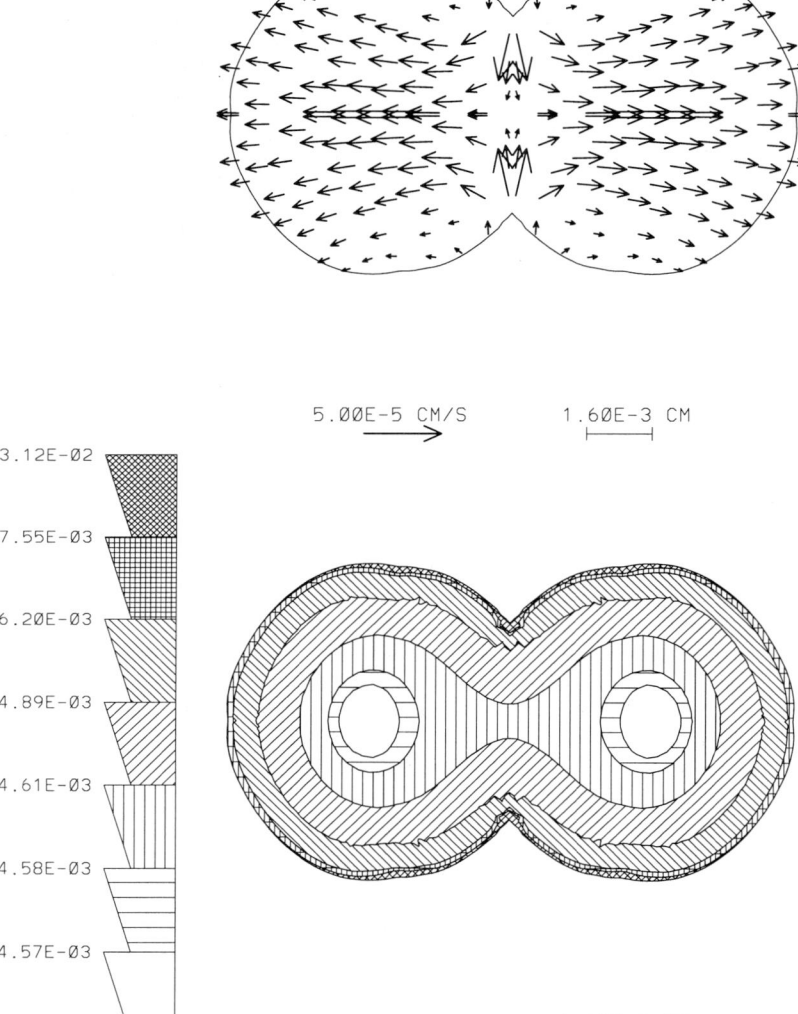

Figure 3: Cell shape and filament network distribution of a sea urchin egg at the middle stage of cleavage ($t = 270$). At the indicated stage the equatorial radius has decreased to a half of its original value but the cleavage furrow is still cutting rapidly into the cytoplasm. Despite the great change in ring circumference and mass, the density and area of the ring when viewed in cross section change are almost constant. This kinematic principle of contractile ring dynamics was first described by Schroeder (1972).

asters are surgically removed. In any event it is clear from such lines of evidence that the continuous, long-term presence of the signal from the asters is not necessary in order to guide the progress of the cleavage plane or to maintain the condensation of the contractile ring.

As a final illustration of the results obtained with the RIF model, Fig. 3 shows the main spatial fields at the middle stage a cytokinesis (furrow diameter = 50% of starting value). The progression of the cleavage by this stage is largely autonomous (i.e. the signal from the asters no longer matters). Although the main result can already be seen from Fig. 1, all other aspects of the predicted geometry at the various transient stages of cleavage are remarkably consistent with the corresponding experimental observations (Hiramoto 1956, 1968).

4 Conclusion

The concept of the continuum has proven versatile for quantitative understanding and analysis in many fields of physical science. For the analysis of amoeboid motility and cytoskeletal dynamics, however, application of continuum mechanics is controversial. Partially this has been because it is difficult to construct experimental systems that behave in a reproducible fashion and that are amenable to quantitative biophysical study. A further problem has been that algorithms for solving realistic continuum theories of the cytoskeleton have been inadequate to the task of making definitive contact with experiment. Nevertheless, in this report we have shown that a large amount of information on the first cytokinetic cleavage of the echinoderm egg can be successfully integrated and analysed using a theory based solidly on the classical language of interacting fields. While this is only a special case, cytokinesis is a ubiquitous and necessary cellular activity. We are therefore encouraged to believe that similar analysis will improve understanding of other instances of amoeboid motion.

Acknowledgement This work was supported by NIH grant RO1-AI21002.

I.8

Shape Behavior of Closed Layered Membranes and Cytokinesis

Saša Svetina, Bojan Božič and Boštjan Žekš (Ljubljana)

The analysis of the mechanism of cell cleavage in general consists in studies of cellular molecular processes which generate forces acting on the cell surface, and in studies of the effect of these forces on cell shape. Knowledge about both these aspects of cytokinesis is rapidly increasing (White & Borisy 1983, Rappaport 1986, Satterwhite & Pollard 1992, Fishkind & Wang 1995). The focus in this contribution is on the mechanical aspects of the movement of the cell's plasma membrane in relation to the shape behavior of closed layered membranes. First we shall review some general properties of shapes of vesicular structures for the case of a laterally homogeneous membrane. Then we shall specify what modifications of the treatment are needed for describing cytokinesis. Some examples of relevant shape transformations will be presented, e.g., it will be shown how cell shapes depend on the forces constricting the equatorial region of an axially symmetrical cell. Finally, a mechanism will be described for the emergence of a laterally inhomogeneous distribution of membrane-embedded and membrane-associated cell components, based on an assertion that such a distribution is caused by variations of membrane principal curvatures over the cell surface.

1 Bending energy and system enthalpy

It is generally accepted that an unsupported flaccid cell or vesicle with its volume (V) smaller than the maximum volume which the membrane of a given area (A) can enclose ($V_s = 4\pi R_s^3/3$ where R_s is the radius of the sphere with the same area) assumes a shape corresponding to the minimum value of the membrane elastic energy. When the layers of a closed-layered membrane behave as two-dimensional liquids, and are in contact but unconnected in the sense that they are free to slide one by the other, the relevant contribution to this energy is the bending energy (W_b) which is the sum of the non-local and local bending terms (Svetina & Žekš 1992),

$$W_b = \frac{1}{2}\frac{k_r}{A}(C - C_0)^2 + \frac{1}{2}k_c \int (c_1 + c_2 - c_0)^2 dA, \qquad (1)$$

where C is the integral of the sum of the principal curvatures c_1 and c_2 over the membrane area, i.e. $C = \int(c_1 + c_2)dA$, C_0 is the equilibrium value of this integral and k_r the non-local membrane modulus. The local bending term is an integral of the local membrane bending energy with k_c being the local bending constant and c_0 the membrane spontaneous curvature (Helfrich 1973). Here it has been taken that the membrane is practically incompressible and that its area is constant. Thus the variational procedure for obtaining the shape with the minimum energy is to be carried out by minimizing Eqn. 1 under the constraints of constant membrane area and constant cell volume. This, the so called generalized bilayer couple model (Heinrich et al. 1993, Miao et al. 1994) contains as its limits the strict bilayer couple model (Svetina & Žekš 1989) corresponding to $k_r = \infty$, and the spontaneous curvature model (Deuling & Helfrich 1976) corresponding to $k_r = 0$.

In the strict bilayer couple model the variational procedure for shape determination involves an additional constraint of a constant integral of membrane curvatures C. In this model the obtained shapes do not depend on the bending constant. Their only determinants are the geometrical parameters which are the relative volume ($v = V/V_s$) and the relative integral of the sum of membrane curvatures ($\bar{c} = CR_s/2$). By varying v and $\bar{c}$ continuously, each of these shapes changes continuously. However, such continuous shape changes can in general occur only within certain regions in the $v/\bar{c}$ phase space. Accordingly, classes of shapes can be defined, where a class is defined to comprise all shapes of the same symmetry characteristics that can be obtained in the described continuous manner. Several of such classes have been characterized in greater detail (Svetina & Žekš 1989, Seifert et al. 1991, Svetina & Žekš 1991) and their stability was analyzed (Heinrich et al. 1993). Analogously, in the spontaneous curvature model the variety of cell shapes and shape classes are obtained for different values of relative vesicle volume v and spontaneous curvature c_0 (Deuling & Helfrich 1976, Seifert et al. 1991). Shapes of freely suspended vesicles and cells are modified under the influence of external forces acting on them. The shape determination in this case is based on varying the free energy of the system which, in general, involves the membrane elastic energy (Eqn. 1) and the potential energies of the external forces.

These general results on cell shape formation and shape transformations can be applied to studies of cytokinesis by taking into consideration some specific features of this system. The essential structural elements of the cell boundary visible by an optical microscope are the plasma membrane and the membrane cortex which is a relatively dense submembraneous layer of the cytoplasm containing an active actin-myosin system. Some actin binding proteins are integral membrane proteins (Bretscher 1993) which indicates that actin is binding to the plasma membrane. The cortex is an elastic entity and can be considered as one of the membrane layers. Cytokinesis occurs at constant cell volume, and in the course of the cleavage process there is an increase of cell surface area which increases to $A = 2^{1/3} A_s$ in an ideal case where two spherical daughter cells are formed out of a spherical mother cell with the area $A_s = 4\pi R_s^2$. Such an increase of the cell surface can only occur if an additional membrane material is recruited from a plasma membrane compartment which does not contribute to the visible cell surface. For example, in the division of the sea urchin egg, the additional membrane material comes from microvilli (Schroeder 1978). Cytokinesis proceeds by energy consumption; therefore it is necessary to include into shape problems the forces generated by the corresponding active processes.

The effect of having membrane distributed in between the two compartments can be treated by assigning to these compartments different values of the surface energy density (σ). At a given value of the difference between these two energy densities ($\Delta\sigma = \sigma^{\text{cell surface}} - \sigma^{\text{storage}}$) the equilibrium shape can be determined by minimizing at constant cell volume the system enthalpy

$$H = W_b + \Delta\sigma A. \qquad (2)$$

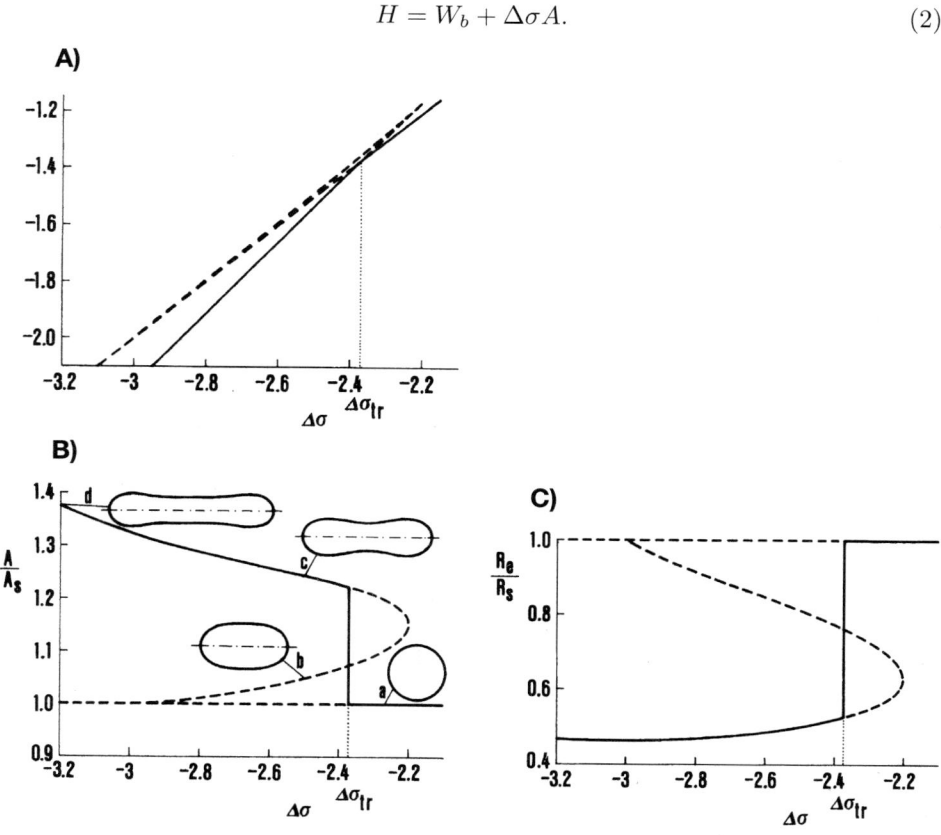

Figure 1: The calculated dependencies of the enthalpy H (in units of the bending energy of the sphere $8\pi k_c$) on the difference between the surface energy densities $\Delta\sigma$ (given in units of $2k_c/R_s^2$) (A), relative membrane area A/A_s (B), and the relative cell equatorial radius R_e/R_s (C), obtained within the spontaneous curvature model by taking $c_0 = 0$. The full lines correspond to stable shapes whereas the broken lines correspond to the unstable shapes. At $\Delta\sigma = \Delta\sigma_{tr}$ there is a discontinuous transition from a sphere to an elongated shape. Cross-sections of some characteristic shapes are depicted in (B).

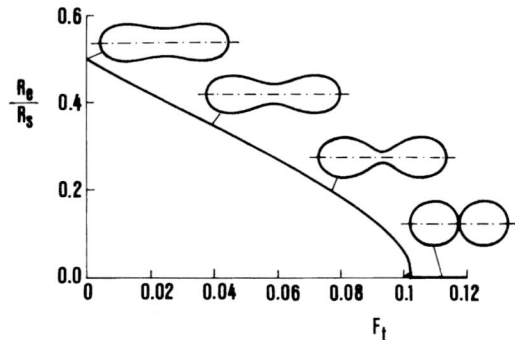

Figure 2: The dependence of the relative equatorial radius R_e/R_s on the normalized constriction force F_t (in units of $4k_c/R_s$) for $\Delta\sigma = -2.5 \times 2k_c/R_s^2$. Some characteristic shapes are also depicted.

In Fig. 1 are shown the calculated dependences on $\Delta\sigma$ of the enthalpy (A), membrane area (B), and the cell equatorial radius (C), obtained within the spontaneous curvature model by taking $c_0 = 0$. Also shown are the cross-sections of some characteristic shapes. At higher values of the surface energy density difference $\Delta\sigma$ the described model predicts a spherical cell. By decreasing $\Delta\sigma$, the system exhibits a discontinuous transition to an elongated shape at $\Delta\sigma = \Delta\sigma_{tr}$.

Fig. 2 illustrates the effect on the cell shape of the equatorial constriction force. The equilibrium state of this system is calculated by minimizing the free energy

$$G = W_b + \Delta\sigma A + F_t 2\pi R_e \qquad (3)$$

where F_t is the force and R_e the equatorial radius. The last term in Eqn. 3 represents the potential energy of the constriction force. The dependence of the equatorial radius on the relative force was determined for a given value of the surface energy density difference $\Delta\sigma$. It is also indicated in Fig. 2 that at a relatively small value of R_e the cell changes its shape discontinuously into the shape composed of two spheres.

The presented analysis indicates that cytokinesis might involve at least two consecutive stages. The results presented in Fig. 1 show that the initial shape changes can occur due to certain processes which affect the surface energies of the plasma membrane in its different compartments while otherwise the membrane remains laterally homogeneous. However, these results also show that the decrease of the difference between surface energy densities does not lead to the fully constricted shape which is a prerequisite for the cell cleavage. On the other hand, as shown in Fig. 2, constriction can be realized by applying the constriction force, i.e. by utilizing energy.

2   Generalizations: effects of principal curvatures

The described examples of the utilization of properties of closed layered membranes for interpreting cytokinesis certainly represent an oversimplified picture of this complex process, and have to be viewed mainly as a starting point for more realistic treatments. A possible generalization which also suggests a mechanism for the formation of the contractile ring is discussed in the following. For every non-spherical cell, the principal curvatures in general vary over the cell membrane. In axisymmetrical cells the two principal curvatures are equal at each pole and at the equator they differ. Specifically, for cells exhibiting an equatorial neck as can be seen in shapes c and d in Fig. 1B, the equatorial principal curvature along the parallels is positive and the corresponding principal curvature along the meridians is negative. It was asserted (Kralj-Iglič et al. 1996) that the distribution of curvatures over the membrane surface may cause an inhomogeneous distribution of membrane components over the membrane surface. By taking into consideration a curvature dependent interaction between the embedded molecule and the membrane matrix, it was shown that the membrane embedded molecules tend to accumulate in membrane regions where the shape of their embedded part corresponds most to the principal curvatures of the cell shape. The cell shape changes in cytokinesis are particularly suited for inducing driving forces for the lateral movement of membrane embedded molecules in the direction of the equator if the intrinsic shapes of these molecules correspond to the membrane principal curvatures of different signs. The concomitant drag of actin filaments in the direction of the equator by actin-binding membrane-integral proteins exhibiting the described property would not only cause the accumulation of these filaments in this region but would also make them properly oriented. Namely, in the equatorial neck, the molecules favoring principal curvatures of different signs would preferentially reside in the two opposite orientations. This could cause actin filaments also to have two opposite orientations which is needed for the actin-myosin system to perform its constricting function. In view of the presented ideas, the formation of the contractile ring and the performance of its function can be considered as consequences of the same basic mechanism.

I.9

Protrusion-Retraction Dynamics of an Annular Lamellipodial Seam

Wolfgang Alt (Bonn)
Robert T. Tranquillo (Minneapolis)

1 Contractile cortex, cytosol flow and pressure gradient

The geometry of single cells (as leukocytes or keratinocytes) spreading on a surface like a "fried egg" (Winklbauer *et al.* I.1 this volume) is characterized by a *dense cortical layer* which surrounds the whole cell body and, near the substratum, forms a more or less pronounced ring, visible e.g. in Fig. 1A of Hinz & Brosteanu (I.2 this volume). Mainly consisting of actin filaments crosslinked via myosin or other proteins, and more or less closely connected to the plasma membrane, this *contractile cortex (CC)* is an obvious candidate for performing two important biomechanical functions of the cell's "statics and dynamics", namely the following counteractions: (1) to provide the force for extending various membrane protrusions (filopods, ruffles, lamellipods or even blebs – having different functions within the processes of adhesion and locomotion) and, (2) to hold or withdraw these protrusions, thereby determining the shape of the cell and guaranteeing its integrity. Indeed, while not undergoing cell division (He & Dembo I.7, Svetina *et al.* I.8 this volume), a typical 3-dimensional cell in its contracted state is nearly ball-shaped, so that for this homogeneous situation we would, according to usual mechanical concepts, presume (1) the existence of a (nearly constant) intracellular *hydrostatic pressure* (P_B) which, by bulging out folds and ruffles within the membrane, provides an effective outward normal force onto the "cell boundary", but also (2) a (nearly constant) "effective surface tension" which prevents the occurrence of "fingering" instabilities with increasing negative (concave) curvature of the cell periphery. However, for a closer understanding of the underlying molecular mechanisms, details of the dynamics within each of the different protrusions (such as filopods, lamellipods, or blebs) have to be investigated and modelled; for a review and further references see Discussion & Open Problems of Chapter I.

In this contribution we will apply the paradigm (Dembo *et al.* 1984, Dembo 1989a, He & Dembo I.7 this volume) that the peripheral cytoplasm constitutes a two-phase "reactive interpenetrating fluid": The highly viscous (F-actin) network phase (due to assembly, disassembly, swelling or contraction) creates and modulates a less viscous flow of the aqueous phase (cytosol) pouring through the network, thereby exerting a drag force onto the network and inducing a gradient of the "effective hydrostatic pressure" P within the whole two-phase flow system. By deriving and analysing an approximating system of differential equations (for the case of lamellipods), we try to argue that the "fine tuning" of pressure and fluid flow can be induced and controlled solely by the reactive and contractive properties of the F-actin network (as a universal dynamical and spatially distributed motor system) and generate the observed phenomena of (steady or rhythmic) protrusion and retraction of lamellae at the cell periphery. Further, an account is made for the modulating role of adhesion to the substratum.

2 Model equations

Here we only consider the dynamics of lamellipodial protrusions which, caused or favored by more or less adhesive connections to the substratum via transmembrane proteins, form a flat, essentially 2-dimensional seam surrounding the contractile cortical (CC) ring. For simplicity we assume that the *"inner CC-ring"* has a circular shape with fixed radial ring distance form the cell center, R_B, fixed ring diameter, δ_B, and variable *F-actin concentration* around the cell, $a_B(t,\alpha)$, $0 \leq \alpha \leq 2\pi$. Furthermore, we assume that the flat lamella seam extending up to its "outer boundary", i.e. the *lamella tip* at radial distance $R(t,\alpha) = R_B + L(t,\alpha)$, $0 \leq \alpha \leq 2\pi$, has constant height and mainly consists of a more or less loose network of actin filaments with concentration $a(t,\alpha)$ and mean (2-dim) flow velocity $v(t,z)$. This network flow experiences drag resistance relative to the (2-dim) flow $w(t,z)$ of the intracellular aqueous phase as well as frictional resistance due to (transient binding) interaction with adhesion proteins beneath the lamella. We neglect any frictional forces or tensions of the plasma membrane itself, in contrast to Svetina *et al.* (I.8 this volume), with the idea that it just serves as a thin loose bag enveloping the *two-phase cytoplasm fluid*. Therefore, at the very tip of the lamella, where vertical curvature and relatively low surface tension of the membrane just determine its height (taken to be fixed in our model), we can assume pressure $p = P_L = 0$ to equal the exterior pressure. On the base of these hypotheses we deal with a (partially free) boundary value problem for the *two-phase flow model system* as it is treated by He & Dembo (I.7 this volume: Table I) and Lendowski & Mogilner (I.11 this volume: Eqns. (4)-(6)). For relatively low F-actin concentration it takes the following form

$$\partial_t a + \nabla \cdot (av) = \eta(a^* - a) \quad (1)$$

$$\nabla \cdot (\frac{1}{\phi a}\nabla p - v) = 0 \quad (2)$$

$$\nabla \cdot (aM\nabla v + I\sigma(a) - Ip) = \Phi av \quad (3)$$

with mean aqueous flow $w = v - 1/\phi a \nabla p$ and various parameters as network assembly rate η, drag resistance ϕ, adhesional friction Φ, viscosity tensor M, and a contractility function $\sigma(a)$, while I denotes the unit tensor.

I.9 Protrusion-Retraction Dynamics

With the assumption of relatively small seam diameter $L \ll R_B$, even smaller diameter of the cortical cell border $\delta_B \ll L$, and moderate network drag ϕ in comparison to Φ and the leading terms in M, we can rescale the seam coordinates and use an approximating finite element scheme in the radial direction (piecewise constant F-actin densities in the CC border and the lamella, $a = a_B(t,x)$ and $a_L(t,x)$, respectively, uniform tangential flow in the CC-ring, $V = V(t,x)$, piecewise linear radial flow components and, finally, quadratic elements for the pressure with mean $P(t,x)$, all quantities parametrized over the normalized peripheral angle $0 \leq x \leq 1$). Then, under further hypotheses mentioned below, from Eqns. (1)-(3) we can derive the following non-linear differential equation system on the unit circle for both F-actin concentrations, a_B and a_L, for the lamella length, L, and for the tangential cortical flow, V:

$$\partial_t a_B + \partial_x \cdot (V a_B) = \eta(a_B^* - a_B) + U_B a_L \tag{4}$$

$$\partial_t (L a_L) + \partial_x \cdot (V L a_L) = \eta L(a_L^* - a_L) - U_B \delta_B a_L \tag{5}$$

$$\partial_t L + \partial_x \cdot (VL) = \frac{1}{A} \cdot \left(\frac{P_B}{\phi} - \psi(a_L) \frac{a_L L^2}{2 + \Phi L^2} \right.$$
$$- (\delta_B a_B + (1 - \frac{\Phi L^2}{4}) a_L L) \delta_B \Psi(a_B) \right)$$
$$+ \partial_x \left(\frac{P_B \rho}{2 a_L} \partial_x (\frac{a_L L^2}{A}) \right) \tag{6}$$

$$\Phi A V = \partial_x \left(2\mu A \partial_x V + a_B \delta \Psi(a_B) + a_L L \psi(a_L) \right.$$
$$\left. - P_B \frac{a_L L^2}{2A} - P_B a_L \partial_x (\frac{\rho}{a_L} \partial_x (\frac{a_L L^2}{A})) \right) \tag{7}$$

with

$$U_B = \Psi(a_B) - \frac{\Phi L^2}{2 + \Phi L^2} \psi(a_L) a_L / a_B \tag{8}$$

denoting the contraction rate of the CC ring in the radial direction, and

$$A = A(t,x) := \delta_B a_B(t,x) + L(t,x) a_L(t,x)$$

the total F-actin mass (distribution per angle) in cortex and lamella at radial section angle x.

To obtain this compressed 1-dimensional PDE system (with two hyperbolic actin mass transport equations, one parabolic geometrical shape evolution equation and one elliptic peripheral force balance equation) as an approximation of the full free boundary value problem for the 2-dimensional PDE system (Eqns. (1)-(3) consisting of one hyperbolic mass transport equation and one elliptic pseudo stationary Stokes system), we suppose that actin filaments in the lamella have a preferential orientation orthogonal to the cell border (Lendowski & Mogilner I.11, Civelekoglu & Geigant I.12 this volume). However, in our bio-chemical/mechanical model this filament alignment, though not modeled explicitly, is thought to be induced (!) by preferential actin polymerization at the lamella tip, where the outward protrusion due to hydrostatic pressure generates new "free space" for assembly from the G-actin pool (assumed to be homogeneously distributed), in concert with their steady viscous flow towards the contractile cortical (CC) border, which is indicated by the steady centripetal motion of ruffles (Hinz & Brosteanu I.2 this volume). Furthermore, crosslinking of actin filaments is thought to be stronger, or different, in the CC ring compared to the lamella region, where the bundling protein filamin might dominate and myosin II might be less or not at all present. In our model we consider all these properties by postulating three hypotheses:

(a) in the prefered radial alignment direction within the lamella, y, the already low filament network drag, $\phi = \phi_y$, is yet stronger than in peripheral direction, so that $\rho := \phi_y/\phi_x$ is of the order R_B/L, meaning that cytosol flow through the anisotropic network is reduced along the fibers;

(b) also in the alignment direction, dilatational viscosity of the network is much larger than its shear viscosity, so that essentially no tangential shear occurs and radial and peripheral components of actin filament transport can be decoupled;

(c) in the tangential direction within in the CC ring, in contrast, shear viscosity dominates dilational viscosity, so that $\mu := \mu_{shear}/\mu_{dilat}$ is relatively large, and also (myosin induced) contractility becomes more important than in the lamella: $\Psi(a) > \psi(a)$. Both functions are relative contractile tensions defined, for example, as $\Psi(a) := \sigma_B(a)/a = \psi_0 a\, exp(-a/a_{max})$ and $\psi(a) := \sigma_L(a)/a = \psi \cdot a$ (as a simple example).

The first two hyperbolic differential Eqns. (4) & (5) describe F-actin assembly and disassembly with rate η and two (eventually different) chemical equilibrium concentrations, a_B^* and a_L^*. In addition, they contain the transport of F-actin from the lamella (a_L) into the cortical border (CC) with rate U_B defined by Eqn. (8): The actin flow into the CC ring associated with CC actin contractility, $\Psi(a_B)$, is diminished by a term that is proportional to lamellar contractility, $\psi(a_L)$, and increasing with adhesional friction in the lamella, $\frac{\Phi L^2}{2+\Phi L^2}$. A similar term appears in Eqn. (6) for the velocity of the free boundary $\partial_t L$: The positive protrusion rate, proportional to the internal cell body pressure, P_B, and inversely proportional to the network drag coefficient, ϕ, is also diminished by a term proportional to $\psi(a_L)$ but now decreasing with adhesional friction, Φ, though monotonically increasing with lamella length. Also, the third term in Eqn. (6) which reduces lamellar protrusion due to contractile transport, $\Psi(a_B)$, is decreasing with Φ. Thus we conclude an overall enhancing effect of adhesion: the more adhesion molecules are slowing down centripetal F-actin flow, the easier is the lamella protruded forward. Notice that in our model the main protrusion force is hydrostatic pressure, P_B, but also swelling of the actin polymer network could contribute by negative values of $\psi(a_L)$ for lower F-actin concentrations a_L at the lamella tip, cp. (Lendowski & Mogilner I.11 this volume). Notice, moreover, that our model also comprehends the case of a non-contracting lamellar network, i.e. $\psi = 0$ due to the lack of myosin II;

then only the CC ring is continuously contracting in the radial direction and just uniformly pulling the highly viscous lamellar network towards the cell body, cp. (Winklbauer et al. I.1 this volume).

Finally, the last terms in the two partial differential Eqns. (6) & (7) are essentially (weighted) second derivatives of lamella length $L(t,x)$ with respect to peripheral angle x. Thus, they represent the curvature of the (free) lamella boundary which being convex (concave) induces a local retraction (extension) of the lamella by decreasing (increasing) the mean effective pressure $P(t,x)$ in the lamella. We emphasize that this effect is induced just by the shape of the lamella, modulating its internal pressure gradient from P_B down to zero, thus altering the corresponding cytosol flow through it, and not by any additional surface tension of the "floppy" plasma membrane.

3 Analysis and simulations

For a possible analysis of the PDE system (4)-(7) we proceed along two different lines of further approximations and simplifications:

(A) In order to represent the slow dynamics withing the CC border we assume a moderate contractility Ψ compared to assembly rate η and "protrusivity" P_B/ϕ, so that the pseudo-stationary Eqns. (5) & (6) for lamellar actin concentration and length yield the approximate identities $a_L = a_L^*$ and $L = L_*(a_B)$, being a decreasing function of a_B. It should be mentioned that roughly the same "principle" has been experimentally observed and analysed by Bereiter-Hahn & Lüers (1994) with keratinocytes and fibroblasts, namely that in radial directions with lower "rigidity" of the CC ring (indicating lower actin concentration) usually more extended lamellae are formed.

Thus, the remaining Eqns. (4) & (7) again constitute a hyperbolic-elliptic PDE system like the original one, but now 1-dimensional, cp. (Tranquillo & Alt 1996). After further assuming that frictional forces dominate shear forces, this system can be reduced to one nonlinear fourth order parabolic differential equation for F-actin concentration, a_B, in the contractile cortical ring with constant diameter δ_B. Analysis and simulations of this "basic morphogenetic model" show the appearance of steady states with one (or two) actin concentration minima, corresponding to lamellar protrusion maxima and being stable up to peripheral shifting. They describe the polarized geometry of the cell and their peripheral CC dynamics, for more details see (Tranquillo & Alt 1996) and also (II.4, this volume).

(B) In order to catch the relatively faster protrusion-retraction dynamics of the lamellar seam as experimentally observed with human keratinocytes (Hinz & Brosteanu I.2 this volume) we neglect the dynamics in the CC ring by considering the system of Eqns. (4)-(7) in the approximative limit for vanishing ring diameter $\delta_B = 0$.

In the *homogeneous case (angle independent quantities)* we obtain a very simple nonlinear differential equation system for total F-actin mass, $A = a_L L$,

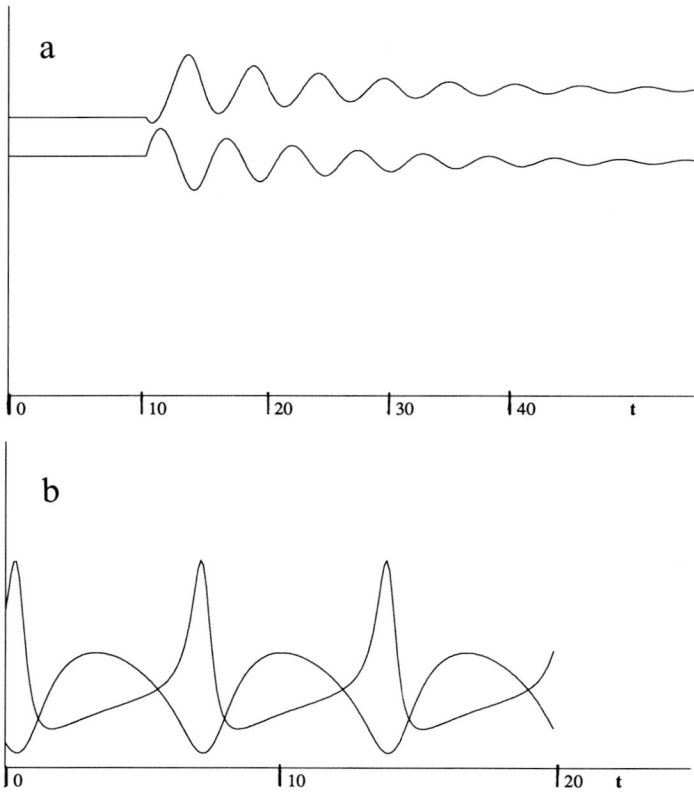

Figure 1: (a) Typical simulation of the homogeneous protrusion-retraction model, Eqns. (9) & (10). Plots over dimensionless time (one time unit = ca. 30 sec); upper plot: F-actin density $a_L(t)$, lower plot: lamella length $L(t)$. For moderate parameters, $\eta = 1$, $a_* = 1$, $\Phi = 0.4$, $P_B = \phi = 1$ and $\psi = 3.5$, the system shows an oscillatory stable steady state. At "t = 10" a step increase of actin assembly, $a_* = 1.1$, and adhesional friction, $\Phi = 0.7$, results in a sudden transient extension of the lamella followed by an increase of F-actin concentration, several gradually damped retraction-protrusion cycles (with a period of about 5 time units) and a final adaptation to the same lamella length, but with increased F-actin level; (b) plots as in Fig. 1a, but with even more increased actin assembly, $a_* = 1.3$, and increased internal pressure $P_B = 1.5$, leading to a sustained protrusion-retraction cycle (with a period of ca. 7 time units). Notice that the relatively rapid length contraction of the lamella at the end of each cycle implies a rapid compaction of F-actin which then rapidly disassembles, thereby inducing a length relaxation; but already during the relatively slow extension of the lamella, F-actin concentration gradually increases due to re-assembly. The cycle period of about 3.5 min lies well within the range of periods observed e.g. in the ruffle dynamics of keratinocytes, ca. 2.5 min (I.2 this volume: Fig. 1E).

and lamella length, L, namely with $a_* := a_L^*$

$$d_t A = \eta(a_* L - A) \tag{9}$$
$$d_t L = \frac{P_B}{\phi A} - \frac{\psi A}{2 + \Phi L^2}. \tag{10}$$

Computation of steady states $A_* = a_* L_*$ with $L_*^2 = 2/(\frac{\phi \psi a_*^2}{P_B} - \Phi)$ reveals the important property that mean lamella length L_* is increased by higher values of adhesion (Φ) but decreased by higher values of polymerization (a_*) in the lamella. Further consequences are discussed in Tranquillo & Alt (II.4 this volume).

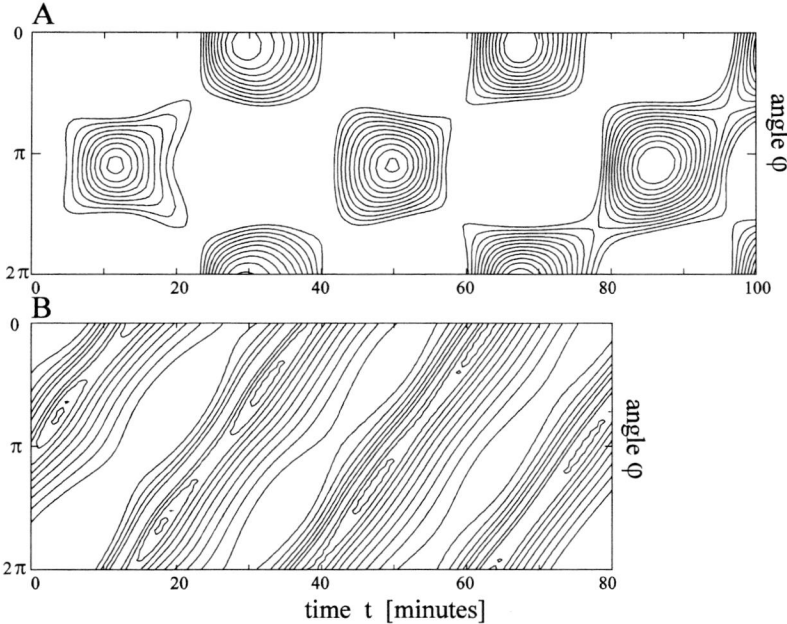

Figure 2: Simulations of the approximating protrusion-retraction dynamics by three partial differential equations for F-actin density, tangential F-actin flow and lamella length $L(t, \varphi)$ over the unit circle with angular coordinate φ. (a) and (b): Topographic plots of protrusion length $L(t, \varphi)$ in a $\varphi - t$ diagram resembling the analogous plots of observed protrusion data in Fig. 3 of (I.2 this volume). (a) Alternating pulsations of two opposite lamellae with a period of about 35 min. Notice that the lamellar protrusions possess a slight tendency to drift counterclockwise, similar as in the experimental plots. (b) For only slightly different parameters (mainly relaxed contractility) there appears a conterclockwise rotating wave of one broad lamella protrusion with rotation time of about 25 min. Notice the overlaid oscillations with a period of ca. 15 min, also detectable in the experimental plots. Reproduced from (Alt et al. 1995; Fig. 11).

Moreover, for moderate values of contractility (ψ) this steady state is (mostly) oscillatory stable, i.e. excitable, meaning that (even small) sudden changes of certain parameters (as contractility, polymerization or adhesion) could induce one or more protrusion-retraction cycles, before the system comes to rest again, see the example plotted in Fig. 1a. Finally, instability of the steady state and subsequent "Hopf bifurcation" to an ongoing, stable protrusion-retraction cycle occurs for values of $\psi a_*^2 > \beta^2/(\beta - a_*)$ with $\beta := P_B \Phi/\phi$, see the example plotted in Fig. 1b.

In the *inhomogeneous case (angle dependent quantities)* the system of Equations (4)–(7) reduces to three partial differential equations which, with slightly different parameter functions, have been investigated previously (Alt & Tranquillo 1995) as a "basic morphogenetic system" modeling shape changes of migrating cells. Linearization around the constant steady state typically provides instabilites for certain wave numbers, usually with oscillatory characteristics, but also non-oscillatory, in particular for the lowest wave number. Accordingly, numerical simulations of the full nonlinear system revealed both stable steady states, representing stationary polarized cells with one leading lamella (Winklbauer *et al.* I.1 this volume), and periodic solutions, namely stationary pulsations or circular traveling waves, just depending on the chosen parameters for contractility and "cortex tension" (Alt 1996). All these spatio-temporal patterns have, in principle, been observed with keratinocytes (Alt *et al.* 1995, Hinz & Brosteanu I.2 this volume): the experimental plots shown there qualitatively resemble the simulated ones shown in Fig. 2.

4 Discussion

With our models for the lamellipodial dynamics of amoeboid cells as keratinocytes, leukocytes or microorganisms as *Dictyostelium* or *Amoeba proteus*, we like to contribute to a more general "biomechanics" of cell shape formation by restricting to only a few biomolecular components of the cytoskeleton (F-actin, myosin) but performing a detailed analysis of the physical quantities and constraints (friction, viscosity, stresses and pressure). In particular, we essentially used the fact that motion of the lamella tip is intrinsically connected to cytosol flow through the filamentous actin network, thereby experiencing a concentration dependent resistance leading to pressure gradients within the lamella. Notice that this concept does not explicitly make use of microtubules that might also serve as cytoskeletal elements withholding the cytosol flow, as has been proposed by Winklbauer *et al.* (I.1 this volume). However, cell surface protrusion and retraction as well as cell locomotion is fairly independent of microtubules, especially in small cells, so that we claim the model to be adequate without them. Furthermore, we want to remark that our "reactive fluid" dynamical model shows pulsations and autowaves of F-actin also without assuming intracellular calcium or other proteins as substantial regulators, and without taking into account the explicit kinetics or diffusion of monomeric G-actin or myosin. However, such additional modelling is desirable, as also suggested by Vicker & Xiang (I.3 this volume) but for a reaction-diffusion model.

By comparing the modeled lamellipodial protrusion-retraction cycle with other periodic patterns in cellular motility, the most prominent example appears to be the contraction-relaxation cycle in the slime mold *Physarum polycephalum*. This also reveals both local, self-sustained pulsations of the ectoplasm, and globally coordinated "excitatory waves" (Teplov *et al.* I.10 this volume). Indeed, the dynamical system presented in the biomechanical model there (I.10 this volume: Eqns. (5) & (6)) possesses strong analogies to the system of Eqns. (9) & (10) here. In particular, the appearance of sustained oscillations relies on the same sequence of events in the contraction – disassembly – relaxation – reassembly cycle, only that the mechanisms providing the restoring forces could be quite different, ranging form hydrostatic pressure over swelling forces (osmotic or capillary) to elastic extension of eventually compressed cytoskeletal elements, e.g. intermediate filaments. Furthermore, there could be different mechanisms preventing instabilites in the free boundary of the cell surface, mathematically guaranteed by a parabolic term in the partial differential equation for the lamella length, see Eqn. (6). Whereas the *Physarum* model in (I.10 this volume) relies on tension forces generated by (contractile actin) fibers in the outer ectoplasmic layer, the lamella model here provides this "smoothing term" just by an effective pressure gradient in the thin lamellipodial seam. Its fine tuning prevents an exceeding cytosol flow into the tip of an eventually far and narrowly extending lamella, but rather facilitates sidewards cytosol flow and subsequent broadening of the lamellipod. By the way, this property of the lamella dynamics does not contradict to the appearance of long and thin filopods which are observed to arise as independent "one-dimensional" protrusions, usually not influencing the dynamics of the lamella seam, rather being slowly overtaken by the proceeding lamella or eventually withdrawn by the retrograde actin flow.

Acknowledgements This work has been supported by the DFG (SFB 256 "Nonlinear Partial Differential Equations": project on Hyperbolic Equations and Systems). It is based on the experiences of many biological colleagues, to whom we express our thanks, in particular to Andrej Grebecki, Graham Dunn, Jürgen Bereiter-Hahn, and Hans Wilhelm Kaiser whose dermatology laboratory enabled the experiments with human epidermal keratinocytes, performed by Boris Hinz. For continuing discussions about theory and models we thank Micah Dembo, who originally initiated the biomechanical approach followed by us.

I.10

Auto-oscillatory Processes and Feedback Mechanisms in *Physarum* Plasmodium Motility

Vladimir A. Teplov (Pushchino)
Yuri M. Romanovsky and Dmitri A. Pavlov (Moscow)
Wolfgang Alt (Bonn)

1 Introduction

The multinucleate plasmodium of the acellular slime mold *Physarum polycephalum* is one of the most suitable organisms for the study of protoplasmic streaming and amoeboid motion. During migration on the surface of agar or any moist surface the plasmodium differentiates into a leading frontal zone and a network of interconnected strands at the posterior region (Fig. 1). The strands, which can be up to 2 mm in diameter, contain the gel-like ectoplasm forming a cortical tube and the sol-like liquid endoplasm representing the core of the tube. The endoplasm exhibits a regular back-and-forth streaming activity throughout the plasmodium with a velocity up to 1 mm/s and a characteristic periodicity for changing the streaming direction of about 2 min (Wohlfarth-Bottermann 1979). This so-called shuttle streaming of the endoplasm is caused by non-stationary gradients of the intracellular pressure, which are generated by periodic contractions of the ectoplasm (Kamiya 1981). Any excised fragments of the plasmodium or even droplets of the endoplasm extruded by puncturing plasmodial strands are capable of rhythmical contractions (Achenbach *et al.* 1979) indicating that both the force-generating ability and the driving oscillator are distributed throughout the plasmodium.

The oscillatory contractile phenomena are intimately connected with plasmodial migration (Baranowski 1976, Beylina *et al.* 1984). In addition to shuttle streaming, the following kinds of autowave phenomena were revealed in *Physarum* plasmodium. In 10-20 min after excision the plasmodial strand suspended in moist air starts local rhythmic contractions, which later become synchronous and synphase over the entire strand (Yoshimoto & Kamiya 1978). Standing waves of higher modes can occur in long isolated strands adhered to an agar surface (Ermakov & Priezzhev 1984). Propagation of peristaltic contractions has also been shown in such strands (Baranowski & Wohlfarth-Bottermann 1982). In moderately small plasmodia (5 cm or less in diameter) there are nearly synchronous and synphasic radial pulsations of all strands in the plasmodial network, contraction phases of the

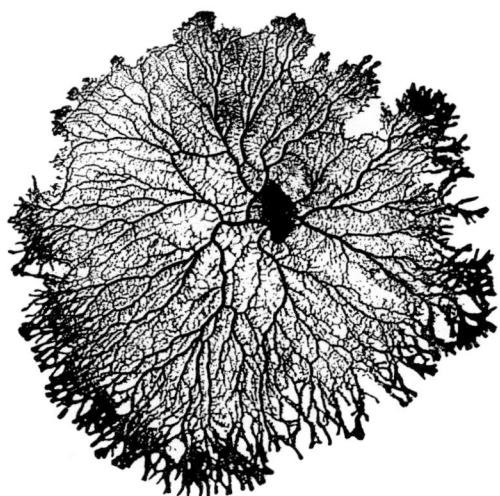

Figure 1: Typical shape of the *Physarum polycephalum* plasmodium during its spreading over the agar surface.

strands being coincident to the expansion phases of the advancing front (Grebecki & Cieslawska 1978a). Autowave motile activity of the plasmodium most visually manifests itself in its frontal sheet. For example, quasi-stochastic oscillations of sheet thickness with their subsequent synchronization as well as wave-like propagation of changes in the thickness can be observed (Baranowski 1978, Beylina et al. 1984, Teplov & Romanovsky 1987). Sometimes, in the frontal zone there is a circulation of such waves. Running waves can be artificially induced by gradients in temperature or chemotactically active factors in the underlying substratum (Hejnowicz & Wohlfarth-Bottermann 1980).

From a theoretical viewpoint the protoplasm of amoeboid cells is a truly unique system for the research of oscillatory self-organization in distributed active media. In this paper we shall present two approaches in describing how autooscillations and autowaves can arise in *Physarum* plasmodia due to properties of the actomyosin filament system. The main goal of modeling is to explain the spatial coordination of contractile activity by considering different possibilities of mechanical and chemical feedback mechanisms: hydrodynamical coupling via the endoplasm and control by Ca^{2+}-ions (Sect. 2), visco-contractile coupling and simple assembly kinetics (Sect. 3). This corresponds to the interest of cytologists in possible controls of state and organization within the actomyosin network (Stockem & Brix 1994). The principal value of *Physarum* as an object for investigating autowave contractile phenomena is the wide range of experimental techniques that can be employed in the examination of this behaviour and, thus, in developing and verifying theoretical models.

2 Autowave behaviour of hydrodynamically interconnected contractile units

The first mathematical models developed for *Physarum* motility described the shuttle flow of a viscous liquid between two volumes coupled by a rigid tube (Romanovsky et al. 1981, Odell 1984, Oster & Odell 1984). Here we shall consider one- and two-dimensional continuum models, which are considerably closer to real organisms and describe contraction waves and endoplasmic shuttle flows in long isolated strands (Teplov & Romanovsky 1987, Romanovsky & Teplov 1988, Teplov et al. 1991) and in plasmodia at early stages of their spreading on a surface (Latushkin et al. 1988, Romanovsky & Teplov 1995). The shaped plasmodial sheet can be presented as a round viscoelastic porous plate whose diameter is much greater than its thickness. The pores, wherein the endoplasm streams, form stochastically distributed channels within the actomyosin containing ectoplasmic gel. We simplify the complex rheology of endoplasm and ectoplasm (Teplov 1988): The endoplasm is assumed to be a Newtonian viscous incompressible fluid, and the ectoplasm is represented as a Voigt element in parallel with an active (sliding filaments) element (Teplov 1989, Teplov et al. 1991, Romanovsky & Teplov 1995). In all cases we exclude longitudinal deformations of ectoplasm in view of strong adhesion of the cell to the underlying substratum. As the Reynolds number for described motions is very small, $Re \sim 10^{-6}$, these endoplasmic flows may be described by the equation for Poiseuille flow in a tube of slowly variable radius and by Darcy's law for flows through a porous medium.

There is a body of convincing evidence that endoplasmic flow is necessary to maintain the synchrony of contraction rhythm throughout the plasmodium (Yoshimoto & Kamiya 1978, Achenbach & Wohlfarth-Bottermann 1981a,b, Baranowski & Teplov 1992). This key circumstance should be taken properly into account for mathematical modeling of the autowave motility in the *Physarum* plasmodium. The other conspicuous characteristic of the plasmodium is its *strain-induced activation* (Kamiya 1968, Yoshimoto & Kamiya 1978, Krüger & Wohlfarth-Bottermann 1978) which suggests the existence of some *positive feedback loop* between a deformation and an active stress.

The structural and biochemical similarities between plasmodial actomyosin and its counterpart in muscle systems immediately suggests a significant role of calcium in the contraction-relaxation cycle (Stockem & Brix 1994). Oscillations of internal Ca^{2+} concentration in *Physarum* plasmodia have been documented (Kuroda et al. 1988) and a requirement of calcium ions for contraction has also been shown (cf. Yoshimoto & Kamiya 1984, Kuroda et al. 1988). The chains of events, by which calcium stored in membranous vesicles is released and controls the chemical state of the actomyosin system, can include inositol-3-phosphate, cyclic ADP-ribose, calmodulin, cAMP, myosin light-chain kinases, etc. whose kinetics might involve coupling to mechanical deformations. As the sequence of chemical events is not determined yet, we consider the dynamics of only one single chemical that might represent the rate-limiting step in the entire chemical chain. For the sake of definiteness let Ca^{2+} itself be such a surrogate variable. For simplicity, we shall not take into account diffusion and drift of the chemical with endoplasmic streaming (because it is thought to be effective only in the ectoplasm). Further, we restrict consideration to elucidating the possibility of oscillation synchronization throughout the plasmodium only via hydrodynamic interactions.

With all simplifying assumptions the following system of equations governing contractions in *Physarum* plasmodium can be obtained (for details see Romanovsky & Teplov 1995):

$$\partial_t u = \nabla \cdot (\nu \nabla (\alpha u + \beta \partial_t u + \gamma p)) \tag{1}$$

$$\partial_t p = k_1 f_1(c)(p_m - p) - k_2 p \tag{2}$$

$$\partial_t c = k_3 f_2(u)(c_m - c) - k_4 c, \ x \epsilon \Omega, \ t \geq 0 \tag{3}$$

$$0 = \mathbf{n} \cdot \nabla(\alpha u + \beta \partial_t u + \gamma p), \ x \epsilon \partial \Omega, \ t \geq 0 \tag{4}$$

where Ω is either an interval $(0, l)$, or a two-dimensional domain with sufficiently smooth boundary, $\partial \Omega$, $\mathbf{n}$ denoting its unit normal vector; $u(t, x)$ describes (small) vertical deformations of the cell surface from its mean position; $p(t, x)$ is the active component of the intracellular pressure generated by actin-myosin interactions in the ectoplasm; $c(t, x)$ is the internal Ca^{2+} concentration; p_m is the maximum possible active component of intracellular pressure which occurs when all myosin oligomers simultaneously participate in force generation; c_m is the maximum Ca^{2+} concentration for the case when all calcium ions are released from their storages. Parameters $\alpha, \beta, \gamma, \nu$ characterize visco-elastic properties of the protoplasm. In the one-dimensional case, for the cylindrical strand, we have $\alpha = EhR_0/16$, $\beta = \eta h R_0/16$, $\gamma = R_0^3/16$, and $\nu = 1/\mu$. Here E and η are the Young's modulus and viscosity coefficient of the ectoplasm for radially symmetric strand deformations, respectively; h is the strand wall thickness; R_0 is the mean radius of the strand in the absence of intracellular pressure; μ is the endoplasm viscosity. In the two-dimensional case, for the plasmodial sheet, we have $\alpha = H_0^2 E$, $\beta = H_0^2 \eta$, $\gamma = H_0^3$, $\nu = m/\mu$ where H_0 is the sheet thickness in the absence of intracellular pressure and m is the porosity coefficient equal to the ratio of pore volume to plate volume ($0 < m < 1$); k_1 and k_2 denote the effective rate constants for formation and dissociation of actomyosin complexes, respectively; k_3 and k_4 are the effective rate constants for Ca^{2+} release and sequestering, respectively. Calcium control of actin-myosin interaction is ensured by entering the dimensionless function of activation, f_1, which is a sigmoid, monotone increasing function, e.g. approximated by either a polynomial of third degree or by an arc tangent function. The dimensionless function, f_2, provides a means for strain-induced activation of the control chemistry. Here the rate of Ca^{2+} influx into fibrillar space is assumed to increase linearly with local deformation u. We restrict our consideration to the no-flux boundary conditions (4).

Typical values of all parameters involved in the model and general conditions for the autowave excitation were discussed earlier (Teplov & Romanovsky 1987,

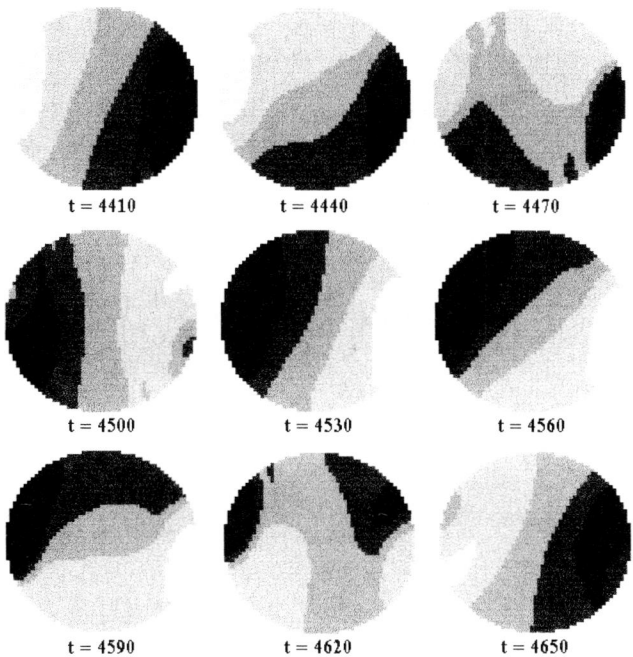

Figure 2: Simulation of the first model system (Eqns. 1-4) on a 2-dimensional disk showing a pulsating, partly rotating wave with period of about 4 min. (Elapsed time is written in seconds, increasing thickness of the plasmodium sheet is represented by lighter zones.)

Teplov et al. 1991, Romanovsky & Teplov 1995), and mathematical problems connected with proofs of existence and uniqueness of solutions were considered by Pavlov & Potapov (1994). Computer simulations of spatially discretized versions of these equations were run for slight perturbations of the stationary state and for various values of parameter k_3. When the linear stability analysis permits excitation of only one or two first Fourier modes, the corresponding numerical solutions of the full nonlinear system are already quantitatively consistent with most of the available data. Specifically, both the amplitude of Ca^{2+} oscillation and phase relation between Ca^{2+} and contraction are in good agreement with experimental data of Kuroda et al. (1988). In this case Ca^{2+} functions as an activator below 100 nM, but by proper adjusting functions f_1 and f_2 it is also possible to model the inhibitory effect of Ca^{2+} on the *Physarum* actomyosin above this dose level. By further increasing parameter k_3 the quasi-harmonic standing autowaves give way to quasi-stochastic regimes. In this case a modulation of oscillations of all variables is observed as a result of nonlinear interactions between the modes.

Periodic contractile activity in the frontal sheet of spreading plasmodia sometimes circulates around the whole cell periphery just as it occurs in the other amoeboid cells (Smolyaninov & Bliokh 1976, Alt 1996). Such a behaviour can also be simulated by the presented model (Fig. 2). Like Alt & Tranquillo (1995) we have observed standing pulsations which at first glance seem to be stable, but finally lose their stability and converge to rotating waves, and vice versa, transient rotating waves coordinating into standing pulsations; see also Alt & Tranquillo (I.9 this volume).

Thus, the only hydrodynamical interaction among identical ectoplasmic contractile units can automatically lead to globally coherent ectoplasmic deformations and vigorous endoplasmic shuttle streaming. Because of the simplifying assumption made that interconversion between endoplasm and ectoplasm does not occur, this model does not address cell migration. Locomotion of polarized amoeboid cells is accompanied by continuous gel-sol transformation of ectoplasm at the posterior end and sol-gel conversion of endoplasm at the anterior, constituting a unidirectional flow of endoplasm towards the leading edge. A family of models of such cytoskeletal dynamics has been formulated using continuum multiphasic mixture theory as a basic language in Dembo et al. (1984), Dembo (1989b, 1994) and in the following section; see also He & Dembo (I.7 this volume).

3 Biomechanics of contractile units with simple assembly-disassembly kinetics

In the early events of freshly exposed protoplasmic droplets one can observe that the droplet expands and acts like a sponge, whereby extracellular water probably is absorbed into and again squeezed out of the developing plasmalemma invagination system (Achenbach et al. 1979, Achenbach & Wohlfarth-Bottermann 1981c), see also the comparable observations with so-called microplasmodia (Stockem & Brix 1984). Similarly, in small plasmodial veins, one can observe that during the contraction-relaxation cycle not only the outer radius $R^+(t,z)$ and the inner radius $R^-(t,z)$ of the ectoplasmic tube are periodically changing, but also the thickness (width) of the tube wall $L(t,z) := R^+(t,z) - R^-(t,z)$ (Grebecki & Cieslawska 1978b).

In order to derive a general biomechanical model for both phenomena, we consider any cross-section (at location z) along a vein, or any volume element of the invagination system in a droplet, with its actomyosin cortex having contact with the plasmalemma, as an autonomous visco-elastic contractile unit of width L. As in the previous section we again represent the unit by a Voigt element, but allow the forces within the unit to induce local contraction of $L(t,z)$, instead of assuming that such forces drive an endoplasmic flow which indirectly enters in Eqn. (1). Then this equation would be locally replaced by a simple ordinary differential equation as

$$\partial_t L = E(L) - \sigma(a) \tag{5}$$

with a nonlinear elasticity function, for example, $E(L) = 1/L + \tau_1$ and an active force $\sigma(a) = \tau_2 a e^{-\lambda a}$ depending on the concentration $a = a(t, z)$ of "active elements", i.e. actomyosin, in the ectoplasm at location z. This function, on the other hand, has to fulfill a mass conservation equation

$$\partial_t a = \eta(1-a) - a \cdot \partial_t L/L \qquad (6)$$

which contains a simple linear assembly and disassembly kinetics ($\eta = 0.5/\text{min}$) and a dilution term proportional to the rate of extension or compression of the volume element of length L.

This ODE system for L and a is already written in dimensionless form, so that only three further parameters appear: a contraction saturation parameter $\lambda = 0.14$, a tension parameter τ_1 describing the biomechanical state of the plasmalemma invagination system, and a connection parameter τ_2 quantifying the amount of connections between actin filaments and plasmalemma (see Alt 1987). Simulations have been performed with parameters reflecting observations (Achenbach et al. 1979), that during the first 4 min after deposition of a droplet the plasmalemma invagination system is built up (τ_1 increasing from -2 to $+1$), whereas within 10–15 min connections to actin filaments are established (τ_2 increasing from 0 to 3). Under this slow change of parameters the dynamical system (5), (6) which originally is only (oscillatory) excitable, passes through a critical Hopf bifurcation, so that after 15 min typical sustained pulsations of contraction and relaxation appear, showing a period of about 3 min (Fig. 3a) consistent with experimental observations (Fig. 3a in Achenbach et al. (1979)).

The oscillatory capacity of this autonomous and local model system is due to disassembly of F-actin in the contracted state and reassembly in the diluted, relaxed state. We could reformulate this geometry-dependent dis- and re-assembly of "contractile elements" again as a *strain-induced* activation that provides a *(positive) feedback* onto the biomechanical elasto-contractile unit.

The *Physarum* droplet model can easily be translated into a one-dimensionally distributed model for *Physarum* plasmodia with developed outer ectoplasmic tube, but then Eqn. (1) should be replaced by two equations for outer and inner radii $R^\pm(t,z)$ similar to Eqn. (5). Now, realizing that Eqn. (2) for the "active pressure component" p, which in the first model is essentially proportional to actomyosin density, might be replaced by Eqn. (6) above, we again obtain a dynamical system that provides a strain-induced positive feedback onto the contractile unit even without the necessity of a chemical control substance as c, thus without Eqn. (3). It would be interesting to perform simulations of this altered model system with the inclusion of global coupling by endoplasmic flow ($\gamma > 0$ in Eqn. (1)) and compare it to the simulation results of the previous section and to Figs. 9–11 in (Teplov et al. 1991). In earlier simulations (Alt 1987) we have treated an alternative model ($\gamma = 0$), where spatial coupling is not globally transmitted via the endoplasm but only locally within the ectoplasm. In addition to elasticity, corresponding to the coefficient α in Eqn. (1), we allowed longitudinal force transduction and slight translocation of actin filaments within the ectoplasmic tube due

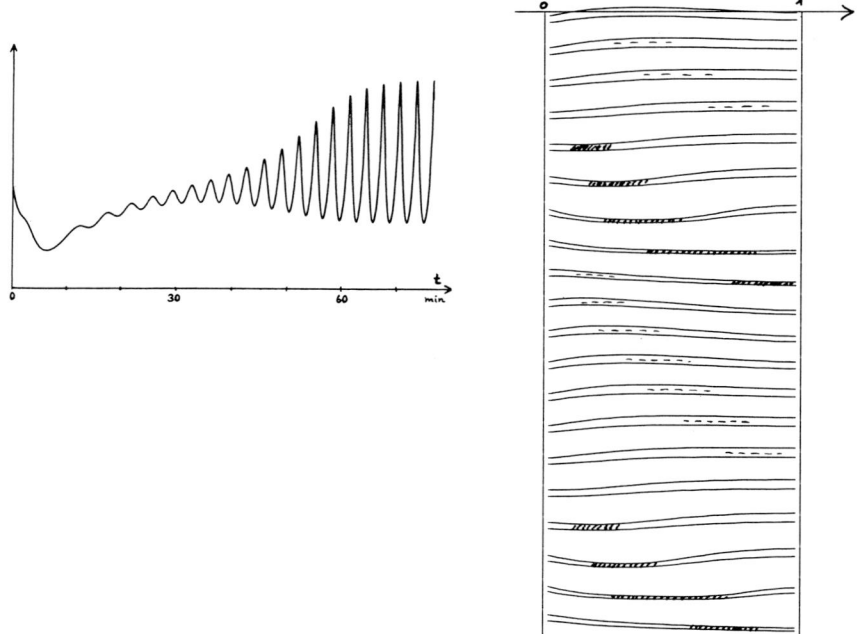

Figure 3: (a) (left) Plot of the resulting tensiometric force in a protoplasmic droplet of *Physarum polycephalum* according to the second contraction-relaxation model: Typical simulation of the ordinary differential equation system (5)–(6) with slowly increasing parameters τ_1 and τ_2 (see text for details); (b) plots of outer and inner radii $R^\pm(z,t)$ of a protoplasmic strand for successive times t (each 6 sec) showing contraction and relaxation waves: Simulation of an altered partial differential equation system for visco-contractile interactions in the ectoplasm (as in (Alt 1987, Fig. 6) with the same biomechanical parameters; longer time records reveal changes of wave direction about every five cycles).

to viscous and contractile interactions between filaments. This was described by an additional elliptic force balance equation for the filament translocation velocity in longitudinal z-direction (for details see (Alt 1987)). Then, suitable choices of parameters revealed simulations of alternating extensions and retractions at two ends of a finite interval, such that the phase gradients could be regarded as contraction resp. relaxation waves (Fig. 3b). However, these waves change their direction about every 5 cycles (Fig. 6 in (Alt 1987)), a phenomenon that is consistent with observations of irregular, only partially synchronous pulsations in ectoplasmic strands with and without endoplasmic fluid (Samans et al. 1984).

4 Discussion

In both presented models of rhythmic motility in *Physarum* it is assumed that plasmodial actomyosin plays an essential role for the cell oscillator and that spatial coordination of contractile oscillations is achieved by mechanical interactions between different regions of the plasmodium. Whereas the second model considers mechanical force transduction via the ectoplasm only, the first model considers global coordination by the endoplasmic flow. Also, both models rely on strain-induced feedback mechanisms: in the second model already the dilution of F-actin during extension triggers re-assembly from G-actin and subsequent contraction, whereas the first model requires a release of Ca^{2+} as a control factor to stimulate an interaction of myosin with existing actin filaments. In this connection the work by Smith & Saldana (1992) should be mentioned, wherein plasmodial myosin is also a principal component of the cytoplasmic Ca^{2+} oscillator which is thought to power shuttle streaming in the plasmodium. Based on data from studies of smooth muscles the authors suggest a mechanism that uses a phosphorylation-dephosphorylation cycle of myosin light chain kinase. However, this model is purely biochemical since it does not take into account the mechanical states of contractile proteins.

A wide-spread opinion that shuttle streaming is driven by locally autonomous chemical oscillations, which function independently of tension generation, is based on the observation that spontaneous Ca^{2+} oscillations continue for some time in the supernatant fraction of homogenized cytoplasm (Yoshimoto & Kamiya 1982). Unfortunately, nobody has reproduced these data yet, nor those of an older experiment on torsion oscillations in glycerol-extracted strands (see Kamiya 1968). Conceivably, oscillations of the glycerol model are due to an intrinsic character of the contractile proteins just as in the cases of ciliary and flagella motion and in the contractions of insect flight muscle.

These difficulties and uncertainties are due to a wide range of cellular processes capable of being involved in generating and regulating sustained oscillations (Wohlfarth-Bottermann 1979). They might be synchronous if common components are available, however, if a purely chemical oscillator were to be locally dominant, there would remain the necessity for explaining the spatial coordination of the contraction-relaxation cycle. The well-known proposal, that coupling is achieved via diffusion of chemicals, has been worked out (see e.g. Oster & Odell 1984), but corresponding simulations typically lead to expressed traveling waves rather than to the observed slight and irregular phase gradients (see Figs. 2 & 3b).

Therefore, we strongly favor the alternative possibility that spatial coordination is directly evoked by mechanical coupling of the stimulated contractile ectoplasm system itself (Baranowski & Teplov 1992). In our contribution we have just offered two biomechanical modeling approaches, both allowing for a "strain-induced activation" of the contractile system, but describing two possible ways in which spatial coordination can be achieved by mechanical interactions: one is by fluid dynamical pressure within the endoplasm, the other is by visco-contractile tension within the ectoplasm. The autowave patterns, which we obtain by simulations of each model, have different characteristics: the first model typically shows standing waves, which can become irregular as the positive feedback increases, and appear to be mixed with phases of transient local traveling waves (see Figs. 10 & 11 in Romanovsky & Teplov 1995). The second model typically produces synchronous pulsations or asynchronous ones with local phase gradients. In an intact *Physarum* plasmodium all these types of spatio-temporal modes can be observed; thus it seems probable that both kinds of postulated mechanisms are relevant for coordination of motility in extended strands.

There remains the problem of modelling interconversion between endoplasm and ectoplasm, in particular, generation of new ectoplasm in the front region of a migrating plasmodium. According to Stockem & Brix (1994) the rate of G-F-actin exchange within the microfilament system of *Physarum* is distinctly correlated with the degree of dynamic activity. Highly dynamic caffeine droplets exhibit a turnover rate between G-actin and F-actin of 2-5 min. In contrast to this case of caffeine droplets, the cortical system of the microplasmodia always remains in close contact with the internal face of the plasma membrane which, however, varies in extension and density according to the contraction-relaxation cycle within minutes. This is consistent with the hypothesis of assembly-disassembly kinetics in Sect. 3. On the other hand, larger alterations of cell surface morphology and rebuilding of thicker actomyosin fibrils proceed on a much longer time scale, with mean turnover time of 10-30 min in active locomoting small microplasmodia and up to 60 min in dumbbell-shaped and vein-like microplasmodia.

Anyhow, the future steps in a theoretical treatment of *Physarum* motility should include "free" moving boundary conditions. Namely, an integration of the model for autowave contractions and endoplasmic streaming with that of ectoplasm self-assembly and pseudopodium extension could allow the construction of a more complete theory of cell locomotion. Such a theory should also try to explain the regulation of cell migration by internal and external factors.

Acknowledgements Many ideas are based on earlier experimental and theoretical contributions, in particular by Zigmunt Hejnowicz, Zbigniew Baranowski, Sofia Beylina, Karl-Ernst Wohlfarth-Bottermann, Wilhelm Stockem and Klaudia Brix to whom we are very grateful. This work has been supported in part by a grant (No. 94-04-12233a) from the Russian Foundation for Basic Research.

I.11

Origin of Actin-induced Locomotion of *Listeria*

Volker Lendowski (Bonn)
Alex Mogilner (Davis)

1 Introduction

Listeria monocytogenes is an intracellular parasite which can cause serious infections (Tilney & Portnoy 1989). It invades a wide variety of cell types including macrophages, fibroblasts, epithelial cells and enterocytes. This rod-shaped, gram-positive bacterium spreads from cell to cell by moving to the peripheral membrane of the host cell and inducing filopodia-like projections on its surface that are subsequently internalized by the adjacent cell. Then, the two host membranes encapsulating the bacteria are lysed and another cycle of spreading starts. Once they penetrate into the host, these parasites spread without being exposed to the extracellular medium, bypassing the humoral immune system of the organism.

At the beginning bacteria become coated with a cloud consisting of a large population of actin filaments. At later stages, the actin cloud rearranges to form a tail-like structure extending outward from one end of the bacterium (Tilney & Portnoy 1989, Theriot & Mitchison 1991, Marchand *et al.* 1995). This transition in actin architecture is required for bacterial motility. The following experimental observations are important for the understanding of this phenomenon of cell locomotion.

The filaments in the tail are often found to be oriented predominantly parallel to the axis of *Listeria*, with their barbed ends pointing towards its direction of motion (Tilney & Portnoy 1989, Zhukarev *et al.* 1995). The tail has cylindrical shape and remains stationary in the cytoplasm (Theriot & Mitchison 1991). The rate of movement of *Listeria* is not correlated with the density of filamentous actin in the tail (Zhukarev *et al.* 1995). This rate is independent of the bacterium's size (J. Theriot, personal communication). *Listeria*'s velocity grows linearly as a function of concentration of monomeric actin at low concentration and saturates at high concentrations (Marchand *et al.* 1995).

The first question we address here is what is the nature of the transition from the actin cloud surrounding the bacteria to the tail. The second one concerns how this so formed tail is involved in bacterial motility. These require microscopic modelling at the molecular level. Then we discuss a macroscopic hydrodynamic-like model for the tail.

2 Dynamics of the actin network near the cell's surface

2.1 Spatio-angular organization of the tail's actin

Theoretical models reviewed in this section are based on the results of articles by Mogilner (1996) and by Mogilner & Oster (1996). Although there are substantial discrepancies surrounding the experimental data, we suggest the following scenario for the tail's microscopic dynamics. Actin is polymerized at the interface between the rear end of the bacterium and the filamentous network of the tail (proteins *ActA* and *profilin* play an important role in this process supplying activated G-actin and/or enhancing nucleation). Actin filaments grow until their lengths reach a few hundred nanometers and their plus ends are capped. Cross-linking of the filaments by actin-binding proteins such as α-actinin is a crucial step in the transition of the actin cloud surrounding the bacterium into the rigid tail.

We suggest that a short filament, nucleating near the rear end of a bacterium, where the concentration of activated monomeric actin is sufficiently high, is bound via a single actin-binding protein to another filament, which is already capped and rigidly cross-linked into the actin meshwork. The short growing filament rotates freely about the single cross-link until it intersects with a free actin-binding domain in the actin meshwork. Then, with some probability, the filament becomes cross-linked into the meshwork for the second time. After that this fiber polymerizes uni-directionally (filaments of lengths typical for *Listeria*'s tail can be modelled as rigid rods (MacKintosh *et al.* 1996)) establishing more bonds with the meshwork through other vacant actin-binding proteins. Dissociation of actin-binding proteins causes breaking of some links; however, estimates show that the actinin concentration is sufficient to provide more than two cross-links per filament, which makes the actin meshwork effectively rigid.

This picture leads to the description of the density distribution $f(\Omega, r, t)$ of filamentous actin in angular and real space (Ω denotes a point on the unit sphere representing the direction of the filament's growth; r is the three-dimensional spatial coordinate of the filament's center-of-mass; t is time). The behavior of this distribution is governed by the following non-linear integro-differential equation (Mogilner 1996):

$$\frac{\partial f}{\partial t}(\Omega, r, t) = -\alpha[f] f(\Omega, r, t) + \int W[f](\Omega, r, \Omega', r') f(\Omega', r', t) \, d\Omega' \, dr'. \quad (1)$$

The first term represents actin depolymerization with density dependent rate α. The second term describes the filament's polymerization at a point r in the direction Ω as a result of the nucleation on a filament located at r' and oriented at Ω'. The function W is the corresponding density-, space- and angle-dependent polymerization rate. The analysis of this equation at biologically feasible parameter values shows that the growing filament binds with high probability to the same stable polymer on which it nucleated. This leads to the creation of a unipolar cylindrical actin bundle. Furthermore, polymerization velocity is the fastest when filaments are oriented parallel to the bacterial axis, with their barbed ends towards the direction of *Listeria*'s motion. This explains the angular organization of the actin in the *Listeria*'s tail. At the same time the tail is cross-linked into the cytoplasmic cytoskeleton, and the actin in the tail remains stationary in the cytoplasm.

2.2 Polymerization-driven bacterial propulsion

The function of the tail is to propel the bacterium forward. Here we demonstrate that polymerization of the actin filaments itself can generate a propulsion force.

Peskin et al. (1993) formulated a theory to account for the force generated by the polymerization process. They proposed that the addition of actin subunits to the ends of the growing filaments rectified the Brownian motion of any diffusing object in front of the filament, and demonstrated that this "ratcheting" of diffusive motion could generate a significant force. However, careful estimates show that bacterium's fluctuations are too slow to account for the observed cell velocities. We generalized the model by assuming that an actin filament's bending fluctuations, which are fast because of the small size of the filaments (Käs et al. 1993b) can be responsible for the force generation.

We described the thermal undulation of the free ends of the actin fibers against the rear surface of *Listeria* with the following Fokker-Planck equation:

$$\frac{\partial P}{\partial t} = D\frac{\partial^2 P}{\partial x^2} - V\frac{\partial P}{\partial y} - \frac{D}{k_B T}\frac{\partial}{\partial x}(F(x,y)P) + \qquad (2)$$
$$\begin{cases} k_{on}M(P(x,y+\Delta) - P(x,y)) + k_{off}(P(x,y-\Delta) - P(x,y)), & x > \Delta \\ k_{on}MP(x,y+\Delta) - k_{off}P(x,y), & x < \Delta \end{cases}$$

Here $P(x,y,t)$ is the probability distribution for a filament with the equilibrium position of its tip at y and the current (thermally bent) position at x (both measured backward from the cell surface). The first term describes the effective diffusion of the filament's tip, D is the corresponding diffusion coefficient. The second term stands for the convection of the equilibrium position of the tip with the bacterial velocity V. The third term describes the drift velocity of the tip under the elastic force F. k_B is Boltzmann's constant and T is the absolute temperature. The last term is responsible for the polymerization/depolymerization. Here M is the actin monomer concentration, k_{on} and k_{off} are the polymerization and depolymerization rates, respectively, and Δ is half the size of an actin monomer. The form of this term is determined by the condition that polymerization cannot occur if the gap between the tip and the cell surface is smaller than Δ, so that a monomer cannot intercalate onto the tip.

With the help of a multiscale perturbation analysis (the polymerization rate is much smaller than the rate of thermal motion) we solved Eqn. 2 and found the probability of the effective gap, p_Δ, sufficient for the intercalation of a monomer onto the polymer's tip. We showed that the velocity, V, could be represented as the free polymerization velocity modulated by this probability: $V \simeq \Delta(k_{on} M p_\Delta - k_{off})$. The probability p_Δ was found to be a decreasing function of the resisting force, thus

$$V \simeq \Delta(k_{on} M p_\Delta(f) - k_{off}). \qquad (3)$$

Furthermore, our estimates demonstrated that *Listeria* moves in a force-independent regime when $p_\Delta(f) \simeq 1$. First, this means that the velocity does not depend on the bacterium's size (which affects only the resistance force f) or the tail's density (which affects the value of the total generated propulsion force). Second, the predicted propulsion velocity is approximately equal to the free polymerization velocity. Substituting the known values of the parameters into Eqn. 3 we obtained the estimate $V \sim 0.3$ μm/s, which is the same order of magnitude as observed.

Finally, the predicted dependence of the propulsion velocity on the actin monomer concentration is linear. However, this conclusion is valid at low concentrations only. At higher concentrations, when the velocity and the viscous resistance force grows, the force-independent limit is not valid. Then the polymerization velocity slows down.

3 The macroscopic model

This section is based on the results of Lendowski (1996). In the situation with *Listeria*, when the biological phenomenon is so complex, a considerable amount of speculation is required for microscopic modelling. Thus a macroscopic phenomenological model may be of value. Therefore, we apply a macroscopic model based on mass and momentum balance equations, the so called RIF (Reactive Interpenetrating Flow) model (Dembo et al. 1984, Dembo & Harlow 1986, Dembo 1994, He & Dembo I.7 this volume). In contrast to the microscopic models introduced above, here we neglect the orientation and length distributions of the actin filaments. Instead, the cytoplasm is regarded as a mixture of an actin network phase a and an aqueous phase $(1-a)$ as volume fraction functions of the independent variables time t and space x. Both phases are considered as Newtonian fluids which may interact. A specific feature of the model is the interpenetrating flow, i.e. the two phases may have completely different velocity fields, denoted by v and w. Furthermore, polymerization kinetics are included and the following interactions between the different phases are assumed. A drag force between the actin network and the aqueous solution is proportional to the difference in phase velocities. Also, there is an internal stress field inside the network due to swelling or contractile forces.

The RIF model consists – under the assumption of relatively low viscosity of the aqueous phase – of the following hyperbolic-elliptic PDE system for the dependent variables, actin concentration a, actin velocity v and pressure p:

$$\partial_t a + \nabla \cdot (av) = R(a; t, x) \tag{4}$$
$$\nabla \cdot (\mu(a)\nabla v) - \nabla p + \nabla \sigma(a) = 0 \tag{5}$$
$$\nabla \cdot (\kappa(a)\nabla p) - \nabla \cdot v = 0. \tag{6}$$

Here R describes the polymerization kinetics, μ denotes the network viscosity, κ is the permeability of the network, and σ describes an internally produced stress in the network, as functions of a, cf. Table I in the contribution of (He & Dembo, I.7 this volume). For known a, v and p, the velocity field w can be calculated as $w = v - \frac{\kappa(a)}{1-a}\nabla p$.

We consider a movable bacterium inside a fixed container filled with cytoplasm. We want to predict the shape of the actin tail and the speed c of the bacterium. In our approach, we give actin-actin interactions special emphasis, which can be described by the internal stress $\sigma(a)$ in (5). A proof for the existence of such stress is clear from magnetic particle experiments, where locally increased actin concentrations lead to a swelling stress and a pushing force on the particle. The molecular origin of the stress fields need not to be specified (see discussion, however). When some contractile or swelling repulsive forces between actin filaments are assumed, we can describe these by a function $\sigma(a)$ with $\sigma'(a) > 0$ or < 0. For the latter case, the simplest approach is to define $\sigma(a) := \sigma_0 a$ with a negative parameter σ_0.

For our application we choose no-slip boundary conditions for the actin velocity v, i.e.

$$v = 0 \quad \text{at the wall of the container and} \tag{7}$$
$$v = c \quad \text{at the surface } S \text{ of the bacterium,} \tag{8}$$

Furthermore, we assume a Neumann condition for the pressure

$$\nabla p \cdot n = 0. \tag{9}$$

The latter should be interpreted as the impermeability of the outer wall and the cell surface for actin network and aqueous phase, respectively. Finally, we assume a balance of forces at the surface of the bacterium which can be calculated by integrating all stress terms multiplied by the outer normal:

$$\int_S \{\mu(a)\nabla v \cdot n - pn + \sigma(a)n\} \, d\Gamma = 0. \tag{10}$$

It can be shown that for a given actin distribution the subproblem (5-10) has a unique solution under reasonable assumptions, in particular the speed c is uniquely determined.

For biological significance we have to specify the parameter function for the polymerization term R in Eqn. 4. The simplest way to do this is to define $R(a; t, x) := \eta(a_{eq}(t, x) - a(t, x))$, where the kinetic equilibrium a_{eq} depends on the current position in the network relative to the bacterial surface: with a high value in the vicinity of the rear of *Listeria* and a decreasing value farther away from it.

Fig. 1 shows a detailed plot of the numerical solution in a 2-D implementation of this problem. The actin concentration is shown as isolines. The high actin concentration at the rear is a consequence of the high polymerization rate, whereas the local maximum at the front is a result of compression due to the movement. The actin velocity field is shown in a moving coordinate system as it is perceived by an observer sitting on the moving bacterium. Therefore, the speed of the bacterium can be read only at the outer boundary of the domain.

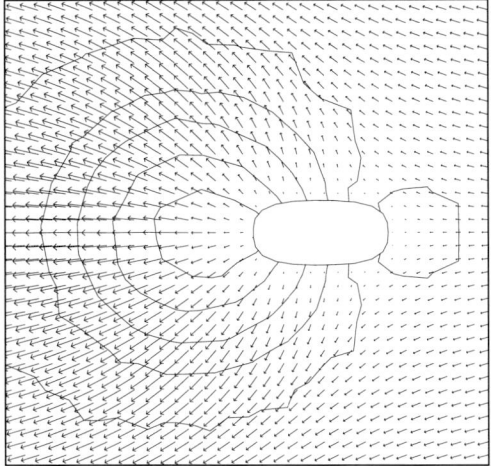

Figure 1: Detailed view of actin concentration (isolines) and actin velocity (arrows) in a moving coordinate frame.

4 Discussion

When we compare the presented macroscopic simulation with macrographs demonstrating the cylindrical shape of the tail, we see an important difference: in our simulation the shape of the tail is more or less round. Hence, the basic macroscopic model that we have sketched must be improved. Reasons for the round shape include simple linear dependencies of swelling stress σ and polymerization kinetics J on the actin concentration a and the isotropic character of the model (neglecting the filament's orientation). To include anisotropy in the macroscopic model is a hard task; it can perhaps be solved with the help of the orientational tensors technique described in the contribution of Mogilner *et al.* (III.1 this volume). However, changes in the parameter functions are simple to implement. One idea is to build in a threshold for swelling only above which swelling takes place. In combination with a nonlinear polymerization kinetics (fast depolymerization above the threshold, slow depolymerization below it) we expect a lengthened shape of the tail. Another idea is to change the boundary conditions. We can prescribe $(v-c)\cdot n$ at the bacterial surface to be equal to the polymerization velocity calculated from the present microscopic model and determine the speed c again by an additional force balance condition at the bacterium.

We presented a microscopic approach to the bacterial motility where the angular organization of actin filaments during polymerization and the propulsion force are due to the thermal undulations of the actin filaments. In this model we ignored the stresses in the actin meshwork and the fluid flow of the cytoplasm, which are considered in the phenomenological model with the emphasis on spatial transport of actin and induced stress fields.

At the present level of modelling we are able to qualitatively explain only some of the basic features of this fascinating biological phenomenon. For further progress an attempt to unify the microscopic and macroscopic models has to be made. The parameter functions J, μ, κ, σ are to be derived from microscopic models. The problem concerning the stress function $\sigma(a)$ is the most challenging one. It may include analyses of the role of entropic writhing of dynamically cross-linked actin fibers (see MacKintosh *et al.* 1995) and a possible role of myosin like motor proteins bound to actin. The final goal should be a 2-D and 3-D model, in which the stress tensor and other quantities are based on microscopic calculations and depend in an anisotropic way on length and orientational distribution of filaments and actin-binding proteins.

Acknowledgements We would like to thank G. Civelekoglu for her valuable contribution, and L. Edelstein-Keshet, W. Alt, G. F. Oster and M. Dembo for helpful remarks and discussions.

I.12

Models for the Formation of Oriented F-actin Structures in the Cytoskeleton

Gül Civelekoglu (Istanbul)
Edith Geigant (Bonn)

1 Actin structure and function

The *cytoskeleton*, the molecular scaffolding of *eucaryotic* cells, enables a cell to adopt a variety of shapes and to carry out various functions including motility. It consists of numerous proteins and actin filaments which act in concert in response to internal and/or external stimuli and may exhibit physical properties similar to a liquid or a solid at different times and at different locations in the cellular cytoplasm.

Actin was discovered in non-muscle cells in the 1960s. It is the most abundant protein in cells. The actin molecule exists both as a monomer *(globular or G-actin)*, and as a polymer *(filamentous actin, F-actin, or microfilaments)* in cells. Microfilaments are an important determinant of the structure and mechanical properties of the cytoplasmic matrix. The slow step of assembly of a few actin molecules provides a nucleus for the formation of an actin filament. A steady state is quickly reached when the actin monomer concentration falls below the so called "critical concentration", in equilibrium with the filaments. In non-muscle cells, actin filaments are highly dynamic on a second to minute time scale.

The orientation of the actin monomers in the double helical structure of the filament provides it with a unique polarity. F-actin may be viewed as a polar macromolecule with a pointed (slowly growing) and a barbed (fast growing) end.

Actin filaments are rarely found solitary, but associate into networks, bundles, or various complex structures which undergo dynamic rearrangement during cell motion, cell division, and other activities of the cell. *Bundles* of actin filaments are e.g. observed in stress fibers and focal contacts whereas *meshworks* or networks of actin filaments are found in motile lamella zones and ruffling membranes (Small et al. 1982, Stossel et al. 1985).

The self-assembly of actin structures is regulated in a remarkably precise manner; it occurs at specified times and locations within the cell. It is now recognized that this control is conferred by actin binding proteins, e.g. Pollard & Cooper (1986). Some actin binding proteins act on single filaments, e.g. nucleation proteins like gelsolin, capping proteins and membrane-binding proteins. Others lead to *interactions between filaments* by linking them together in various ways. *Cross linking* proteins, like ABP-50 and filamin, promote the formation of nearly orthogonal meshworks and *bundling proteins*, like villin, fascin, fimbrin and α-actinin, promote the alignment of filaments in bundles. The types, the amounts, and the affinities of the attachment proteins affect the type of actin structure being formed, and are thus of great importance to cellular function. However, the precise mechanism by which a variety of filament structures form or switch from one to another in a cell when all actin binding proteins are acting in concert is still unclear.

There is a wealth of literature which focuses on actin and its dynamics. Recent cellular automaton simulations by Dufort & Lumsden (1993a) have provided dynamic visual images of the actin cytoskeleton and its interactions with many binding proteins. In the mechanical approach, the effects and the balance of forces in and outside the cell are considered while the microscopic interactions and their influences on the mechanical properties of the cytoskeleton are neglected (see Sherratt & Lewis 1993). Other models focus on the dynamic contractile behaviour of actin-myosin gels, e.g. Oster *et al.* (1985) and Oster & Odell (1984). The actin-myosin meshwork can also be viewed as a creeping viscous fluid with negligible elasticity (Alt 1987, Dembo 1989a).

The importance of the key structural elements in these phenomena, the actin binding proteins, has been noted in the above papers. However, the interactions between the actin filaments and the actin binding proteins and the consequences of these interactions have not been included in most of these models.

2 Modelling interactions of F-actin with actin-binding proteins

We will review three different models for the formation of microfilament structures, based on molecular level interactions. We will briefly describe the formulation, and the analysis of the models and outline their similarities and differences from a mathematical point of view.

2.1 Modelling assumptions

In the following models the hypothesis that *molecular interactions between actin filaments and the actin associated proteins alone, may lead to the formation of various microfilament structures and the switch between them* has been investigated. It is proposed that the formation of these structures as well as the transitions between them can occur even in the absence of external mechanical forces. One or several types of actin binding proteins, including bundling proteins, crosslinking proteins, and nucleation proteins, have been considered to interact with microfilaments in these models. It is shown that biochemical properties and interactions between the molecules at the individual level, described by differential equations, suffice to determine the collective behaviour of the ensemble.

With these models, the sensitivity of the actin cytoskeleton dynamics to parameters such as kinetic rate constants, concentrations, and affinities of the various intermediates has been investigated. The importance of this issue is proposed in Wachsstock et al. (1994). Recent interest has arisen in the comparison of cytoskeletal structures found in different species of organisms whose actin associated proteins have slightly different kinetic rates.

The actin binding proteins which act on single filaments, e.g. capping proteins, or between filaments and other structures such as the cell membrane, e.g. membrane-binding proteins, are not included in these models. Therefore, membrane cortex interactions are not considered. Actin polymerization was only incorporated in first versions of the models, and omitted in later versions where a "treadmilling" case (where G-actin and F-actin concentration is in equilibrium) has been treated.

Our models focus only on the orientational distribution of filaments in cytoskeletal structures and neglect the spatial variations, but see Mogilner et al. (III.1 this volume). The models have been initially formulated to account for a 2-D geometry, and later extended to full 3-D geometry. Each filament is associated with a unique direction (since it has a barbed and pointed end), and viewed as a rigid rod. Thus, the models only account for the actin cytoskeletal structures formed by relatively stiff and short filaments, yet long enough to allow binding of more than one actin binding protein.

2.2 Modelling strategies

In this section, we will outline the repertoire of microfilament dynamics considered in the three different types of models.

In the first model, Civelekoglu & Edelstein-Keshet (1994) distinguish between free and bound filaments. Free filaments are the solitary filaments which can undergo rotational diffusion and bind to other filaments. Bound filaments are linked to one or more others via actin associated proteins, and are assumed to be fixed in angular space as long as they remain bound. The bounds are assumed to be dynamic, dissolving at a constant rate. A free filament can bind to another free or bound filament, instantly turn into a new direction as it binds, and become a bound filament itself. Its new direction is determined by the geometry of the actin binding protein involved in binding and by the direction of the filament it binds to. The geometry of the actin binding protein is reflected by the specific form of the binding kernel. The equations of this model can be found in the box below, Eqns. 1.

While Civelekoglu and Edelstein-Keshet use *binding probabilities* to describe the interaction of filaments, Geigant et al. (1997) and Geigant (1997) use *turning probabilities*, see the box, Eqn. 2. They consider a single population of filaments which can interact by means of actin associated proteins. Binding of the actin associated protein causes the filaments to change their relative orientation and to turn into new directions determined by the geometry of the actin-binding protein.

Mathematical modelling of actin aggregation

Here, we will first define the variables and the parameters used in the models, then present the equations and outline the main mathematical features of functions and terms.

In all the models below, θ_1, θ_2 are unit vectors, i.e. $\theta_1, \theta_2 \in S^{n-1} = \{x \in \mathbb{R}^n \mid |x| = 1\}$, $n = 2$ or $n = 3$, denoting filaments pointing in that direction, and $\angle(\theta_1 - \theta_2)$ is the angle between θ_1 and θ_2.

System of partial integro-differential equations. Let $F(\theta, t), B(\theta, t)$ denote the densities of free and bound actin filaments, respectively, oriented in direction θ at time $t \geq 0$. The kernel function, $k(\angle(\theta - \theta_i))$ ($k : \mathbb{R} \to \mathbb{R}_0^+$ 2π-periodic), is the probability that two filaments bind (by means of an actin binding protein) when their relative angle is $\angle(\theta - \theta_i)$ (the index i stands for *interacting*). Then, the following system of equations captures the dynamics of the two populations of actin filaments described above (Sect. 2.2):

$$\begin{aligned}\frac{\partial F}{\partial t} &= \delta \triangle_\theta F - \beta_1 F \cdot k * F - \beta_2 F \cdot k * B + \gamma B \\ \frac{\partial B}{\partial t} &= \qquad\qquad \beta_1 F \cdot k * F + \beta_2 B \cdot k * F - \gamma B\end{aligned} \quad (1)$$

where $k*F(\theta) := \int_{S^n} k(\angle(\theta-\theta_i)) F(\theta_i) d\theta_i$, $k*B$ is defined analogously. Here δ denotes the rate of rotational diffusion of free filaments, β_1 and β_2 the binding rate of free to free and free to bound filaments, and γ the dissociation rate of bound filaments. The convolution integrals such as $-\beta_2 F \cdot k * B$ represent the rate at which free filaments, F, turn away by binding to and aligning with other bound filaments, B. The kernel k is an even function, $k(\angle(\theta - \theta_i)) = k(\angle(\theta_i - \theta))$, to avoid chirality, i.e. interactions to the left and to the right are equivalent.

Ordinary integro-differential equation. In the second model, $F(\theta, t)$ denotes the density of actin filaments (free and bound) oriented in direction θ at time $t \geq 0$. The function

Note that in the first model, reorientation is caused *only* by rotational diffusion of free filaments or by binding of a free filament to a bound filament to which it turns. In the second model this corresponds to spontaneous or induced turning, respectively. In both models, the binding rate is proportional to the filament density at a given orientation, meaning that filaments tend to bind to existing bundles, promoting the formation of clusters. However, in the second model biasing the interactions alone cannot cause bundling: biased turning is a necessary requirement. Also a term accounting for the dissociation of the cross-links is included in both equations.

$w(\theta_o, \theta_n, \theta_i) \geq 0$ denotes the probability that a filament with "old" direction θ_o turns to a "new" direction θ_n whilst being bound to an "interacting" filament in θ_i via an actin associated protein. And $d(\angle(\theta_o - \theta_n))$ denotes the probability of spontaneous turning by an angle $\angle(\theta_o - \theta_n)$. Then, the following equation conveys the dynamics of actin filaments described in Sect. 2.2 as the second approach:

$$\frac{\partial F}{\partial t} = -(\gamma + \beta M) F + \gamma\, d * F + \beta \int_{S^n} \int_{S^n} w(\theta_o, ., \theta_i)\, F(\theta_i)\, F(\theta_o)\, d\theta_i\, d\theta_o, \quad (2)$$

where the constant $M := \int F(\theta)\, d\theta$ is the total mass of filaments in the system. The term $-\gamma F(\theta, t)$ accounts for the loss of filaments from direction θ by dissociation of filaments due to spontaneous turning of filaments, thus γ is the dissociation rate of filaments, and $-\beta M F(\theta, t)$ accounts for the filaments which turn away from θ by binding to other filaments, thus β is the binding rate of actin binding protein to filaments. The remaining positive terms give the rate at which filaments turn to the (new) direction θ spontaneously or by interaction with other filaments and actin binding proteins. As above, the turning probability d is even to avoid any chirality. Also, the function w is invariant with respect to turnings and reflections for the same reason.

Model for gradual turning and alignment of filaments. Here, $F(\theta, t)$ denotes the density of actin filaments oriented in θ at time $t \geq 0$, and $k(\angle(\theta - \theta_i))$ is the strength of the interaction between two filaments with relative angle $\angle(\theta - \theta_i)$. The following equation describes the dynamics of filaments in a given direction by gradual turning and alignment, and by rotational diffusion:

$$\frac{\partial F}{\partial t} = -\beta\, \nabla (F \cdot k * F) + \delta\, \triangle_\theta F, \quad (3)$$

where β is the binding affinity and δ, the diffusion rate. The first term in Eqn. 3 represents a continuous drift with angular velocity $k * F$. Assuming symmetry of turning to left and right, the kernel k, in this case, is an odd function.

The models above are only appropriate for the case of rapid (nearly instantaneous) interactions and reorientation (e.g. alignment) of filaments and cannot describe cases in which actin filaments are gradually pulled into alignment. The interactions of myosin with actin are of this type: myosin binds to two actin filaments and gradually pulls them into an aligned (anti-parallel) configuration. The model in Civelekoglu (1994) differs from the previous models in having drift terms (in angle). There, the *turning rate*, or the drift velocity, of an actin filament is assumed to be influenced by the total density of actin filaments which can interact with it via actin binding proteins, see the box, Eqn. 3. This represents the gradual reorientation of filaments which are attached via actin binding proteins, towards the "correct" configuration determined by the geometry of the actin binding protein.

3 Analytical and numerical results

By analysis of the models described above we wish to answer the following questions:

What is the behaviour of such systems on short time scales, e.g. can the homogeneous steady state be disturbed by small perturbations? If so, what would be the growing pattern, and what are the control parameters?

What is the long time behaviour of the above systems? Is there angular aggregation in form of bundles or networks, if so, which angles are favored?

Can transitions between various patterns (bundles or networks) occur? And if so, how sharp are these transitions and do they necessitate the breakdown of the original structure?

3.1 Formation of bundles and orthogonal networks

With linear stability analysis of all three models, criteria (commonly referred to as dispersion relations) determining the stability of the homogeneous steady state are derived.

It is found in each case that, *if* the "attracting forces" in the system *dominate* the "diffusive forces", small perturbations of the homogeneous distribution lead to the formation of bundles or orthogonal networks. Large binding rates, such as β_1, β_2 in the first model and β in the second and third models, mean strong attraction, while large diffusion coefficients δ and/or large dissociation rates γ counteract cluster formation. Furthermore, the sensitivity of the model to the parameters can be seen in the dispersion relation: some parameters appear in quadratic, and others in linear terms.

The dispersion relation predicts also the destabilizing mode number which is the number of clusters in the growing pattern. The type of the structure that forms is determined by how the filaments bind to each other, more precisely by the properties of the binding probability functions, k in the first and third model, and w in the second model (Civelekoglu 1994, Civelekoglu & Edelstein-Keshet 1994, Geigant 1997). For example, if filaments are more likely to bind at small angles (first model) or are attracted to each other's orientation (second and third models), bundles of filaments form, or if the binding occurs at nearly orthogonal angles, or if filaments repulse each other to such configurations, networks of filaments form. This analysis only reveals the relative orientation of the clusters. The actual direction of these aggregates depends on the initial distribution of the filaments and/or the orientation of pre-existing small bundles which act as aggregation nuclei.

Precise statements, e.g. on variance or mass of aggregates (the size of the clusters) and on the system's long time behaviour, can be made for parameters near

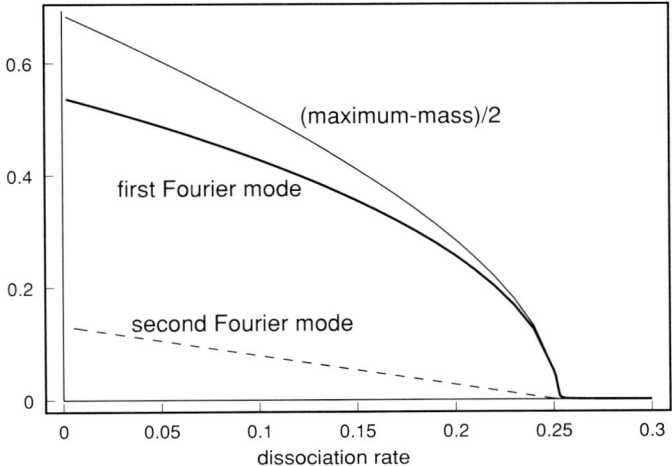

Figure 1: Bifurcation diagram for the second model. If the dissociation rate, γ, is too high the filaments are randomly oriented (all modes are zero). At the phase transition, i.e. near the critical dissociation rate (which can be calculated with linear stability analysis), the solution looks like a weakly oscillating cosine. This means that a loose bundle of equally oriented filaments results from the interplay of depolymerization and attraction. With decreasing dissociation rate the bundle tightens (higher modes are added to the dominant first mode) until a limiting thinness is reached for small (and zero) dissociation (the modes converge to limits smaller than 1; the maximum of the solution is finite). The first Fourier mode of the solution F bifurcates like $\hat{F}_1 \approx c_1(\gamma_{bif} - \gamma)^{1/2}$, the second like $\hat{F}_2 \approx c_2 \hat{F}_1^2 = c_2 c_1^2 (\gamma_{bif} - \gamma)$. In this example $w(\theta_o, \theta_n, \theta_i) = g_{0.1}(\theta_o - \theta_n - \frac{1}{2}(\theta_o - \theta_i))$ and $d(\theta_o, \theta_n) = g_{0.1}(\theta_o - \theta_n)$, where $g_{0.1}$ are periodic Gaussians with deviation 0.1.

phase transitions from unstructured to structured states (Mogilner & Edelstein-Keshet 1995, Geigant 1997). It can be shown that, in all models, on long time scales, the aggregates forming bundles (or orthogonal networks) will attain an equal mass. This bifurcation analysis uses eigenfunction expansion, a synergetic approach and/or asymptotic expansion. Fig. 1 shows a typical bifurcation diagram (calculated for the second model) using the dissociation rate γ as bifurcation parameter and an attracting interaction mechanism. With decreasing polymerization rate the meshwork of randomly oriented filaments changes into a loose bundle of equally oriented filaments. For smaller rates the bundle becomes tighter and tighter until a limiting width is reached where the diameter depends solely on the characteristics of turning.

For the third model explicit steady state solutions can be determined if the interaction kernel k is of sufficiently simple form (Civelekoglu 1994). These solutions agree with the results of both the linear and the nonlinear bifurcation analysis; i.e. the integer mode number that destabilizes the homogeneous steady state coincides with the number of peaks appearing in the non-homogeneous steady state solution.

It is possible that small deviations from the homogeneous state initially lead to meshworks with three, five or more bundles with almost equal angles between them. For this to occur, the interaction range has to be very small. The system's long time behaviour is then determined through "communication" between the peaks – in contrast to the individually based short time behaviour. In general this is rather complicated to analyse, however it is possible to carry out the analysis for some limiting cases, e.g. in the absence of diffusion in the first model or for small and/or exactly directed turning in the second model (Mogilner & Edelstein-Keshet 1996, Geigant et al. 1997). It is found that in most of these cases, the solution is a sum of delta peaks, i.e. infinitely small clusters. By perturbation analysis, it is shown that these peaks flatten due to the effect of weak diffusion or less exact turning properties.

All modelling approaches lead to the conclusion that microfilaments tend to organize either in orthogonal networks or bundles depending on the relative "effectiveness" of the orthogonal and parallel actin binding proteins, and the total mass of actin available. "Effectiveness" represents both relative binding rates and affinities as well as concentration of binding proteins when they co-exist in a mixture.

3.2 Transitions between structures

A minute change in some parameter(s) can result in a drastic change in the dispersion relation and may lead to a sharp switch of the dominant mode number, meaning a transition from one aggregation pattern to another. If e.g. bundling proteins are present in an assay as well as crosslinking proteins then small changes in their concentrations can cause a sharp transition from bundles to orthogonal networks or vice versa. Numerical simulations are in total agreement with this prediction (Civelekoglu & Edelstein-Keshet 1994, Geigant 1997).

In all models a switch between two structures is possible only under certain conditions, for example if the dissociation rate (of cross links) is high enough, or the diffusion rate is very high. In numerical simulations it has been observed that during the transitions from bundles to orthogonal networks two new clusters develop in the middle of the two existing ones. In the process of reformation half the filaments are dissociated from the bundles, reorient and bind anew. In contrast, during transitions from orthogonal networks to bundles, the initial structure is *fully dissolved*, all four clusters "disappear", then two new ones are formed, usually at or near the positions of the former ones.

4 Conclusion

Results of the above models indicate that the phenomena modelled here are robust to the modelling technique. The details of the mathematical approach, the number or type of equations, the specific forms of functions or kernels do not heavily influence the predictions of the models. Only, some general symmetry properties of the functions, representing the geometry of the molecules involved are essential.

As a final remark we would like to note the similarities of the above models from a mathematical point of view. The equations in all three models share the following properties:
a) They are mass conserving.
b) They have the same symmetry properties, namely all operators are $SO(n)$ - equivariant ($n = 2, 3$), this means that the solutions may be rotated (in space).
c) They all have quadratic non-linearities.
These similarities enable to use the same mathematical methods for analysis and also explain some of the similarities in their results, such as the transition between different types of structures.

There are more connections between the models which are not yet fully explored. We speculate for example that, in the case of fast diffusion and large binding rates, turnings are fast like jumps. Therefore some models of the second type may be derived from the equations of the first model. To do this mathematically one uses boundary layer methods and introduces a fast time scale for turnings of free filaments and a slow time scale for bound filaments (Mogilner& Geigant, unpublished). Also the transport-diffusion equation of the third model can be formally derived from the second model in the case of small turnings assuming suitable approximation arguments (Ladizhanski 1994). The reason is that several very small jumps look like a continuous movement.

Discussion & Open Problems

While it is clear that biophysicists and biomathematicians will need to work alongside cell and molecular biologists if we are eventually to obtain a reasonably complete explanation of the mechanisms of cell motion, it is not yet clear how this joint enterprise will most effectively operate. Various modelling proposals have been presented here to describe some of the phenomena and to explain some of the possible mechanisms of cytoplasmic motion and cell shape formation: the two interconnected topics of this Chapter. The discussions that took place in different sessions of the workshop revealed the differences but also emphasized the complementarity of the two classical methodologies: on the one hand, the statistical analysis of data (here of cell images) using new modelling ideas and, on the other hand, the analytical and numerical treatment of mathematical models (here of differential and/or integral equations) implementing recent estimates of parameter values from experimental data.

The difficulties and relative effectiveness of these two approaches are illustrated by the particular problem of pseudopodial/lamellipodial dynamics. Former statistical analysis of cell shape had been restricted to assigning, at each time instant, characteristic shape factors, such as elongation and dispersion, to a two-dimensionally mapped cell image (Dunn & Brown 1990, Hartman *et al.* 1993a) or even generalized to three dimensions (Inkielman & Doroszewski I.6). Now, the question of characterizing "pseudopods" as extending protrusions compared to stationary or retracting parts of the cell periphery, can be handled by either investigating the temporal changes extracted from the image series, or by distinguishing regions of protrusion or polymerization activity with the help of other criteria such as local dry mass density (based on optical path measurements) or immunofluorescent staining (Dunn *et al.* I.5, Coates *et al.* 1992, Hartman *et al.* 1993b, Keller & Bebie 1996). Some interesting approaches that should stimulate further experimental research and model refinement are the angular-temporal correlations between (an averaged) protrusion direction and the translocation vector as well as the typical oscillations that occur at both levels with periodicities in the range of several minutes (Vicker & Xiang I.3, Dunn *et al.* I.5, Brosteanu 1994, Hartman *et al.* 1994, Ehrengruber *et al.* 1996). It is becoming clear from these observations that periodic events and stochastic fluctuations in the peripheral activity (here particularly in the F-actin dynamics) determine the stochastic shape and path characteristics of locomoting cells.

The question at hand is how are modelling counterparts to these statistical results currently developing? So far, one class of models has been emphasizing the molecular aspect of F-actin assembly (and disassembly) such as the "Brownian ratchet" model (I.11), the diffusion-limited growth theories (Evans 1993) or models containing a complex of chemical and physical submodels, the most recent by Skierczynski *et al.* (1994). While these investigations also compute protrusion forces and speeds, a further class of dynamical models directly concentrates on the biomechanical aspect of F-actin networks. In particular, the mathematical analysis of cortical F-actin dynamics has been successfully explored and applied to various scenarios of cell shape deformation such as cell division, lamellipodial protrusion or pulsatile motion of plasmodia (He & Dembo I.7, Alt & Tranquillo I.9, Teplov *et al.* I.10). We are immediately struck by the power of these viscous fluid-like models of the cytoplasm and its interactions with the cortical plasma membrane complex, since they capture the main phenomena of self-organized motion, even though they consist of simple systems of partial differential equations that neglect many details of possible biochemical regulations. These approaches could possibly stimulate further solutions of the rather old problem of modelling the "budding of membranes" (Oster & Moore 1989) by covering the whole spectrum of scales from microscopic fluctuations of the plasma membrane alone, as in vesicles (Käs *et al.* 1993a), to the successive re-structuring of cortical statics and dynamics by intramembraneous and membrane-bound proteins (cf. Svetina *et al.* I.8).

Of particular importance are the cases of cortical interaction with integral membrane proteins mediating cell-substrate or cell-cell contacts, both of which are central modelling topics of Chapters II and III, respectively. Here we briefly mention the first problem of substrate adhesion within our context of actin dynamics. As experimental, quantitative data on the temporo-spatial dynamics of adhesion are gradually becoming available (see e.g. Schmidt *et al.* 1993, Dembo *et al.* II.2), so mathematical simulation models are slowly developing that include possible mechanisms linking the actin "motor" to its "feet", e.g. (DiMilla *et al.* 1991, Dickinson & Tranquillo 1993, Skierczynski *et al.* 1994, Li *et al.* 1994, Vereycken *et al.* 1995) and, most recently (Bottino 1996, Tranquillo & Alt 1997). So far, all attempts to combine the intracellular dynamics of the cytoskeleton with receptor kinetics (of diffusion and binding) still suffer from being quite phenomenological. Of particular interest with respect to the involvement of the cytoskeleton in signal transduction, however, is the emerging concept that adhesion-dependent signal transduction is mediated by contraction of the actomyosin meshwork which itself is moderated by the presence of microtubules (Bershadsky *et al.* 1996). Thus it may turn out that the mechanical and contractile properties of the actin-based meshwork are responsible not only for determining the cell's motility but also for regulating many other activities related to the growth of the cell. This raises the possibility that the induction of external stress fields, transmitting tension via the extracellular matrix, is a fundamental aspect of intercellular signalling – but this is a topic for the next Chapter.

Chapter II

Dynamics of Cell Interaction with the Environment

Coordinator: Robert T. Tranquillo

> *Die Mechanik lehrt den Ingenieur gewisse Gesetze kennen, die er bei seiner Tätigkeit in Anwendung zu bringen hat, wenn er mit möglichst geringem Material eine möglichst große oder aber bei gegebenem Material die größte mögliche Wirkung (Festigkeit etc.) erzielen will. Wenn wir also morphologische Gebilde, die mechanische Funktion erfüllen, derart organisiert antreffen, daß sie wie der Ingenieur die genannte Minimum-Maximum-Aufgabe lösen, daß sie wie von einer Intelligenz ausgeführt erscheinen, nennen wir sie mechanisch angepaßt, mechanisch zweckmäßig.*
>
> *Hans Driesch* (1891)

Introduction

Enormous progress in molecular and cell biology over the past fifteen years has led to the identification of many molecular components involved in cell motility. However, an understanding of the interrelationships between adhesion, traction, locomotion, and directed migration (each being a functional prerequisite to the next, with cell motility, the other prerequisite to traction, being the subject of Chapter I) is still a major challenge, even for cells on planar surfaces yet alone within complex tissue environments. In particular, an understanding of the physics of these processes, as well as their regulation by intracellular "second messenger" chemicals originating in extracellular signals, is far from complete. While by no means comprehensive, this Chapter contains a collection of papers that deal with aspects of each, using an assortment of approaches that span the spectrum from experimental to theoretical as appropriate for the specific objectives. A brief description of each contribution follows.

The paper by Grebecki (II.1) reviews the adhesion properties of *Amoeba proteus* and observations about their relationship to the pattern of pseudopod activity and cell locomotion on a rigid surface. The following paper by Dembo, Oliver, Ishihara & Jacobson (II.2) summarizes a quantitative methodology to map the pattern of traction exerted by a cell exhibiting locomotion on an elastic surface. While not explicitly revealing the interrelationships between adhesion and locomotion, such traction maps provide the data that, in conjunction with other measurements (e.g. the pattern of pseudopod activity and cell locomotion), will enable their elucidation. Vicker's paper (II.3) describes a series of experiments aimed at assessing the relative contributions of chemokinesis and chemotaxis to the directed migration of *Dictyostelium discoideum* in response to cAMP, depending on whether the cAMP is presented to the cells mainly as a wave-like pulse or a developing spatial gradient. The paper by Tranquillo & Alt (II.4) explicitly models the connection between the kinetics of the cell's sensory apparatus (stochastic receptor binding and diffusion at the cell surface) and the dynamics of its contractile cortex (morphogenetic assembly and flow of the actomyosin network in a peripheral cortical ring). Particular consideration is given to the mechanism and role of adaptation for the chemotactic response in temporal and spatial gradients, and also of adhesion. Contact guidance, the tendency of cells to align and move bidirectionally along a principal direction in an anisotropic medium, is the subject of a mathematical model presented by Dickinson (II.5, technical details to be found in Box II.6). In contrast to the mechanistic chemotaxis model, the contact guidance model provides a framework for characterizing cell tracks, although it can be connected to a mechanistic model as he has previously shown for haptotaxis and chemotaxis. Finally, the paper by Tranquillo & Barocas (II.7) incorporates the anisotropic diffusion model for contact guidance (and an anisotropic stress tensor for cell traction, accounting for another effect of contact guidance) into a model for fibroblast-driven dermal wound contraction, wherein the contraction induced by the fibroblast traction leads to the anisotropy (i.e. aligned tissue fibers).

II.1

Cell-Substratum Interactions of *Amoeba proteus*: Old and New Open Questions

Andrzej Grebecki (Warszawa)

1 Mechanisms of adhesion

Little is known about the mechanism of adhesion of free-living amoebae. Even though much has been learned about the motor functions of various proteins inside these cells, it remains conjectural whether there are specific molecular bridges binding their outside surfaces to the substrata, as well as which non-specific physical forces are involved in the interaction of both surfaces. In spite of the incomplete knowledge about the basic mechanism of adhesion, much more is known about relations of the motor functions of giant fresh water amoebae to their attachment to solid substrata. For example, in the case of *Amoeba proteus*:

1. Although the cortical actin cytoskeleton moves uniformly rearward in unattached cells, in adhering ones it is centripetally retracted toward the cell-substratum attachment sites (Grebecki 1984).

2. The absence of locomotion in enucleated amoebae has previously been attributed to a failure to adhere (Lorch 1969, Grebecka 1977). New observations indicate that the cortical cytoskeleton of enucleated amoebae is disorganized, but it can be repaired by cell attachment to highly adhesive substrata (Grebecka *et al.* in preparation).

3. Adhesion to the substratum is needed for phagocytosis, but it is missing during cation-induced pinocytosis (Opas 1981). Recently, it was clarified that the absence of adhesion is not a necessary condition but a result of pinocytosis (Klopocka *et al.* 1996).

Rough, low magnification pictures of *A. proteus* adhesions to glass were provided long ago by the side-view observations of Bell & Jeon (1963) and Grebecki (1976). The attachment sites appeared as blunt knobs, termed contact or supporting pseudopods. These were found to have transparent hyaline caps, similar to those previously seen in the tips of advancing pseudopods at the cells' front. In the expanding tips of advancing pseudopods, the actin cytoskeleton is incessantly withdrawn and restored beneath the frontal edge (Grebecki 1990). But the contact pseudopods have not yet been examined closely enough to determine if they present the same distribution and movements of actin.

The adhesions made by the tips of contact pseudopods have also been studied by reflection interference microscopy (Haberey 1971, Opas 1978), although the distances separating cell surfaces from the substrata were not measured. It therefore seems premature to claim the homology between these contacts in *A. proteus* and the focal contacts formed by many metazoan cells. However, with the same optical technique the cell-substratum gap was estimated to about 20 nm at the close contact sites of the soil amoeba *Naegleria gruberi*, and an analogy to focal contacts has been suggested (Preston & King 1978).

It should be emphasized that *A. proteus* and other giant fresh water amoebae differ from metazoan motile cells in several respects. Some of their special features may account for the differences in their adhesion and adhesion-dependent motor functions.

1. The absolute velocity of movement is much higher in *A. proteus*, than in all metazoan tissue cells. The relative locomotion rates may be similar (as in the case of neutrophils), but more often they are inferior in the tissue cultures. A fibroblast needs 1.5-2 hours to cross a distance equal to its own length, whereas an amoeba covers it 40 times quicker, within 2-3 min. Thus, the amoeba must more promptly establish and more quickly destroy contacts with the substratum.

2. Cell size of the *Proteus*-type amoebae exceeds by 5-10 times the size of motile metazoan tissue cells. This probably changes also the geometry of contact with the substratum, and the total area of close adhesion.

3. The glycoprotein coat on the surface of *Amoeba proteus* is extremely thick (> 200 nm). This always raised doubts whether the markers attached to the surface reflect movements of plasma membrane or only of the mucus. For example, location of concanavalin receptors, in the membrane of amoebae or in the coat, is still subject to discussion.

4. Fresh water amoebae migrate in nature on non-protein substrata, usually on fresh or degraded plant material. Neither are proteins present in the surrounding solution, in natural or artificial media. On the contrary, the adhesion of cultured metazoan cells depends on serum proteins and is usually studied in their presence.

5. The environment of all fresh water amoebae has a very low ionic strength. In the Pringsheim culture medium it reaches only 5.5 mM. It is 25 times lower than in the natural milieu and culture media for the motile cells of higher animals.

The absence of proteins in the medium and its low ionic strength contribute to the unexpected behaviour of amoebae interacting with substrates that differ in wettability and electrostatic charge. Most metazoan cells, in the presence of serum albumins, adhere better to polystyrene substrata bearing increased number of hydrophilic groups on their surface (Harris 1973, Curtis et al. 1986). Kowalczynska & Kaminski (1991) have produced styrene/methyl methacrylate copolymers and modified their sulfonation method, which made possible a calibration of wettability of the surfaces. The density of hydrophilic sulfonic groups is regulated by the proportion of styrene in the copolymer, because only styrene units undergo sulfonation. In fact, the wettability of these surfaces and, consequently, the number of L1210 cells (lymphocytic leukemia, mouse) adhering to them, linearly increased with the increase of styrene to methyl methacrylate ratio in the copolymers.

We have tested 4 types of these substrata: sulfonated and nonsulfonated copolymers containing 5% or 50% styrene units. It is known that floating cells of *A. proteus* have a radial symmetry, but recover the tail-front polarity and unipolar pseudopodial activity a few minutes after settling on the surface of glass. Cell morphology, rates of shape readjustment and locomotion resumption, as well as the eventual velocity of progressive movement, were the same on all copolymers, regardless of the great differences in density of highly hydrophilic sulfonic groups on their surface (Kolodziejczyk et al. 1995).

This indifference of *A. proteus* to the variable wettability of modified polystyrene substrata conforms well with old observations that this cell can move over solid paraffin and a more recent demonstration that *Naegleria* migrates on the fluorocarbon oil/water interface with a similar speed as on glass and untreated polystyrene (King et al. 1981). It is unknown why amoebae fail to react to the variable wettability of substrata.

On the contrary, *A. proteus* promptly and differently reacted in our experiments to a positively charged polylysine-coated substratum and to a negatively charged surface of gelatin gel (Kolodziejczyk et al. 1995). Coating of glass with polylysine raises its adhesiveness for amoebae, as for the cultured metazoan cells. All amoebae in a population adhere to the polylysine already after 3 min, whereas they attach only after 13 min to the untreated glass. The recovery of locomotive morphology on polylysine is complete, although the cells are flat and move slowly. On the other hand, amoebae never adhere to the gelatin gel. Only 25% of them regulate their shape and undertake directional pseudopodial activity, but their growing pseudopods fail to adhere and they slip against the substratum, so that they cannot move the whole cell.

The enhanced adhesion of *A. proteus* to polylysine and its failure to adhere to gelatin probably depend on the low ionic strength of the medium. In very dilute solutions, nonspecific repulsion or attraction forces extend a considerable distance from the charged surfaces into the medium and prevent or promote close contact of cells with substrata. Such effects of low ionic strength media on adhesion were earlier observed in fibroblasts (Curtis 1964), *Dictyostelium* (Gingell & Vince 1982) and *Naegleria* (King et al. 1983).

On the other hand, however, a problem arises of how amoebae adhere to and move over other negative substrata, such as glass and polystyrene. It may depend on competition between the nonspecific repulsive force and a specific molecular cell-to-substratum binding as was theoretically anticipated (Bell *et al.* 1984) and empirically confirmed in other cells, but not in amoebae. The expected molecular bridges connecting amoebae to the substratum are not yet revealed.

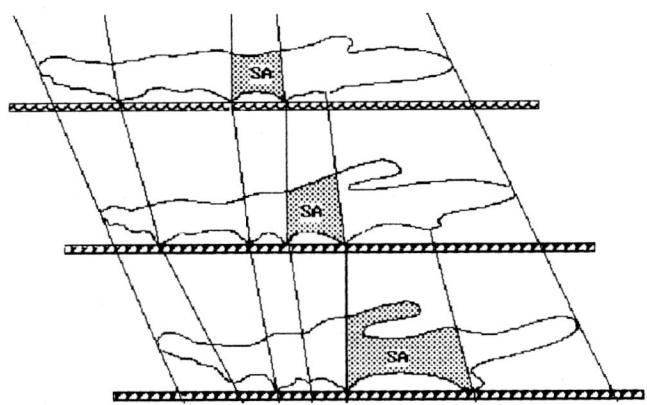

Figure 1: Schematic side view of the contact pseudopodia in *Amoeba proteus*. Note the stable adhesion zone (SA) where the pseudopodial tips do not change their position relative to the substratum, development of new contact pseudopods beneath the advancing frontal pseudopod, and gliding of old contact pseudopods behind the stable attachment zone.

2 Motile dynamics relative to attachment zones

Also the old side-view observations of *A. proteus* moving on glass raised a number of still unresolved problems. They do not directly concern the mechanism of attachment, but are more related to motor phenomena that accompany adhesion.

Contacts established by *A. proteus* with the substratum become stable for 30-60 sec and function during that time as footholds for the moving cell. This stable adhesion zone, where contact pseudopods do not change their position relative to the external reference points (Fig. 1), is usually located at about 1/3 of the body length behind the front (Grebecki 1976). It seems significant that vinculin, characteristic of focal contacts in metazoan cells, was also found at the same body region of *A. proteus* (Brix *et al.* 1990). Location of the attachment zone at 1/3 body length behind the front explains why, in average, just 2/3 of the amoeba's body shrinks and retreats, while the anterior 1/3 expands (Grebecka & Grebecki 1975).The ectoplasmic cylinder, which includes the submembraneous actin network, behaves in the same way: it is stationary for about 1/3 of the cell length, but moves at the front and at the rear (Grebecki 1976).

Behind the stable attachment zone only a few contact pseudopods may remain fixed to the substratum until they are broken by the retracting tail. Usually, the posterior contact pseudopods of quickly moving specimens start freely gliding over the substratum (Fig. 1). A similar distinction between stable and gliding contacts has been made by reflection interference microscopy (Opas 1978).

New pseudopods at the anterior cell region are unattached and extend forwards or obliquely upwards. Most often these new extensions arise beneath the older ones. Subsequently they bend toward the substratum. Gravitation is not responsible for their falling down, since they bend toward the substratum even if amoebae crawl in the upside down position (Nowakowska & Grebecki 1978). We do not know what other force or signal orientates growing pseudopods toward the substratum. The ce of 5-10 μm, is too large for the involvement of electrostatic forces. More credible is an asymmetric contraction of the new pseudopod. The cell cortex between the existing adhesion sites is probably under tension created by isometric contraction of cytoskeletal actin. One can then imagine a contraction wave propagated from older adhesion sites forward along the ventral side of a new pseudopod, which results in its bending toward the substratum.

On the upper side of amoeba, on the contrary, the contour of the front seen in side-view, moves upward. It occurs because new pseudopods develop underneath the older ones and they are not immediately inhibited by making contact with glass. The new pseudopods leaning against the substratum continue growing, so that they lift the frontal part of amoeba up (Grebecki 1977). Almost the same picture was drawn by Cheng (1992) as explanation of ruffling and retraction of lamellipodia in fibroblasts. Cheng observed "minipodia" which develop under the lamellipodia, lift them up and eventually replace them. It seems that in spite of morphological differences between amoebae and fibroblasts, their fronts obey the same rule: the new frontal structures arise and grow underneath the earlier leading edge.

In new contact pseudopods of amoebae the cortical cylinder is intensely retracted, which results in a strong fountain streaming of cytoplasm. It produces a picture strikingly resembling a frontal pseudopod making contact with prey. Later on, pseudopods leaning against the substratum start contracting between their attachment sites and the cell body. Simultaneously, they produce new free unattached tips which continue expansion directed forwards (Grebecki 1977). In reality, however, we do not know, when the tip of a contact pseudopod actually adheres to the substratum: at the moment of making first contact, or after stopping growth downward? Consequently, the temporal and causal relations between adhesion and contraction have to be explored.

3 Open questions

This brief review of interactions of free-living amoebae with the substratum may be summarized in the form of a list of some still pending questions:

1. Why are amoebae indifferent to the wettability of the substratum?
2. How do amoebae attach to the negatively charged surface of glass or polystyrene?
3. What sort of adhesion molecules bind amoebae to these substrata?
4. What force or signal orientates growing pseudopods toward the substratum?
5. Is adhesion established when the pseudopod makes first contact or after its growth stops?
6. How is the actin cytoskeleton organized and how does it function in the tips of contact pseudopods?
7. Is contraction a prerequisite of adhesion, or adhesion a signal to contract?

Acknowledgement Our recent results presented above were obtained under the provisions of grant 0453/P2/93/04 from the Committee of Scientific Research (KBN).

II.2

Imaging Traction Stresses

Micah Dembo (Boston)
Tim Oliver, Akira Ishihara and Ken Jacobson (Chapel Hill)

1 Introduction

In studies of cell locomotion and motility, a major goal has been to correlate the kinematic observations available by direct cytological methods and the underlying processes of cytoskeletal force generation (see Lauffenburger 1991, Lee et al. 1993, Evans 1993, Condeelis 1993, Cramer et al. 1993, Sheetz 1994, Lee et al. 1994). An important contribution to this program, Harris (1981), was the basic idea of observing the reactions of elastic substrata to the presence of locomoting or adhering cells. To implement this idea Harris and his coworkers developed a procedure for fabricating highly compliant yet stable films made of silicone rubber. They subsequently demonstrated that cells could induce large lateral displacements and wrinkling when placed on such substrata. Recently, some refinements to the basic method of Harris have been introduced in order to produce more reproducible and quantitative measurements of the elastic deformation (Oliver et al. 1995, Lee et al. 1994).

Measurements of substratum deformation are merely secondary reflections of the underlying biologically generated traction forces. Therefore, in order to deduce the actual traction stresses being exerted on a substratum at each location by cytoskeletal activity it is necessary to process the information inherent in displacement measurements by some sort of statistical procedure. Our approach to the calculation of such "traction images" starts by approximating the distribution of traction stresses acting on the substratum as a superposition of elementary "delta function" influences. Next, given a set of noisy displacement observations, we use standard methods of data fitting to find the most likely amplitudes and locations of the elementary influences. In this chapter we give only an abbreviated summary of needed techniques and the results obtained (a more detailed description appears elsewhere (Dembo et al. 1995)).

2 Theory

Consider a infinite flat membrane or plate composed of a uniform isotropic elastic material and introduce Cartesian coordinates $[x_1, x_2, x_3]$ such that the center of the membrane is coincident with the plane $x_3 = 0$. Presuming that the only external loads acting on the membrane consist of tangential tractions, T_α, on its upper surface, then one may consistently reduce the dimensionality of the problem by invoking a standard "plane stress" approximation (Landau and Lifshitz 1986, Dembo et al. 1995). If we consider that the membrane has some degree of elastic compliance, then as a result of the stresses it will deform slightly from its reference state as a result of the applied loads. In particular, a point in the midplane of the membrane with initial position x_α will be displaced to a new equilibrium position $x'_\alpha = x_\alpha + d_\alpha$. The coupled second order partial differential equations for the components of the displacement vector in the midplane are:

$$\frac{E_s}{1-\nu^2}\partial_1\partial_1 d_1 + \frac{E_s}{2(1+\nu)}\partial_2\partial_2 d_1 + \frac{E_s}{2(1-\nu)}\partial_1\partial_2 d_2 = -T_1$$
$$\frac{E_s}{1-\nu^2}\partial_2\partial_2 d_2 + \frac{E_s}{2(1+\nu)}\partial_1\partial_1 d_2 + \frac{E_s}{2(1-\nu)}\partial_2\partial_1 d_1 = -T_2 . \quad (1)$$

There are two constant moduli in these equations; ν the Poisson's ratio and E_s the *surface* Young's modulus. Although the latter will usually be a constant for any given film its actual value must be determined empirically by separate calibration measurements. This is because in addition to linear dependence on the membrane thickness, E_s can be a sensitive function of the degree of cross-linking and also of the amount of prestress. The surface Young's modulus of the standard elastic film preparation used in our studies of keratocyte locomotion was found to be 54 ± 15 dynes/cm by the pinch technique; see Dembo et al. (1995).

In the plane stress approximation, a constitutive law for the thickness averaged stress in the membrane can be easily obtained by combining the 3-D form of Hooke's law with the plane stress requirement and the condition of incompressibility;

$$\sigma_{\alpha\beta} = \frac{E}{(1+\nu)}\epsilon_{\alpha\beta} + \frac{E\nu(\epsilon_{11}+\epsilon_{22})}{(1-\nu^2)}\delta_{\alpha\beta} + \bar{\sigma}\delta_{\alpha\beta} . \quad (2)$$

In this expression, $\epsilon_{\alpha\beta} \equiv 0.5(\partial_\alpha d_\beta + \partial_\beta d_\alpha)$ is the strain tensor associated with this deformation field, and $\bar{\sigma}$ is the prestressed drumhead tension of the membrane. The latter is presumed to be constant. Although the drumhead tension has no direct bearing on the observable deformation produced by distributed tractions, it is still conceptually important to realize that any given film has some prestress. Doubly so because the amount of prestress can be controlled by changing the conditions of manufacture and because, if the prestress is small or negative, then one would expect an elastic film to become unstable towards the various wrinkling modes. For purposes of quantitative analysis, the latter are to be avoided, since they represent a violation of both linearity and the plane stress conditions.

For certain boundary conditions and traction distributions, elegant solutions for the fields of displacement that satisfy Eqn. 1 are known (Timoshenko 1934). For example if the elastic medium is *infinite* and if there are simple stress boundary conditions $\sigma_{\alpha\beta} \to \bar{\sigma}\delta_{\alpha\beta}$ as $|\mathbf{x}| \to \infty$ then the components of the displacement field can be expressed in integral form:

$$d_\alpha(\mathbf{x}) = \frac{1}{E_s} \int (g_{\alpha 1}(\mathbf{x},\mathbf{f})T_1(\mathbf{f}) + g_{\alpha 2}(\mathbf{x},\mathbf{f})T_2(\mathbf{f})) \, d\mathbf{f} \qquad (3)$$

where

$$g_{\alpha\beta} \equiv \frac{(1+\nu)^2}{4\pi} \left[\frac{(x_\alpha - f_\alpha)(x_\beta - f_\beta)}{|\mathbf{x}-\mathbf{f}|^2} + \delta_{\alpha\beta} \frac{(3-\nu)}{(1+\nu)} \ln\left(\frac{1}{|\mathbf{x}-\mathbf{f}|}\right) \right]. \qquad (4)$$

Physically, the nondimensional functions $g_{\alpha\beta}(x_1, x_2, f_1, f_2)$ can be thought of as giving the displacement in the α direction at location (x_1, x_2) induced by a delta function traction density acting in the β direction at location (f_1, f_2). In other words the $g_{\alpha\beta}$ are the components of a so called displacement Green's Function. Chapter 5 of Timoshenko (1934) gives more details.

Substitution of some special choices for the traction density reveals that for many loadings the integral required by Eqn. 3 will not exist. Physically this should not be surprising since it is certainly conceivable that under some conditions an unbounded elastic membrane with stress boundary conditions will slip tangentially and undergo an infinite displacement without reaching equilibrium. To avoid such pathology and thereby assure existence of the integral in Eqn. 3 it is sufficient that the traction density field should have bounded support and satisfy a constraint of global force balance;

$$\int T_1(\mathbf{f}) \, d\mathbf{f} = \int T_2(\mathbf{f}) \, d\mathbf{f} = 0. \qquad (5)$$

Fortunately it is easy to see that the traction field generated by a freely locomoting cell, normal to the gravitational field, and having no contacts with the external world except through the elastic substrate, will always satisfy Eqn. 5. Therefore, in the subsequent discussion we will concentrate strictly on the analysis of such freely locomoting cells. The important generalization to the case of cells locomoting against external loads requires additional work and will be dealt with in a subsequent publication.

3 Image analysis

We now turn to the question of what can be learned about the traction field acting on an elastic film from observations of the displacement field. In the interests of specificity we presume that a digitized image, with pixel resolution s, has been obtained so as to record the centroid location of a finite number of small marker particles imbedded in the undisturbed elastic material. We further presume that at some later time, a cell wanders into the observation field causing a small disturbance. The particle locations are then observed for a second time and for each particle the displacement relative to the undisturbed position is computed using image processing software. We let $\hat{\mathbf{d}}^p = (\hat{d}_1^p, \hat{d}_2^p)$ and $\mathbf{x}^p = (x_1^p, x_2^p)$ be the resulting *experimental* measurements of displacement and initial location of the $p-th$ particle $(p = 1, 2 \ldots P)$. Finally, we assume that the theory of the preceding section is applicable, that both E_s and ν are known and that the presence of particles in the film does not contribute a significant perturbation to its elastic behavior.

Since displacements are computed by subtracting a reference image from a disturbed image, it is naturally very important that the two images should share the same origin and orientation with respect to a set of fixed reference markers. The systematic error introduced into the measured displacements by misalignment of the two images is called a "drift correction". If we include both random error and translational drift error, then the observed displacement of the $p-th$ particle is given as the sum of three vectors;

$$\hat{\mathbf{d}}^p = \mathbf{d}^p + \mathbf{w}^0 + \mathbf{r}^p. \tag{6}$$

Here $\mathbf{d}^p = \mathbf{d}(\mathbf{x}^p)$ is the systematic displacement that would be obtained by a perfect measurement (this term could be computed using Eqn. 3 if we knew the exact traction density field being applied to the substratum by the cell). The second term to the right of Eqn. 6 is the drift correction and the third term is the random error. In our experience, translational drift is always present and must be carefully accounted for even though the actual size of the misalignment may be only a fraction of a pixel. This extreme sensitivity arises because the translational drift is the same for all the particles and therefor introduces a cumulative systematic error that gets progressively worse as the number of particle observations is increased. In contrast to the drift corrections, the r^p_α are independent random numbers which tend to cancel. We will henceforth assume that the r^p_α are all sampled from a single Gaussian distribution. We will further assume that the mean value of this distribution is zero and the standard deviation is equal to the pixel radius.

We now construct a space of finite dimension from which to select candidate traction images. For this purpose it is important to formally recognize the common sense idea that cells can apply tractions to a substratum only at points where they make some direct contact. This means that in constructing the space of possible traction images it is safe to assume that all the tractions occur within the boundaries of some sharply delimited region, Ω. Usually but not always Ω will correspond to the projected image of the locomoting cell. As a second condition on the candidate traction images it is also necessary to eliminate any image violating the constraint of global force balance, Eqn. 5.

Aside from such physical constraints, the T_α could conceivably be any continuous functions. To efficiently represent this large space of possibilities we will assume that the interior of Ω is paved by a mesh of quadrilaterals. To smoothly interpolate the T_α over our mesh we then define so called nodal "shape" functions, $S^k(\mathbf{f})$, all with the usual finite element property of C_0 continuity. We also require that each and all of the S^k should satisfy the "localization" conditions;

$$S^k(\mathbf{f}^j) = \delta_{kj} \quad \forall j = 1, 2, \ldots \mathsf{N}. \tag{7}$$

In the last equation, N is the number of nodes, $\mathbf{f}^j = (f_1^j, f_2^j)$ is the location of the j-th node, and δ_{kj} is the Kronecker delta. Finally, the total traction density distribution over the mesh is written as a sum over nodal tractions;

$$\mathbf{T}(\mathbf{f}) = \sum_{k=1}^{N} \mathbf{w}^k (S^k(\mathbf{f}) - \frac{A^k}{A_T}), \qquad \mathbf{f} \in \Omega . \tag{8}$$

Here

$$A_T \equiv \int_{\Omega} d\mathbf{f} , \tag{9}$$

is the total mesh area,

$$A^k \equiv \int_{\Omega} S^k(\mathbf{f}) \, d\mathbf{f} , \tag{10}$$

is the area surrounding the k−th node. The $\mathbf{w}^k = (w_1^k, w_2^k)$ are the mesh-associated degrees of freedom that implicitly encode the discretized traction image. In view of the definition of the A^k, one may easily verify by substitution into Eqn. 5 that the constraint of global force balance is exactly satisfied by Eqn. 8 for any choice of the $\mathbf{w}^k$; $k = 1, \ldots$ N.

Note that in our notation, the components of the drift correction vector, $\mathbf{w}^0 = (w_1^0, w_2^0)$, are equivalent to degrees of freedom associated with an imaginary zeroth node of the mesh. Consistency then requires that we define the null area and the null shape function as also being associated with the zeroth node. Accordingly, whenever necessary we will henceforth implicitly assume that it is legitimate for the nodal index to take on a value of zero and moreover that $S^0(\mathbf{f}) = 0$ and $A^0 \equiv 0$.

Finally, to avoid any possible confusion we should take this opportunity to point out that according to Eqns. 7 & 8, the traction density vector at the n-th mesh node is

$$\mathbf{T}^n \equiv \mathbf{T}(\mathbf{f}^n) = \mathbf{w}^n - \sum_{k=1}^{N} \mathbf{w}^k \frac{A^k}{A_T} = \mathbf{w}^n - \bar{\mathbf{w}} . \tag{11}$$

Thus in general $\mathbf{T}^n \neq \mathbf{w}^n$.

Let $\{X\}$ be the set of all possible outcomes of an experiment and suppose that the particular result X is obtained in an actual realization of the experiment. A "test hypothesis", H, consists of some guess for the unknown information necessary to generate a complete prediction of this experimental outcome. In the current context such a hypothesis consists of a definite choice of the w_α^k for $k = 0, 1, \ldots, N$, and, $\alpha = 1, 2$ (i.e. 2 N + 2 real numbers). Now, suppose that before conducting our experiment we have recorded an estimate of the probability that the test hypothesis is true. Letting $P(H)$ be this initial probability, we will symbolically denote the corresponding quantity *after* obtaining the experimental result X by $P(H \,|\, X)$. If $\{H\}$ is the universe of all the theories to be tested, then the hypothesis that affords the best explanation of the experiment, is defined by the element of $\{H\}$ that maximizes $P(H \,|\, X)$.

To compute $P(H\,|\,X)$ for arbitrary H and X we utilize the well known rule of inductive logic due to Bayes (Bernardo & Smith 1994). According to this result the post-experimental probability of a hypothesis is expressed as the product of three factors;

$$P(H\,|\,X) = P(X|\,H)\,P(H)S_X^{-1}\,. \tag{12}$$

The first quantity appearing on the right in this equation is interpreted as the probability of obtaining the experimental outcome, X, on the assumption that the test hypothesis is true. It is simply common sense to assert that when this quantity is large, our belief in the underlying hypothesis should increase in a more or less proportional fashion. Eqn. 12 also implies that the post-experiment probability of a test hypothesis will be proportional to the pre-experiment probability. Once again this is common sense since in practice our opinion of a test hypothesis prior to an experiment should be only partially modified as the result of a single experimental experience. Finally, the quantity S_X appearing on the right of Eqn. 12 is just a normalization constant chosen so that the integral of $P(H\,|\,X)$ over all $H \in \{H\}$ will be exactly equal to one.

The probability of a particular experimental outcome, given a test hypothesis, is identical to the probability of obtaining certain actual values of the error vectors by random sampling from an appropriate distribution. In the case of the current experiments, the errors in the particle displacement measurements are assumed to be Gaussian with mean of zero and standard deviation equal to the pixel radius. Thus, except for an irrelevant normalization constant, the probability of the observed experiment is

$$P(X|H) \propto \exp\left[-\mathcal{X}^2\right]\,, \tag{13}$$

where

$$\mathcal{X}^2 \equiv \sum_{p=1}^{P}\left|\frac{\mathbf{r}^p}{s}\right|^2 = \frac{1}{s^2}\sum_{p=1}^{P}\left|\hat{\mathbf{d}}^p - \mathbf{w}^0 - \mathbf{d}^p\right|^2 \tag{14}$$

is the familiar "chi-squared" statistic.

In view of the fact that the $\mathbf{d}^p$ must be evaluated from Eqns. 3, 4 & 8, actually computing the value of $P(X|H)$ using Eqn. 13 can require considerable mathematical effort. Nevertheless, at least the basic idea of Eqn. 13 is unambiguous and objective. In contrast, evaluating the initial probability of a particular test hypothesis, constitutes a more slippery and difficult topic. This is because $P(H)$ necessarily embodies various personal beliefs and biases, because these latter are not usually subjected to objective scrutiny and because there is in any case no absolute standard to apply.

To rhetorically lance the subjectivity of a given individual's prior belief, the classical method is to confront him with the imaginary or the simulated viewpoints of some small collection of trusted and neutral authorities. After first carefully explaining the biases and virtues of these ideal observers we faithfully demonstrate what each concludes from any given experiment and we invite consideration of these conclusions. We then appeal to the empathic facility of our species. If possessed of this ability, an individual should be able to extrapolate from the situation of the imaginary observers and thereby draw his own conclusions. For simplicity we will only consider the viewpoint of one ideal observer in the present discussion (hereafter this observer is designated O_1).

The first and most important characteristic of O_1 is that he believes in a form of Occam's principal. By analogy with Eqn. 13, this means that he always tries to rank the intrinsic likelihood of images according to the inverse of a positive functional called "complexity";

$$P(H) \propto \exp\left[-C^2\right]. \tag{15}$$

The specific definition of complexity used by O_1 is as follows;

$$C^2 = \int_\Omega \left(\frac{T_1(f_1, f_2)}{\lambda}\right)^2 + \left(\frac{T_2(f_1, f_2)}{\lambda}\right)^2 \frac{df_1\, df_2}{A_{tot}}. \tag{16}$$

According to this definition, the image of minimum complexity is the trivial or null image wherein the traction density is everywhere exactly equal to zero. Other images are then ranked according to the average magnitude (or L_2 norm) of the traction density. Readers who think that this definition of complexity is not to their liking are advised that a comparison of the images generated by O_1 with the results of an independent observer is presented elsewhere (Dembo et al. 1995).

The quantity λ, hereafter called the "complexity scale", is a positive real number with the same dimensions as the traction density. In essence λ is just O_1's a-priori "order of magnitude" estimate of the rms traction density of the minimal field that can be expected to produce a detectable effect on the particles under observation. An astute computation of the required quantity can be derived from a dimensional analysis:

$$\left[\frac{E_s s}{\lambda}\right]^2 = \sum_{m=1}^{N} \sum_{p=1}^{P} \frac{N}{2P} \sum_{\alpha=1}^{2} \sum_{\beta=1}^{2} \left[\int g_{\alpha\beta}(\mathbf{x}^p, \mathbf{f})(S^m(\mathbf{f}) - \frac{A^m}{A_T}) d\mathbf{f}\right]^2. \tag{17}$$

The final traction image obtained by maximizing $P(H|X)$ using this equation in conjunction with Eqn. 16 is the maximum likelihood hypothesis according to O_1. Note that according to Eqn. 17, λ depends on the mesh geometry and also on the undisturbed locations of the beads or particles being observed. The complexity scale does not depend on the actual motions of the beads and is thus a true a-priori estimate. Elsewhere we describe the computational deteails of the actual maximization proceedure and we describe Monte Carlo techniques in order to study the characteristic errors and biases of O_1's maximum likelihood image (Dembo et al. 1995).

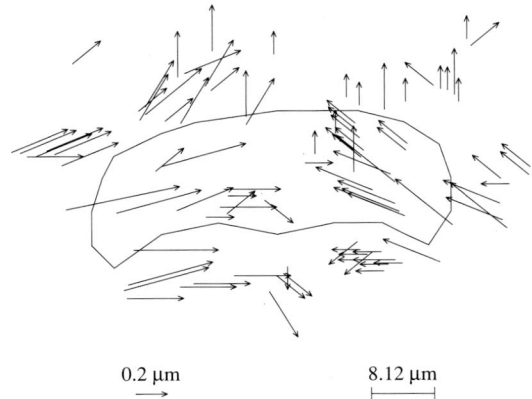

Figure 1: The domain of the cell-substratum contact for a fish epidermal keratocyte locomoting on the standard elastic silicone rubber film; see (Oliver et al. 1995) for details of experimental methodology. Observations of bead displacement are shown by arrows both inside and outside the domain of cell-substratum contact (n=95). The reference for undisturbed bead location was taken from analysis of this same location on the film several seconds before the cell appeared in view. The cell is moving in the upward direction at approximately 0.3 μm per second. The area of the cell-substratum contact is 656 μm^2.

4 Illustrative traction map

Details of the experimental approaches used for preparation of keratocytes, for preparation of elastic films and for collection and recording of data on bead displacements and cell shape have all been described previously (Lee et al. 1994, Oliver et al. 1995).

Fig. 1 shows the mesh representation of the image of a typical keratocyte moving in a straight path towards the positive Y-direction on a standard elastic film. The total area of the cell-substratum contact defined by the interior of this mesh is 656 μm^2. Experimental observations of particle displacements, (n=95), in the field surrounding the moving cell are shown by arrows. Fig. 2 shows the traction density field reconstructed by O_1 based on these data. The tractions of Fig. 2 are fully calibrated based on the best available estimate of E_s for our standard elastic film preparation.

The virtual observer reports tractions applied beneath the cell in a symmetric, pincer-like pattern, and orthogonal to the direction of locomotion. Traction densities are strongest at the two lateral margins of the cell. They diminish and approach zero at the cell centre. A consistent pattern of tractions acting in a rearward direction, appears just behind the nucleus. Relative traction densities at centre-front of the lamella were small and oriented somewhat randomly, so it is difficult to interpret them with confidence. The lateral-front edges of the cell are characterized by forward-directed traction densities. No forward directed tractions

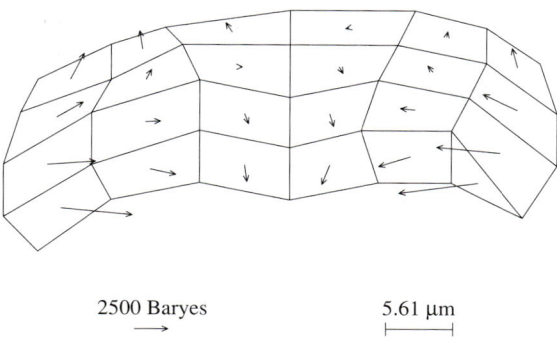

Figure 2: Traction image of a locomoting keratocyte. Mesh of 24 quadrilaterals defining the descretization of the domain of the cell-substratum contact for a fish epidermal keratocyte locomoting on the standard elastic silicone rubber film. At the center of each quadrilateral a vector is drawn indicating the maximum likelihood traction image reconstructed by virtual observer O_1. This reconstruction is based on the data of Fig. 1.

were observed at the rear of the cell. Since the traction fields of Fig. 2 are calibrated we are able to compute some interesting descriptive statistics. For example, the rms average of the cell-substratum tangential stress is 2.2×10^3 dynes/cm^2.

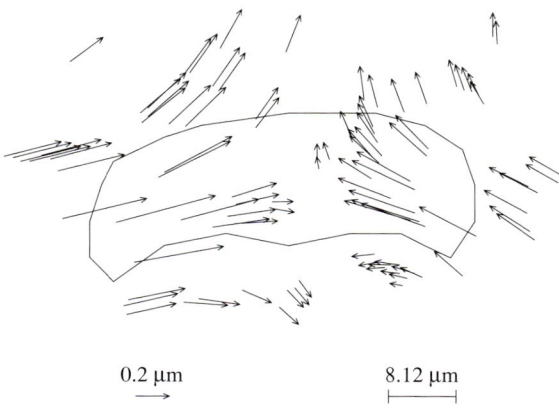

Figure 3: Maximum likelihood bead displacements computed according to Eqn. 3. The rms discrepancy between these theoretically predicted displacements and the displacements actually observed (see Fig. 1) was ± 0.093 μm. This error is consistent with the resolution of the optics.

Fig. 3 shows the theoretically predicted displacements of the particles computed using Eqn. 3 in conjunction with the maximum likelihood image of O_1. The rms error between theoretical displacement vectors in Fig. 3 and the experimental vectors in is ± 0.094 μm. For comparison the pixel resolution at the magnification used in this experiment is ± 0.092 μm. Thus the fit between theory and experiment is very close to the theoretical limit of the optics. In addition, we could detect no spatial correlations of either the directions or magnitudes of the error vectors (data not shown). We should emphasis that in our hands, this excellent fit between theory and experiment is quite typical. A more detailed discussion of the errors and nuances of traction images has appeared elsewhere, (Dembo et al. 1995).

5 Conclusion

It is of interest to check our current estimate of the gross area-averaged traction stress between a keratocyte and its substratum, with prior measurements of similar quantities. One such measurement is provided by the microneedle retraction study of Felder and Elson (1990). These authors reported that a stress of $\sim 5 \times 10^2$ dynes/cm^2 was exerted by a the leading lamella of a fibroblast in the action of retracting a calibrated microneedle adherent to its upper surface. Another approach to measurement of average traction stress is provided by the data of Usami et al. (1992) on the motion of a leukocyte into a capillary tube against counter-pressure. Since the contact area between the leucocyte and the capillary wall is about 2 times the cross sectional area of the capillary in Usami's study, the average tangential stress against the wall must be about the 1/2 the stall pressure. The result is then an estimated traction stress for the leucocyte of $\sim 8 \times 10^3$ dynes/cm^2. Remarkably, both these estimates of traction stress are in order-of-magnitude agreement with our own estimate, ($\sim 2 \times 10^3$ dynes/cm^2). One is therefore encouraged to believe that there is some broad consistency between these various determinations of cell generated traction stress despite big differences in the experimental details.

According to our results a locomoting keratocyte exerts a characteristic traction force of $\sim (2.2 \times 10^3$ dynes/cm$^2) \times (6.5 \times 10^{-6}cm^2) = 1.4 \times 10^{-2}$ dynes. For comparison with our total force estimate, recent measurements indicate that the force exerted by single myosin motor is only 10^{-7} dynes, (see Ishijima et al. 1991). Therefore, it would seem that $\geq 10^5$ myosin motors would be required to produce the tractions we observe. Actually this is probably a rather conservative lower bound since the force generating efficiency of myosin in realistic practice is apt to be very low. In any event, based on studies of amoebae (Clark & Spudich 1974, Warrick & Spudich 1987) it seems to exist adequate conventional myosin (half of total protein mass) to generate the required force.

Acknowledgement Work was supported by NIH grant GM 35325 (to K.J.) and NIH grant AI21002 (to M.D.).

II.3

Chemotaxis and Chemokinesis of *Dictyostelium* Amoebae: Different Accumulation Mechanisms Induced by Temporal Signals and Spatial Gradients of Cyclic AMP

Michael G. Vicker (Bremen)

1 Introduction

Chemotaxis, a primary feature of eukaryotic cells, also determines cell behaviour throughout the development of *Dictyostelium discoideum* (Alcantara & Monk 1974, Darcy & Fisher, 1990). The reaction in post-aggregative-stage cells is regulated by soluble signals of cyclic AMP (cAMP), which are rhythmically emitted as ca. 1 min pulses each 5–8 min (Gerisch & Hess 1974). Reception of the cAMP signal induces the cell to autocatalytically synthesize and emit cAMP (signal relay) and to simultaneously release a specific cAMP phosphodiesterase, which rapidly destroys the cAMP signal (Shaffer 1975). Rhythmic signal relay and the cell's oriented locomotory response lead to the self-organization of dynamic 2-D waves of cell locomotion during cell aggregation, e.g. during culture upon an agar surface, and later within the slug. The two and three-dimensional wave patterns resemble the oscillatory concentric and spiral scroll-waves, respectively, generated in chemical reaction-diffusion systems, notably the Belousov-Zhabotinsky reaction (e.g. Müller 1988). The present contribution addresses the mechanisms of chemotaxis, slug patterning and locomotion, and proposes an alternative to models which assume the involvement of a cAMP spatial gradient in chemotaxis.

Intracellular cAMP induces chemokinesis and chemotaxis in *Dictyostelium* cells. Kinesis refers to the effect of the cAMP concentration [cAMP] on the amplitude of cell motility parameters, e.g. velocity, directional persistence, etc. (Fraenkel & Gunn 1961). Chemotaxis is an oriented turning-and-movement response of a cell to an oriented chemical signal, and is a qualitatively different behaviour from chemokinesis: each form of behaviour being regulated by a qualitatively different cAMP signal (Swanson & Taylor 1982, Vicker 1994). The nature of the induced cell behaviour and the identity of the inducing signal each requires careful investigation. Thus, cAMP diffusion generates two possible signals: a spatial and a temporal concentration gradient, which may be depicted as $\partial[cAMP]/\partial x$ and $\partial[cAMP]/\partial t$, respectively, in order to concretely describe their qualitative and quantitative difference with respect to time and signal molecule concentration at the cell (Vicker 1989). Furthermore, cell accumulation may be induced by mechanisms other than taxis. Unfortunately, cell behaviour and signal quality are usually conflated into an undifferentiated "chemotactic gradient" (e.g. Futrelle 1982, Fisher et al. 1989). The evidence usually presented for spatial gradient-reading mechanisms of taxis is weak and unacceptable, because it indiscriminately involves either non-tactic accumulation behaviour or responses to signals which contain both temporal and spatial concentration gradients. The experiments reported here attempt to differentiate a) the temporal and spatial components of developing gradients and b) the tactic and kinetic elements of cell behaviour.

2 Methods

Dictyostelium discoideum NC-4(H) cells were cultured 48 h on SM agar with *Escherichia coli* as nutrient. Cells were washed free of bacteria in potassium phosphate buffered salts solution (PPBSS, pH 6.5) and incubated 16 h at 6°C in order to develop cAMP sensitivity at the early aggregation stage (Vicker et al. 1984). 5×10^5 cells in 0.5 ml PPBSS were subsequently seeded onto each flat surface of a micropore filter (5 μm pore size, 150 μm thick, 13 mm diameter, which supports free cell locomotion). The filter was held between two chambers in a plexiglass block, which was placed horizontally while being loaded with cells and later set upright (Vicker 1994). If the law of diffusion is to operate as the sole determinant of gradient development and turbulence be suppressed, the gradient needs to propagate in a high viscosity medium, a gel or a micropore filter (Vicker 1981). Experiments began with an even population distribution, induced by infusing the filter for 30 min at 21°C with PPBSS and then for 30 min with 1 nM cAMP in PPBSS including 2 mM dithioerythreitol (DTE) to inhibit cAMP hydrolysis by cAMP phosphodiesterase.

In order to generate a positively developing cAMP gradient, i.e. a signal pulse, across these filter-bound populations, they were washed free of and deadapted to cAMP by incubation in PPBSS including 2 mM DTE and 2 mM caffeine (to inhibit cAMP relay, Siegert & Weijer 1989) for 20 min (5 min being about the nominal deadaptation time in vivo) and then cAMP was added to the left filter side for 10 s before being washed away. The cells were fixed in 5% formalin after 10 min incubation in PPBSS alone. Cells in other filters were pre-incubated in isotropic [cAMP] for 30 min and then exposed to a negatively developing gradient, which was generated by replacing the cAMP on one side of the filter with PPBSS. Spatial gradient stability was maintained by infusing PPBSS, with or without cAMP, respectively, into each compartment for 60 min (Vicker 1994). Spectroscopy of analogous neutral red gradients demonstrated that gradient steepness became virtually linear within 10 min at a ratio of larger than 100 : 1 between the dye in the left chamber and the PPBSS in the right chamber and approached a steepness of 6.7% across a 10 μm length, which is equivalent to the length of a cell (Vicker 1989). Filters were fixed in formalin, stained in Giemsa and mounted in Permount (manufactured by Fisher, Philadelphia). Cell distributions were analysed using 18.75 μm-deep optical sections, beginning ca. 8 μm beneath the filter surface at five to six fields/filter. The median position of the cell population in each filter was calculated from the 1-D cell distribution in grouped form. Its significance was tested by the Wilcoxon signed rank test where Xnorm is less than -2.56 (larger than 2.56) indicates a shift to the left, i.e. up-field, (right, i.e. down-field) at the 1% level of significance.

Cells in 2 μl of PPBSS containing various [cAMP] were spread between a glass slide and coverslip separated by a ca. 50 μm Parafilm ring. Video images of 5-15 cells were digitized every 6 s for 5 min only if the cells were motile, did not leave or enter the field or collide. Grey-value centroid tracks of 33-47 cells were evaluated at each cAMP concentration in each of three separate experiments. For each cell and cell ensemble, the reciprocal root mean squared (*rms*) displacement per step was plotted versus the reciprocal of the step duration. The *rms* directional persistence time (P) and speed (S) were then calculated from a least-squares fit line to this plot. Its slope estimates 1/S and its x-intercept estimates -1/(6P), the average cell diffusion rate or *rms* motility $R = 2 \times P \times S^2$ (Dunn 1983).

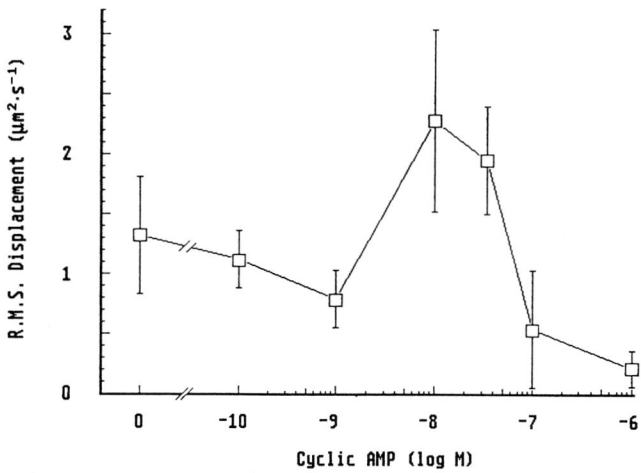

Figure 1: *Dictyostelium* cell motility in various isotropic cAMP concentrations. Motility (dispersal) is expressed as $R = 2 \times P \times S^2$, with P, persistence, and S, speed. 33-47 cells were measured at each cAMP concentration. The vertical bars indicate s.e. (from Vicker 1994).

3 Results and discussion

Cell motility shows a triphasic dependence on [cAMP] to which the cells were adapted, with a primary optimum between 10-30 nM isotropic cAMP (Fig. 1). The optimum is similar to values reported by many others (e.g., Fisher *et al.* 1989). The migratory behaviour of cell populations was examined after exposure to stable spatial gradients of these [cAMP]. However, a brief temporal gradient is also generated during spatial gradient development by diffusion, either positive or negative, depending on how the spatial gradient is produced. Delivery of an absolutely time-invariant spatial gradient is virtually impossible. But in order to eliminate the effects of positive cAMP pulses and assure that cells experienced only spatial concentration gradients without such pulses, the populations were exposed to gradients which developed negatively. Negative gradients are generated naturally as the "back of the cAMP wave" during cell aggregation and, thus, do not interfere with the perception of the chemotactic signal in the next wave a few minutes later. Both positively and negatively developing gradients become linear within less than 30 s in micropore filters (Vicker 1989). The effects of a) an isotropic or b) spatial concentration gradient or c) a single, 10 s pulse of cAMP may be compared on cell behaviour (Fig. 2). The reaction to 10 nM cAMP, which is very near the concentration for optimal cell motility, was different depending on how the cAMP was delivered. Spatial gradients induced slow, down-field accumulation, but a solitary cAMP pulse induced quite rapid accumulation up-field.

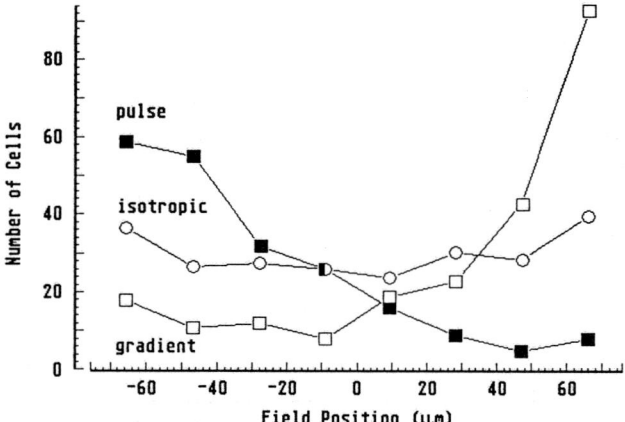

Figure 2: Cell distributions in a micropore filter. Cells were exposed for 60 min to 10 nM cAMP infused either as an isotropic field (circles) or spatial gradient (open squares), which had developed negatively with the sink on the right. Cells were treated with a cAMP pulse (closed squares) from the left for 10 s then washed and allowed to wander in cAMP-free medium for 10 min. The initial cell distributions were even (from Vicker 1994).

Perhaps the most surprising new finding of all was that negatively-developed spatial gradients induced cell accumulation either up- or down-gradient or no accumulation at all depending on the actual [cAMP] applied up-field (Table 1). 100 nM and 1000 nM cAMP induced no accumulation, confirming previous observations (Vicker et al. 1984). 1000 nM is thought to be the [cAMP] in the natural wave (Tomchik & Devreotes 1981). Gradients of 1 nM induced up-field accumulation, but cells in 10 nM gradients accumulated particularly strongly down-field.

		[cAMP] (nM):			
	control	1	10	100	1000
gradients:					
real					
median (μm)	−3.01	−32.10	53.81	3.03	−12.02
Xnorm	−0.80	−4.21	−3.11	0.53	−1.07
simulated					
median (μm)	–	−14.48	8.68	−25.19	−3.06
Xnorm	–	−8.07	6.75	−9.51	−2.99

Table 1: Effects of spatial gradients of cAMP on *Dictyostelium* cell motility. The median position of the cell population distribution in cAMP spatial gradients, which were developed negatively, is compared to that in simulated gradients. The simulated gradients were constructed from motility data of cells in isotropic [cAMP] (Fig. 1). The highest [cAMP] is positioned on the left. Negative median values indicate accumulation to the left, i.e. up-field. The control cell population was not exposed to cAMP. The data for the latter was derived from that in Fig. 2. Xnorm values larger than |2.56| indicate that the median differs from null at the 1% level of significance. The gradient and simulated gradient populations were treated identically for these calculations. Values are in micrometres.

In order to test whether the cell distribution in spatial gradient fields is a sole function of the ambient [cAMP], each axis of the data in Fig. 1 was replotted as its inverse in analogy to Fig. 2. Thus, the ordinate cell "motility" (random diffusion or *rms* motility, R) is expressed as cell "accumulation" (1/R) and the abcissa [cAMP] is expressed within a micropore field position in a simulated linear gradient (Fig. 3). The results demonstrate a strong qualitative resemblance to the original cell behavioural data in gradient fields (Table 1, cf. the effect of 10 nM cAMP gradients in Figs. 2 & 3C). Although 100 nM cAMP appears as an exception, this particular gradient is the only one that contains the cell motility optimum of ∼10 nM cAMP close to mid-field. Therefore, the displacement of the population is likely to be especially sensitive to the relative field position of this concentration. The results in Table 1 indicate that the mechanism of accumulation or non-accumulation, respectively, in cAMP gradients between 1 to 1000 nM is chemokinetic, because cells meander longer (i.e., are "trapped") in [cAMP] regions

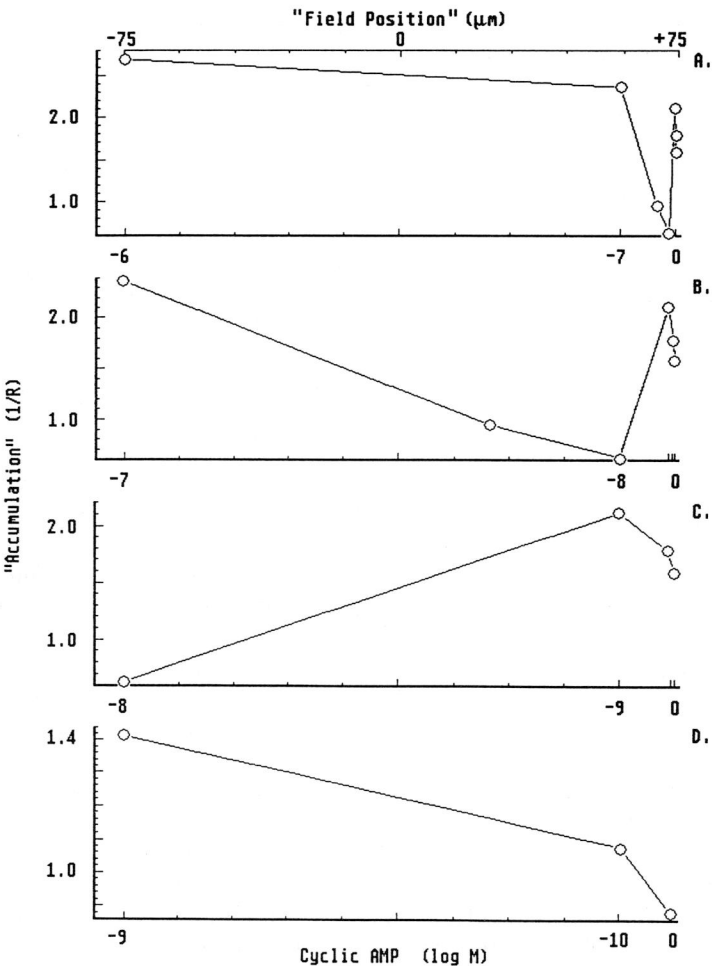

Figure 3: Simulated cell distributions in linear spatial cAMP gradients. The data are taken from Fig. 1 and plotted as in Fig. 2. The ordinate represents "cell accumulation" and is derived from inverted motility (1/R) from Fig. 1. Each of the 4 curves, A-D, is constructed from an inverted segment of the curve in Fig. 1 selected from one point between 1 to 1000 nM cAMP and 0 nM. Thus, "B" is the "expanded", lowest tenth part of curve "A" and so forth. The [cAMP] field is idealized as a spatial gradient in a micropore filter (from Vicker 1994).

where their motility is relatively low and they tend to rapidly evacuate regions where their motility is relatively high. Although any ongoing cell accumulation must necessarily generate some degree of cell orientation in the population, the results provide strong support for the conclusion that only random locomotion, not oriented turning, is involved and that cells react to the ambient [cAMP], but the spatial gradient remains imperceptible to them and is therefore not an orienting signal. Thus, chemotaxis is a qualitatively different cell reaction, because it specifically requires a sharp, oriented cAMP pulse. This conclusion is supported by 13 experiments recording the distributions of nearly 1.4×10^4 cells (Vicker 1994).

Reasons for the difference between reactions to pulses and spatial gradients of cAMP may exist at the molecular level of the cell locomotor. A cAMP pulse induces a transient excursion followed by a rapid but brief oscillation in both the intracellular F-actin concentration (McRobbie & Newell 1984) and in F-actin autowave propagation (Vicker et al., submitted; Fig. 3 in Vicker & Xiang, I.3 this volume). These adaptive processes are neither induced by isotropic fields nor spatial gradients of cAMP. Cells are, rather, fully adapted to the ambient [cAMP] in static fields. Only sharp temporal cAMP signals effect the intrinsic F-actin assembly oscillations and the derivative cell surface oscillations by inducing an initial phase resetting of the oscillatory cell shape modes to mode 0, i.e. the cell rounds up (Killich et al. 1993, 1994). We have suggested that this F-actin reaction follows the cAMP wavefront as it sweeps across the cell and that a wave of recovered capacity for F-actin-dependent pseudopodium generation will follow close behind. Our interpretation of actin behaviour is that the chemotactic pseudopodium will appear on the side of the cell which had reacted first to the initial contact with cAMP, slightly ahead of the distal cell volume (also Vicker et al. 1984).

Timing principles, i.e. adaptive reactions and oscillations, are intrinsic to the chain of molecular reactions leading to chemotaxis. Many systems gain advantages and reduce signal reading errors by communicating periodically rather than through molecular concentration alone (Li & Goldbeter 1992, Harrison & Lacalli 1993). Thus, hormonal communication in various systems depends on temporal signalling (Belchetz et al. 1978, Dietl et al. 1993). Non-linear signalling mechanisms, based on biological time, assure that cells bear a full complement of sensitive receptors coordinated with optimal secretion frequency and optimally (read "economically") efficient signal concentrations. Biological time (Winfree 1990) means that cellular reactions are geared to the intrinsic, system-specific dynamics of intracellular processes, rather than being simple, numerical products of receptor-ligand binding kinetics and signal concentrations. Thus, the motility and intracellular F-actin assembly rates of *Dictyostelium* and neutrophils are both relatively higher than those in fibroblasts (Zigmond 1993).

II.4

Receptor-mediated Models for Leukocyte Chemotaxis

Robert T. Tranquillo (Minneapolis)
Wolfgang Alt (Bonn)

1 Introduction

It is well established that leukocytes like polymorphonuclear neutrophils (PMNs) exhibit chemotaxis in a spatial concentration gradient of a chemotactic factor (CF), that is, biased extension of lamellipods in the gradient direction leading to biased "turning" and orientation during locomotion. Two central and inextricably related questions remain unanswered: what are the molecular mechanisms by which a PMN senses and responds to a spatial gradient, albeit imperfectly, and how does the PMN retain the ability to do so over a large range of concentrations as it migrates in a meandering fashion up a gradient? While it is clear that the answers ultimately lie in the emerging details of receptor-mediated messenger generation and consequent modulation of the actin-based motility system of the cell, they will require an integrated understanding of these biochemical and biophysical phenomena. Our goal here is to examine these questions together from the viewpoint of mathematical models which facilitate such integration. The reader is directed to Tranquillo (1990) for an overview of models focused on stochastic receptor signaling (as the origin of the stochastic cell track during chemotaxis) in order to elucidate the sensing-response mechanisms and to Alt (1994) for more discussion of the pseudo-spatial mechanism in particular (see also below) and other possible sources of stochasticity.

Observations of PMN behavioral and biochemical responses to various types of CF concentration gradients, namely, continuous temporal gradients and step temporal gradients (being spatially uniform macroscopically in both cases), and quasi-steady spatial gradients are highly informative (see Tranquillo (1990) for a review of relevant observations and further references). As shown by Sally Zigmond and colleagues (e.g. Zigmond & Sullivan 1979), step temporal gradient observations clearly indicate the property of behavioral adaptation: when subject to a step increase (decrease), the polarized cell morphology is lost due to transient extension (retraction) of lamellipods from the entire cell surface before being recovered as they become limited to an emerging leading lamella. The capacity for biochemical adaptation was shown by Larry Sklar and colleagues (e.g. Sklar & Omann 1990): when subject to a continuous temporal gradient, intracellular markers of stimulation become elevated if the temporal gradient is above a threshold. Taken together, these observations support the hypothesis that PMN chemotaxis in a spatial gradient reflects an adapting pseudo-spatial mechanism: lamellipods that extend and experience a sufficient increase in CF concentration continue to extend up to some limit determined by cytomechanics, otherwise the rate of adaptation exceeds the rate of stimulation and they are retracted. "Pseudo-spatial" refers to the consequence that the cell senses a spatial gradient by a mechanical "competition" between lamellipods separated in space, each lamellipod being regulated by competition between stimulation and adaptation processes largely, but not necessarily entirely, within itself (Lackie 1986).

Stimulation is associated with the "receptor-measured temporal gradient" (which could involve a spatial component within the lamellipod) as the lamellipod extends, depending on the orientation of the lamellipod relative to the spatial gradient, the rate of lamellipod extension, and receptor binding kinetics. A key receptor event leading to stimulation proceeds through formation of a ternary complex of CF-receptor G-protein and subsequent activation of protein enzymes (and subsequent generation of second messengers) by subunits of the dissociating G-protein. The identity and mechanisms of the adaptation processes are unknown, although there are many candidates such as reversible conversion of receptors from a signaling to non-signaling state (e.g. by sequestration of receptors from G-proteins or down-regulation of the total number of surface receptors by an increased rate of receptor endocytosis).

2 Receptor-mediated adaptation models

While a complete model for receptor-mediated adaptation has not been validated for PMN (see Lauffenburger & Linderman (1993) for an assessment of the specific case of the well studied chemotactic peptide receptor based on the research of Larry Sklar and colleagues), a generic model based on a receptor interconverting between two free and two bound states that yields any desired degree of adaptation has been proposed, where the adapting signal is defined as a weighted sum of the four states (Segel et al. 1986) (the two "free" states might represent in the PMN case a free receptor and a receptor precoupled to G-protein). This model did not account for spatial nonuniformity that is relevant for chemotaxis, nor has it been appropriately extended (e.g. accounting for diffusion and/or convection of the various species) and incorporated into a model that predicts a chemotactic response.

Another adaptation scheme that has been incorporated into chemotactic response models was proposed by Zigmond (1980), wherein a bound receptor rapidly forms an activator, converting an inactive enzyme to an active state, and more slowly forms an inhibitor, which binds to the active enzyme rendering it inactive. By further assuming that the activator was nondiffusible and the inhibitor was highly diffusible, with all the species being in a quasi-steady-state, Tranquillo & Lauffenburger (1986) derived an adapting signal, $s^* = b/\oint b$, where b is the local density of bound receptors and the integral is over the entire cell. Clearly, in a uniform CF concentration s^* is constant independent of the concentration (i.e. complete adaptation after a transient in s^*), whereas in a spatial gradient of CF, a spatial gradient of s^* is induced within the cell. It was shown that the use of s^* rather than the simple nonadapting signal, b, altered the predicted cell response both in the presence and absence of receptor binding fluctuations. However, the response model was only probabilistic, based simply on the existence of a threshold spatial difference in b or s^* that must be exceeded in order to elicit cell orientation with no account for lamellipod extension; it was not stochastic, which allows for continuous time realizations of the response, as in more recent models (Tranquillo & Lauffenburger 1987, Moghe & Tranquillo 1994, 1995), which have also considered mechanisms for adaptation at the receptor and second messenger levels.

3 Pseudo-spatial cell motion models

The phenomenological model for a pseudo-spatial mechanism proposed by Alt (1994) is based on an equation for the rate of change of lamellipod length $L(t, x)$ oriented with angle x with respect to the gradient direction:

$$\partial L/\partial t = (local)\ protrusion\ rate - (local + global)\ retraction\ rate \qquad (1)$$

where the parenthetical qualifiers indicate the suspected origin of the rate regulation. The protrusion rate is assumed to be a (generally nonlinear) function of the experienced (temporal) CF gradient on the lamellipod. Moreover, a probability distribution for lamellipod extension relative to the cell polarity direction is specified, reflecting the observed decrease in lamellipod frequency away from the leading lamella. The integral of L weighted by this distribution over all directions would then capture the competition between lamellipods. We seek here to relate this phenomenological model to a stochastic cytomechanical model we have developed (Tranquillo & Alt 1996, referred to as TA in this paper). The cytomechanical part of this model appears as the limiting case of a more general cytomechanical model derived and analyzed in this volume (Alt & Tranquillo I.9, referred to as AT in this paper, Eqns. (4)–(7)). The notation of TA is used below (where cortical actin is defined as the variable a rather than a_B as in AT)

We modeled the local concentration (or volume fraction) of cortical actin, $a(t, x)$, for each angular location, x, around the cell periphery with the mass balance equation

$$\frac{\partial a}{\partial t} + \frac{\partial (av)}{\partial x} = \eta(a^* - a) \qquad (2)$$

where a^* is a chemical equilibrium concentration around which a varies with first-order rate measured by the assembly/disassembly rate constant, η, and v is the velocity of cortical flow (tangentially) around the cell periphery determined by a mechanical force balance (see Eqn. (4) of TA, or Eqn. (7) of AT). In TA, we analysed how stochastic fluctuations in the density of bound receptors around the periphery of the cell surface induced fluctuating patterns of the distribution of cortical actin in the case of a uniform CF field. Our present interest is to relate these dynamic patterns to cell shape changes associated with lamellipod extension, to assess the effect of a spatial gradient of CF, and to take account of adaptation. An in-depth study of the first two topics, including associated cell movement, will appear elsewhere (Tranquillo & Alt 1997).

Rather than model the intrinsic dynamics/mechanics of lamellipod extension from the cortex in angular direction x (i.e. use the general mechanistic model of AT), we assume that lamellipod extension is simply a linearly decreasing function of a (this is a simplification of the *ansatz* in AT that lamellipods are relatively longer near the leading lamella where the cortical actin concentration is relatively higher). Then we have:

$$\frac{\partial L}{\partial t} \propto -\frac{\partial a}{\partial t} = \eta(a - a^*) + \frac{\partial(av)}{\partial x}. \qquad (3)$$

Assuming that a^* is some function of the (stochastic) density of bound receptors on the lamellipod, $b(t, x)$, that accounts for the stimulation and adaptation processes associated with receptor binding, then a linearization of this function accounting for small deviations of b from its expected value $<b>$ yields (see Eqn. (16) of TA):

$$\frac{\partial L}{\partial t} = \eta(a - [<a^*> + \gamma(b - <b>)]) + \frac{\partial(av)}{\partial x}. \qquad (4)$$

Assuming the signal transduction parameter γ is negative, meaning that an increase in bound receptors leads to longer lamellipods and therefore a depletion of cortical actin per the *ansatz*, then Eqn. (4) can be rewritten as

$$\frac{\partial L}{\partial t} = (\eta(a - \gamma(b - <b>))) - \left(\eta <a^*> - \frac{\partial(av)}{\partial x}\right) \qquad (5)$$

where all terms are positive in the region of the leading lamella for positive fluctuations in b, see Fig. 4 of TA showing that in this region the mean net assembly is negative, at least in a pseudo-steady state approximation, and thus by Eqn. (2) also the term $\partial(av)/\partial x$. Comparing Eqn. (5) with Eqn. (1), and noting that b is a function of lamellipod length in a spatial gradient, we have thus identified the desired relation: the first bracketed term in Eqn. (5) is the (local) protrusion rate of Eqn. (1), which depends on $L(t, x)$ and thus the (temporal) gradient experienced by the lamellipod, whereas the second bracketed term in Eqn. (5) is the (local + global) retraction rate, which involves a local cortical actin disassembly term and a global cytomechanical term.

Simulated shape changes based on Eqn. (5) for the case of a uniform CF concentration field are presented in Fig. 1a. (Here a biphasic functional $L(a)$ was used with a maximum of L at an intermediate value of a, rather than a simple linear decreasing dependence as assumed in writing the proportionality in Eqn. (3),

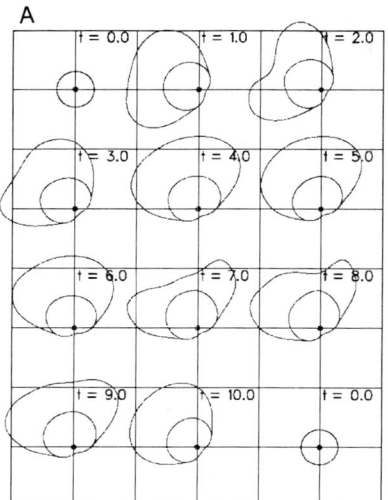
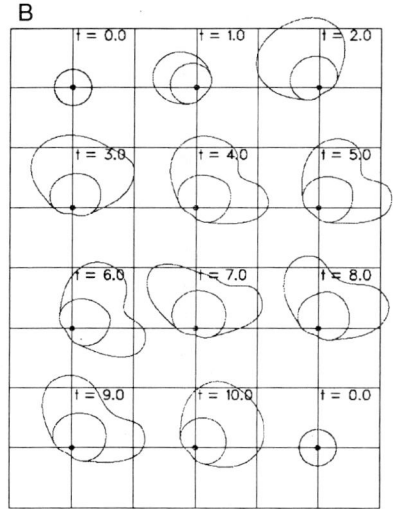
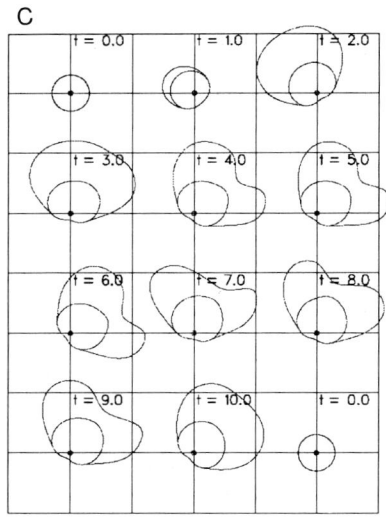

Figure 1: Simulated shape changes of a "model cell" whose cortical F-actin dynamics, according to Eqn. (2) and a simplified version of Eqn. (7) of AT, is altered by stochastic fluctuations in the density of bound CF receptors around the cell periphery, leading to changes in the mean lamella length $L(t,x)$ according to Eqn. (5) for (A) & (B), or Eqn. (6) for (C). Starting with a quiescent cell ($t = 0$) the time series show the outlines of the cell body (inner contour) and the mean lamellar extension (outer contour) for different times after exposure to the receptor fluctuations. They exhibit varying regions of lamellar protrusion activity, (A) in a homogeneous CF field and, (B) & (C) in a uniform spatial CF gradient to the right, obviously inducing preferred extension of lamellae in the positive gradient direction, thus chemotactic orientation of cell polarity.

reflecting the idea that cortical actin flow away from the center of the leading lamella reduces its length there; this argument relies on the term $\partial(VL)/\partial x$ on the left side of Eqn. (6) in AT. This choice of $L(a)$ explains the transient occurrences of two leading lamella, that is, it was not due to a bimodal cortical actin distribution $a(t, x)$, although regions of parameter space do admit this, as demonstrated in TA). Fig. 1b shows the effect of a spatial gradient of CF increasing to the right (i.e. the simulation was repeated using the same sequence of random deviates as inputs to the SDE system that models the fluctuating distribution $b(t, x)$), which can be seen to induce a chemotactic orientation of the cell.

From the earlier discussion about adapting signals, it is now obvious that the mechanistic model can be used to introduce an element of adaptation beyond that conferred by the cytomechanics, as discussed above. For example, $s^* = b/\oint b$ can be substituted for b in Eqn. (5):

$$\frac{\partial L}{\partial t} = \left(\eta(a - \gamma \left(\frac{b}{\oint b} - < \frac{b}{\oint b} > \right)) \right) - \left(\eta <a^*> - \frac{\partial(av)}{\partial x} \right). \quad (6)$$

The simulation of Fig. 1b was repeated with this substitution and is presented in Fig. 1c (the value of γ used in the simulations of Fig. 1a,b was adjusted to account for the integral value of b over the cell periphery that appears in s^*). While the cell shape appears similar at all time points, there are discernible differences (e.g. at t = 7.0).

Thus, both signals (b and s^*) appear to confer positive chemotaxis in the context of the cytomechanical model (which itself is consistent with a pseudo-spatial sensing mechanism, as introduced above). Of further interest would be a comparison of the signals for a spatial gradient of the same steepness but at different absolute values of the CF concentration, c. This would contribute toward answering the second central question posed in the Introduction. Unfortunately, this raises an additional issue that precludes a definitive comparison. Namely, the dependence of $<a^*>$ on c (equivalently, the dependence of $<a^*>$ on b) must be prescribed for the b signal case (there would be no dependence for the s^* signal by definition). This also precludes a comparison of the two signals in the simulation of polarity development when a step temporal gradient is imposed.

Alternatively, a heuristic adapting signal compatible with Eqn. (5) is

$$\frac{\partial L}{\partial t} = \left(\eta \left(a - \gamma \frac{\partial b}{\partial t} \right) \right) - \left(\eta <a^*> - \frac{\partial(av)}{\partial x} \right) \quad (7)$$

meaning that deviations in a^* from $<a^*>$ depend linearly on the rate of CF binding rather than on the amount of bound CF (consistent with the observations of stimulation in a continuous temporal gradient noted above). While no mechanistic justification can be made for this signal, it was instructive to use it in a comparative simulation. In contrast to the b and s^* signals, only a relatively small degree of lamellipod extension occurred, which developed early in the simulation and then diminished.

4 Discussion

This implies that the cytomechanical model used here, which does not admit the rapid extension/retraction of individual lamellipods that can experience large binding rate changes, might be inadequate to capture an important element of adaptation that occurs in a pseudo-spatial mechanism. (As seen in Fig. 1, the model only admits length changes of a continuous lamellar region, since the local length is determined solely by the cortical actin concentration which itself varies smoothly around the periphery.) Therefore, it is necessary to consider how adaptation might regulate lamellipod extension at the level of an individual lamellipod. As a suitable candidate the more general cytomechanical model derived in AT (I.9 this volume) ought to be analysed in this context.

Already the local lamellipodial dynamics simulated there in a simplified situation (AT, Fig. 1a) reveal its intrinsic capacity for adaptation (also generally applicable to the pseudopodial dynamics of other small amoeboid cells): Starting from a resting state with no (spontaneous) lamellar protrusions, an exposure to a *step temporal CF gradient* by which an increase in bound CF receptors would stimulate actin polymerization and, simultaneously, increase the number of bound adhesion sites (e.g. by being identical to them), induces a sudden but transient protrusion of a lamellipod, with successive adaptation of lamellipod length to the resting level (depending on the parameters, the number of (damped) protrusion-retraction cycles can even be reduced to one). Obviously, a sufficiently steep *continuous temporal gradient* would overcome this adaptive regulation by enhancing a lamellar extension that has started once, until a non-adapted mechanical equilibrium is reached. Thus, a "pseudo-spatial response" could similarly work in a *steady spatial gradient*: Spontaneous local lamellar protrusions (periodic as in AT, Fig. 1b, or stochastic due to imposed receptor binding fluctuations as in TA) would be enhanced only when they extend sufficiently fast in a significantly positive gradient direction. In the light of "intrinsic timing principles in non-linear signalling mechanisms" as formulated by Vicker (I.3 this volume) before, we note that the extending lamellipod/pseudopod "uses" a self-made temporal signal "produced" during the extension phase of its mechano-chemical protrusion-relaxation cycle, therefore inducing an (optimal?) nonlinear enhancement and amplification of the signal (e.g. the number of bound CF/adhesion receptors – particularly easy to realize with substrate bound CF, a case that appears prevalent in living tissues). Finally, we like to emphasize that the time characteristics and adaptive properties of the intracellular motor dynamics, though being autonomous in its intrinsic oscillatory capacities, at least within our modelling framework (Alt 1996), are thought to be influenced by (and possibly to have co-evolved with) spatio-temporal characteristics of the coupling to the environment, such as the mechanics or kinetics of binding to extracellular matrix proteins which essentially determine the adhesion dynamics of migrating cells (see the subsequent contributions in this Chapter).

II.5

A Model for Cell Migration by Contact Guidance

Richard B. Dickinson (Gainesville)

1 Introduction

Contact guidance is a bias in cell migration in response to a structural anisotropy, such as aligned grooves etched into a surface (Matthes & Gruler 1988) or in a three-dimensional matrix of aligned fibrils (e.g., in a collagen gel (Dunn & Ebendal 1978, Dickinson et al. 1993)). On these substrata, cells show a bi-directional orientation bias (Dunn & Brown 1986, Guido & Tranquillo 1993) and move preferentially along the axis of alignment. The contact guidance response can be quantified by measuring cell motility coefficients (i.e. diffusion coefficients) parallel and perpendicular to the axis of alignment, μ_{xx} and μ_{yy}, respectively. For example, we have shown that migrating fibroblasts in aligned collagen gels have an increase in μ_{xx} and a corresponding decrease in μ_{yy} as a function of fiber alignment (Dickinson et al. 1993).

How the cell path is modified by the anisotropy to yield a net bi-directional migration bias has not be studied in detail. However, unidirectional biased migration in response to a stationary gradient of chemoattractant or adhesion molecules has been analyzed (Alt 1980, Dickinson & Tranquillo 1995) to find that net drift in the direction of a gradient of stimulus can theoretically result from a number of different modifications of the cell path (Tranquillo & Alt 1990), including a bias in turning toward the gradient direction (tropotaxis), a direction-dependent cell speed or random turning frequency (orthotaxis and klinotaxis, respectively), or a spatially-dependent cell speed or turning frequency (orthokinesis and klinokinesis, respectively). The relative contributions of these cell path modifications can be predicted theoretically by examining the long time behavior of the underlying random walk model (Dickinson & Tranquillo 1995). The purpose of this paper is to perform an analogous analysis for bi-directional movement by contact guidance.

2 Model

As depicted in Fig. 1, the cell path is assumed to be described by two stochastic variables: the cell position, $r(t)$, and an orientation vector, $\Theta(t)$, determined by the direction of cell polarity. We have shown elsewhere (Dickinson & Tranquillo 1995) that if fluctuations in cell rotational and translational velocities relax on a time scale that is short relative to the time scale of directional changes, the time-dependent joint probability density, $p(\Theta, r, t)$, can be written as

$$\partial_t p(\Theta, r, t) = Lp(\Theta, r, t) - \partial_r \left(v(\Theta) p(\Theta, r, t) - B(\Theta) \partial_r^T p(\Theta, r, t) \right), \quad (1)$$

where v is the mean cell velocity in direction Θ, B is a diffusion tensor resulting from fluctuations in the cell velocity which lead to a dispersion in cell position, and L is a time-independent operator that governs the Θ-dependence of $p(\Theta, r, t)$. For two-dimensional movement ($\Theta = [\cos\theta \ \sin\theta]^T$, $r = [x \ y]^T$, $v = [v_x \ v_y]^T$, $\partial_r = [\partial_x \ \partial_y]$) with continuous turning, L can be given by

$$Lp(\theta, r, t) = -\partial_\theta \left(\omega(\theta) p(\theta, r, t) - D(\theta) \partial_\theta p(\theta, r, t) \right) \quad (2)$$

where $\omega(\theta)$ and $D(\theta)$ are the rotational drift velocity and rotational diffusion coefficient, respectively.

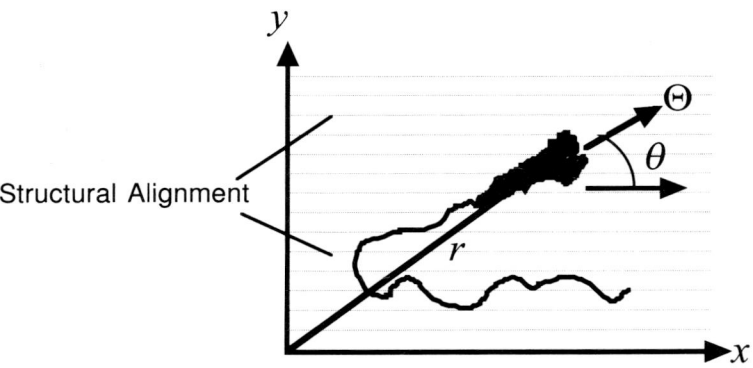

Figure 1: Depiction of a cell crawling on a two-dimensional substratum with a structural anisotropy in the x-direction. The cell path is described by the position vector, $r(t)$, and the orientation vector, $\Theta(t) = [\cos\theta(t) \sin\theta(t)]$.

If the time scale is chosen such that $t \gg (sup(D(\theta)))^{-1}$, then $Lp \gg \partial_t p$, and θ approaches a stationary distribution; i.e. $p(\theta, r, t) \approx p_s(\theta) p(r, t)$, where $p_s(\theta)$ is the solution to $Lp_s(\theta) = 0$. On this time scale, Eqn. 1 asymptotically approaches the diffusion equation (Dickinson & Tranquillo 1995)

$$\partial_t p(r, t) = \partial_r (\mu \partial_r^T p(\theta, r, t)), \quad (3)$$

where μ is the random motility tensor, given by

$$\mu = \int_0^\infty <v(\theta(t'))v(\theta(0))^T> dt' + <B(\theta)>. \quad (4)$$

The mean implied by $<*>$ is under the stationary density, $p_s(\theta)$, such that for any functions $f(\theta)$ and $g(\theta)$, we have

$$<f(\theta)> = \int_0^{2\pi} f(\theta)p_s(\theta)d\theta \tag{5}$$

and

$$<f(\theta(\tau))g(\theta(0))> = \int_0^{2\pi}\int_0^{2\pi} f(\theta)g(\theta_0)p(\theta,\tau|\theta_0,0)p_s(\theta_0)d\theta d\theta_0. \tag{6}$$

Here, the conditional probability density, $p(\theta,t|\theta_0,0)$, is the solution to

$$\partial_\tau p(\theta,\tau|\theta_0,0) = Lp(\theta,t|\theta_0,0) \tag{7}$$

with the initial condition, $p(\theta,0|\theta_0,0) = \delta(\theta-\theta_0)$.

To proceed, we need the θ-dependence of w, D, B, and v. Let ε be a parameter that reflects the magnitude of influence of the structural anisotropy on cell movement. Based on the bi-directional symmetry we assume the following expansions apply for small ε:

$$\begin{align}
w(\theta) &= -w^{(1)}\varepsilon\sin 2\theta + O(\varepsilon^2) \tag{8}\\
D(\theta) &= D^{(0)} + D^{(1)}\varepsilon\cos 2\theta + O(\varepsilon^2) \tag{9}\\
B(\theta) &= (B^{(0)} + B^{(1)}\varepsilon\cos 2\theta)\Theta\Theta^T + O(\varepsilon^2) \tag{10}\\
v(\theta) &= (S^{(0)} + S^{(1)}\varepsilon\cos 2\theta)\Theta + O(\varepsilon^2). \tag{11}
\end{align}$$

To simplify the analysis to follow, we have assumed in Eqns. 10 & 11 that the cell velocity is confined to the direction of cell orientation, thus allowing v and B to have only components proportional to the vector Θ and the dyadic $\Theta\Theta^T$, respectively. The $\cos 2\theta$ and $\sin 2\theta$ dependencies to lowest order of ε in above expansions result from symmetry about the x-axis requiring $f(\theta) = f(\theta+\pi/2)$ for any function, $f(\theta)$. Furthermore, because the symmetry in turning toward the x-axis, $w(\theta) = -w(-\theta)$, thus w must only depend on $\sin 2\theta$. Conversely, $|v|$, D, and $||B||$ are symmetric about $\theta = 0$, resulting in the dependence on $\cos 2\theta$ only.

Introducing Eqns. 8 through 11 into Eqn. 2, we have

$$Lp(\theta) = D^{(0)}\partial_\theta^2 p(\theta) + \varepsilon\partial_\theta\left(w^{(1)}\sin 2\theta p(\theta) + D^{(1)}\cos 2\theta\partial_\theta p(\theta)\right) + O(\varepsilon^2) \tag{12}$$

from which $p_s(\theta)$ can be obtained:

$$p_s(\theta) = \frac{1}{2\pi}\left(1 + \varepsilon\frac{w^{(1)}}{2D^{(0)}}\cos 2\theta\right) + O(\varepsilon^2). \tag{13}$$

Eqn. 13 implies that, to $O(\varepsilon)$, only the turning bias toward the alignment axis (as reflected in $w^{(1)}$) results in a non-uniform stationary orientation distribution.

Fourier transformation of Eqn. 7 yields the following set of coupled ordinary differential equations for the Fourier coefficients,

$$p_m(t|\theta_0) \equiv \frac{1}{2\pi} \int_0^{2\pi} p(\theta, t|\theta_0, 0) e^{-im\theta} d\theta :$$

$$d_t p_m(t|\theta_0) = -m^2 D^{(0)} p_m$$
$$+ \frac{\varepsilon}{2} m \left[\left(\omega^{(1)} - D^{(1)}(m-2) \right) p_{m-2} - \left(\omega^{(1)} + D^{(1)}(m+2) \right) p_{m+2} \right]$$
$$+ O(\varepsilon^2). \tag{14}$$

Given $p(\theta, 0|\theta_0, 0) = \delta(\theta - \theta_0)$, the transformed initial condition for Eqn. 14 is $p_m(0|\theta_0) = \frac{1}{2\pi} e^{-im\theta_0}$. Eqn. 14 can be solved to $O(\varepsilon)$ by expanding p_m into powers of ε, i.e. $p_m = p_m^{(0)} + p_m^{(1)} \varepsilon + O(\varepsilon^2)$, introducing this series into Eqn. 14 then collecting and sequentially integrating terms of like order. The resulting solution is

$$p_{\pm 1}(t|\theta_0)$$
$$= \frac{1}{2\pi} e^{-D^{(0)} t} \times$$
$$\times \left(e^{\pm i\theta_0} + \frac{\varepsilon}{2} \left[\left(\omega^{(1)} + D^{(1)} \right) e^{\pm i\theta_0} t - \frac{\omega^{(1)} + 3D^{(1)}}{8D^{(0)}} e^{\mp 3i\theta_0} \left(1 - e^{-8D^{(0)} t} \right) \right] \right)$$
$$+ O(\varepsilon^2) \tag{15}$$

$$p_m(t|\theta_0)$$
$$= \frac{1}{2\pi} e^{-m^2 D^{(0)} t} \times$$
$$\times \left(e^{-im\theta_0} + \frac{\varepsilon}{2} m \left[\frac{\omega^{(1)} - D^{(1)}(m-2)}{4(m-1)D^{(0)}} e^{-(m-2)i\theta_0} \left(1 - e^{4(m-1)D^{(0)} t} \right) \right. \right.$$
$$\left. \left. + \frac{\omega^{(1)} + D^{(1)}(m+2)}{4(m+1)D^{(0)}} e^{-(m+2)i\theta_0} \left(1 - e^{-4(m+1)D^{(0)} t} \right) \right] \right) + O(\varepsilon^2)$$
(for $m \neq 1, -1$). \tag{16}

Eqn. 15 can now be used to find the components of μ. Noting Eqn. 11, the velocity autocorrelation functions in the x- and y-directions can be written as

$$< v_x(\theta(t)) v_x(\theta(0)) > = (S^{(0)^2} + \varepsilon S^{(1)} S^{(0)}) < \cos\theta(t) \cos\theta(0) > + O(\varepsilon^2) \tag{17}$$

$$< v_y(\theta(t)) v_y(\theta(0)) > = (S^{(0)^2} - \varepsilon S^{(1)} S^{(0)}) < \sin\theta(t) \sin\theta(0) > + O(\varepsilon^2), \tag{18}$$

where we have used the identities,

$$\cos 2\theta \cos\theta = (\cos\theta + \cos 3\theta)/2 \quad \text{and} \quad \cos 2\theta \sin\theta = (-\sin\theta + \sin 3\theta)/2,$$

II.5 A Model for Cell Migration

and noted from Eqn. 15 that terms involving cross-correlation functions of the form $<\cos 3\theta(t)\cos\theta(0)>$, $<\sin 3\theta(t)\sin\theta(0)>$, etc., are $O(\varepsilon^2)$. Off-diagonal terms, $<v_x(t)v_y(0)>$ and $<v_y(t)v_x(0)>$, are in a similar manner proportional to $<\cos\theta(t)\sin\theta_0(0)>$ and $<\sin\theta(t)\cos\theta_0(0)>$, respectively. Introducing Eqns. 13 & 15 into Eqn. 6 yields

$$<\cos\theta(t)\cos\theta(0)> = \int_0^{2\pi} \cos\theta_0 \pi (p_1(t|\theta_0) + p_{-1}(t|\theta_0)) p_s(\theta_0) d\theta_0 + O(\varepsilon^2)$$

$$= \frac{1}{2}\left(1 + \frac{\varepsilon}{2}\left[\frac{\omega^{(1)}}{2D^{(0)}} + \left(\omega^{(1)} + D^{(1)}\right)t\right]\right)e^{-D^{(0)}t} + O(\varepsilon^2) \quad (19)$$

and

$$<\sin\theta(t)\sin\theta(0)> = \int_0^{2\pi} \cos\theta_0 \pi i (p_1(t|\theta_0) - p_{-1}(t|\theta_0)) p_s(\theta_0) d\theta_0 + O(\varepsilon^2)$$

$$= \frac{1}{2}\left(1 - \frac{\varepsilon}{2}\left[\frac{\omega^{(1)}}{2D^{(0)}} + \left(\omega^{(1)} + D^{(1)}\right)t\right]\right)e^{-D^{(0)}t} + O(\varepsilon^2). \quad (20)$$

By similar integration, $<\cos\theta(t)\sin\theta_0(0)>$ and $<\sin\theta(t)\cos\theta_0(0)>$ (and thus $<v_x(t)v_y(0)>$ and $<v_y(t)v_x(0)>$) can be shown to be $O(\varepsilon^2)$.

We can now define the *directional persistence tensor*,

$$P \equiv \int_0^\infty <\Theta(t)\Theta^T(0)> dt,$$

to measure the direction-dependent characteristic time of cell persistence. The xx- and yy-components of P can be calculated from Eqns. 19 & 20:

$$P_{xx} \equiv \int_0^\infty <\cos\theta(t)\cos\theta(0)> dt = \frac{1}{2D^{(0)}}\left(1 + \frac{\varepsilon}{2D^{(0)}}\left(\frac{3}{2}\omega^{(1)} + D^{(1)}\right)\right) + O(\varepsilon^2) \quad (21)$$

$$P_{yy} \equiv \int_0^\infty <\sin\theta(t)\sin\theta(0)> dt = \frac{1}{2D^{(0)}}\left(1 - \frac{\varepsilon}{2D^{(0)}}\left(\frac{3}{2}\omega^{(1)} + D^{(1)}\right)\right) + O(\varepsilon^2). \quad (22)$$

The (scalar) *directional persistence time*, P_t, can then be defined as the trace of P, i.e., $P_t \equiv Tr(P) = P_{xx} + P_{yy}$. Eqns. 21 & 22 imply $P_t = 1/D^{(0)} + O(\varepsilon^2)$.

The contribution to μ from velocity fluctuations is obtained from the stationary mean of B:

$$<B> = \frac{1}{2\pi}\int_0^{2\pi} \left(B^{(0)} + B^{(1)}\varepsilon\cos 2\theta\right)\Theta\Theta^T\left(1 + \varepsilon\frac{\omega^{(1)}}{2D^{(0)}}\cos 2\theta\right)d\theta + O(\varepsilon^2)$$

$$= \begin{bmatrix} \frac{1}{2}B^{(0)}\left(1 + \frac{\varepsilon}{4}\omega^{(1)}P_t\right) + \frac{\varepsilon}{4}B^{(1)} & 0 \\ 0 & \frac{1}{2}B^{(0)}\left(1 - \frac{\varepsilon}{4}\omega^{(1)}P_t\right) - \frac{\varepsilon}{4}B^{(1)} \end{bmatrix} + O(\varepsilon^2). \quad (23)$$

Finally, by combining the above into Eqn. 4, we obtain the principle xx- and yy-components of the motility tensor, μ:

$$\mu_{xx} = \left(S^{(0)^2} P_{xx} + \varepsilon S^{(1)} S^{(0)} P_t\right) + \frac{1}{2} B^{(0)} \left(1 + \frac{\varepsilon}{4}\omega^{(1)} P_t\right) + \frac{\varepsilon}{4} B^{(1)} + O(\varepsilon^2) \quad (24)$$

$$\mu_{yy} = \left(S^{(0)^2} P_{yy} - \varepsilon S^{(1)} S^{(0)} P_t\right) + \frac{1}{2} B^{(0)} \left(1 - \frac{\varepsilon}{4}\omega^{(1)} P_t\right) - \frac{\varepsilon}{4} B^{(1)} + O(\varepsilon^2). \quad (25)$$

We have shown that off-diagonal elements of μ are $O(\varepsilon^2)$ because $< v_x(t)v_y(0) >$ and $< v_x(t)v_y(0) >$ are $O(\varepsilon^2)$. For no bias ($\varepsilon = 0$), Eqns. 24 & 25 reduce to $\mu_{xx} = \mu_{yy} = \frac{1}{2}\left(S^{(0)^2} P_t + B^{(0)}\right)$, which is essentially the well-known result by Dunn (1983) and Othmer et al. (1988) for an unbiased persistent random walk, but modified here to account for fluctuations in cell speed about the mean at any fixed direction.

3 Discussion

As has been acknowledged for chemotaxis (Wilkinson 1988), we show here that mechanisms other than a bias in directional orientation may contribute to an overall contact guidance response. Eqns. 24 & 25 imply that, to $O(\varepsilon)$, the net contact guidance response can also result from an enhanced speed ($S^{(1)}$), persistence time (P_{xx}), and fluctuations in velocity (reflected in $B^{(1)}$) in the direction of alignment, with a corresponding decrease of these quantities in the perpendicular direction. Furthermore, an orientation bias can lead to an increase in the contribution of velocity fluctuations in the alignment direction.

It is interesting to note differences between cell migration responses to unidirectional vs. bi-directional stimulus fields. In our previous analysis of haptotaxis (Dickinson & Tranquillo 1995), $w(\theta)$, $D(\theta)$, $v(\theta)$, and $B(\theta)$ were approximated as functions of $\varepsilon \sin \theta$ and $\varepsilon \cos \theta$, rather than of $\varepsilon \sin 2\theta$ and $\varepsilon \cos 2\theta$ as done here. In that case, the motility tensor, μ, was found to vary with ε^2, which suggest that, to lowest order response, the motility coefficients for chemotaxis and haptotaxis are independent of the gradient steepness. In contrast, as we have shown here, the lowest order contact guidance response varies with ε.

Although the analysis shown here assumes two-dimensional movement, it is straightforward (albeit more complex) to generalize to three dimensions by expanding the functions of Θ into series of spherical harmonics as we have done previously for a unidirectional haptotaxis response. Furthermore, the assumption of continuous turns is inessential. Any turning process can be assumed as long as the orientation process, $\Theta(t)$, can be approximated as a Markov process, and the θ-dependence of L can be expressed in terms of $\varepsilon \sin 2\theta$ and $\varepsilon \cos 2\theta$. This assumption is sufficient to yield equations for the Fourier coefficients of $p(\theta, t|\theta_0, 0)$ that are mathematically equivalent to Eqns. 15 & 16. For example, $\Theta(t)$ can be assumed to be a directional jump process with Poisson-distributed waiting times and a bi-directional dependence for both the turning probability and the new direction.

II.5 A Model for Cell Migration

In principle, all parameters in Eqns. 24 & 25 can be derived from an underlying mechanistic model of a single cell that can predict cell turning, cell speed and associated fluctuations as functions of the cell orientation, as have been developed previously for unidirectional migration bias due to chemotaxis or haptotaxis (Tranquillo & Lauffenburger 1987, Dickinson & Tranquillo 1993). This analysis therefore provides a framework to predict and interpret observed "macroscopic" cell migration behavior (i.e. cell transport and cell movement statistics) in terms of the molecular and physical properties of a single moving cell.

II.6

Derivation of a Cell Migration Transport Equation from an Underlying Random Walk Model

Richard B. Dickinson (Gainesville)

Cell movement in an anisotropic environment has often been modeled by considering the cell velocity, $v(t) \equiv [v_x(t)\ v_y(t)\ v_z(t)]^T$, to be a Markov stochastic process in time, t; i.e. the future statistics of $v(t)$ are assumed dependent only on the current value, not on its history. The statistics of the motion are then determined by the joint probability density, $p(v,r,t)$, where $r(t) \equiv [x(t)\ y(t)\ z(t)]^T$ is the cell position. However, because of its direct analogy with cell density, the probability density of r alone, $p(r,t)$, often is of more interest on a longer time scale where the instantaneous value of v is less important than its long-time stationary distribution. We show here how an approximate differential equation for $p(r,t)$ can be derived if the time scale of interest is much larger than the relaxation time of v. This analysis adopts the projector operator formalism used in Gardiner (1983, Chapter 6) and in Kubo et al. (1991).

If $v(t)$ is a Markov process, the general equation for the time evolution of $p(v,r,t)$ is the differential Chapman-Kolmogorov equation (Gardiner 1983, Chapter 3) which can be written as

$$\partial_t p(v,r,t) = \eta L_1 p(v,r,t) + L_2 p(v,r,t) \quad (1)$$

$$\text{where} \quad L_2 p(v,r,t) \equiv -\partial_r (v p(v,r,t)) \quad (2)$$

and operator L_1 governs changes in the v-dimension, with the following general form for both discrete and continuous changes in v:

$$L_1 p(v,r,t) \equiv -\partial_v (A(v) p(v,r,t)) + \partial_v (B(v) \partial_v^T p(v,r,t))$$
$$+ \int (\phi(v,v') p(v',r,t) - \phi(v',v) p(v,r,t)) dv'. \quad (3)$$

$A(v)$ and $B(v)$ are the drift vector and diffusion tensor for continuous changes in v, and $\phi(v,v')$ is the turning kernel, i.e. the probability per unit time of making a discrete velocity jump from v' to v. r and t have been scaled such that η is a large parameter which reflects the fast relaxation of v on an appropriately long time scale. (Here we adopt the notation that $\partial_r \equiv [\partial_x\ \partial_y\ \partial_z]$ and $\partial_v \equiv [\partial_{v_x}\ \partial_{v_y}\ \partial_{v_z}]$.) In general, A, B, and ϕ may also depend on r (Dickinson & Tranquillo 1995), but here we examine the simpler case where the statistics of $v(t)$ are independent of r.

Let $p_s(v)$ be the pseudo-stationary density of v, independent of r, defined as the solution to $L_1 p_s(v) = 0$, such that the pseudo-stationary mean of any function $f(v,r)$ is $< f(r) > \equiv \int f(v,r) p_s(v) dv$. Then L_2 can be divided into components corresponding to its "mean", $\bar{L}_2 \equiv -\partial_r <v>$ and deviation from the mean, $L_2' \equiv -\partial_r (v - <v>)$.

Now define a projection operator P to project any function, $f(v,r,t)$, into the subset of functions proportional to $p_s(v)$, i.e. $Pf(v,r,t) \equiv p_s(v) \int f(v,r,t) dv$.

Defining $\psi \equiv Pp(v,r,t)$ and $\psi' \equiv (1-P)p(v,r,t)$, and operating on Eqn. 1 by P and $(1-P)$ yields
$$\partial_t \psi = P(\eta L_1 + L_2)(\psi + \psi') = PL_2'\psi' + P\bar{L}_2\psi \qquad (4)$$
$$\partial_t \psi' = (1-P)(\eta L_1 + L_2)(\psi + \psi') = (\eta L_1 + (1-P)L_2)\psi' + (1-P)L_2\psi \qquad (5)$$
where we have used the fact that $PL_1 = L_1 P = 0$ and $P\bar{L}_2(1-P) = 0$. Upon taking the Laplace transform of Eqns. 4 & 5 ($\tilde{\psi}(s) \equiv \int_0^\infty e^{-st}\psi(t)dt$), setting $\psi'(t=0) = 0$, and combining to eliminate $\tilde{\psi}'$, we have
$$\begin{aligned} s\tilde{\psi} - \psi(t=0) &= P\bar{L}_2\tilde{\psi} + PL_2'[s - \eta L_1 - (1-P)L_2]^{-1}(1-P)L_2\tilde{\psi} \\ &= P\bar{L}_2\tilde{\psi} - \eta^{-1} PL_2'L_1^{-1}(1-P)L_2\tilde{\psi} + O(\eta^{-2}). \end{aligned} \qquad (6)$$

Upon inverse transformation, this becomes
$$\partial_t p(r,t) = \bar{L}_2 p(r,t) - \eta^{-1} PL_2'L_1^{-1}(1-P)L_2 p_s(v) p(r,t) + O(\eta^{-2}). \qquad (7)$$

To make the right hand side of the above recognizable, we can formally write
$$L_1^{-1}(1-P) = -\int_0^\infty d\tau\, e^{L_1\tau} \qquad (8)$$
where $e^{L_1\tau} f(v_0, 0)$ is the solution to
$$\partial_\tau f(v,\tau) = L_1 f(v,\tau) \qquad (9)$$
with initial condition, $f(v_0, 0)$, i.e.
$$f(v,\tau) = e^{L_1\tau} f(v_0, 0) = \int dv_0' p(v, \tau|v_0, 0) f(v_0, 0) \qquad (10)$$
such that
$$L_1^{-1}(1-P)f(v_0, 0) = -\int dv_0 \int_0^\infty d\tau\, p(v, \tau|v_0, 0) f(v_0, 0). \qquad (11)$$

Upon introduction of the initial condition, $f(v_0, 0) = L_2 p_s(v_0) p(r,t)$, and combining Eqns. 7 & 11, we have
$$\partial_t p(r,t) = \bar{L}_2 p(r,t) + \eta^{-1} \int dv \int dv_0 L_2' \int_0^\infty d\tau\, p(v, \tau|v_0, 0) L_2 p_s(v_0) p(r,t) + O(\eta^{-2}) \qquad (12)$$
which can be written as
$$\partial_t p(r,t) = -\partial_r (<v> p(r,t)) + \partial_r (\mu \partial_r^T p(r,t)) + O(\eta^{-2}) \qquad (13)$$
where
$$\begin{aligned} \mu &\equiv \eta^{-1} \int_0^\infty d\tau \int dv_0 \int dv\, p(v, \tau|v_0, 0)(v - <v>)v_0^T p_s(v_0) \qquad (14) \\ &= \eta^{-1} \int_0^\infty d\tau <v(\tau), v(0)^T>. \end{aligned}$$

Eqn. 13 is equivalent to a diffusion equation for a non-interacting cell population of density, $p(v,t)$, with drift velocity, $<v>$, and diffusion tensor μ (Note that Eqn. 4 in (Dickinson, II.5 this volume) is a special case of Eqn. 14, where the first term corresponds to the integral of the autocorrelation function of the mean velocity at fixed orientation direction, and the second term corresponds to that of the fast-relaxing fluctuation around this mean).

II.7

A Continuum Model for the Role of Fibroblast Contact Guidance in Wound Contraction

Robert T. Tranquillo and Victor H. Barocas (Minneapolis)

1 Introduction

Wound repair is a basic and vital process for homeostasis. It is also complex: in full-thickness cutaneous wounds, where loss of dermis has occurred, there are characteristic phases of immediate inflammation, short-term granulation tissue formation (angiogenesis, the ingrowth of blood vessels, and fibroplasia, the infiltration of fibroblastic cells), and long-term wound matrix remodeling. Wound contraction, wherein the wound edge and surrounding skin move inward toward the wound center, occurs during fibroplasia. Fibroblastic cells secrete collagen and other extracellular matrix (ECM) components needed to transform the weak fibrin clot into a collagenous scar with mechanical strength. They are also responsible for the forces underlying wound contraction. The reader is referred to the books edited by Clark & Henson (1988) and Cohen et al. (1992) for an overview of the biology of wound repair.

There is great motivation for understanding the interplay between the complex chemical, cellular, and mechanical phenomena that result in wound contraction, as it can be either a beneficial or deleterious feature of wound repair. If a predictive mathematical model of wound healing that accounted for the salient phenomena was available, it would provide a rationale for the design and application of pharmacological modulators of cell function in order to augment or mitigate wound contraction as appropriate for the nature of the wound being managed. A mathematical model of dermal wound contraction based on the continuum-mechanical theory by Oster & Murray (Oster et al. 1983) for cell-ECM mechanical interactions was first proposed by Tranquillo & Murray (1992). It focused on showing that certain inflammation-mediated mechanisms, such as chemotactic accumulation of fibroblasts and growth factor induced enhancement of fibroblast traction, were consistent with data reported for wound contraction. This model was subsequently extended to examine the role of fibroblast contact guidance in the contraction of linear dermal wounds (Tranquillo et al. 1992), accounting for the interrelated phenomena of biased cell movement in aligned tissue fiber networks and the formation of such networks during wound contraction. However, the extension was not generalizable to other geometries and did not account for another aspect of contact guidance, namely, that net cell traction is greater in the direction of fiber alignment due to the associated cell alignment.

Olsen et al. (1995) recently elaborated the model proposed in Tranquillo & Murray (1992) to include explicitly the dynamics of the inflammatory mediator and interconversion of fibroblasts and traction-enhancing myofibroblasts, as well as more complex kinetics for the matrix synthesis and degradation. In addition to characterizing the predicted rate of wound contraction for their model, Olsen et al. characterized conditions for a sustained extent of wound contraction, termed contracture. Olsen et al. (1996c) subsequently derived a caricature model to investigate conditions for prediction of excessive levels of fibroblasts and inflammatory mediators in response to wounding that may be related to pathological outcomes such as keloids and hypertrophic scars, wherein excessive collagen production occurs within or beyond the wound edge, respectively.

2 Model

The anisotropic biphasic theory for cell-matrix mechanical interactions proposed by Barocas & Tranquillo (1994, 1996) extends the original Oster & Murray theory by accounting for the fiber network and interstitial matter of the ECM as distinct phases, and the interrelated phenomena of inhomogeneous network deformation (i.e. anisotropic strain), fiber alignment, and cell contact guidance (i.e. cell alignment, therefore also traction and migration preferentially in the direction of fiber alignment). We have used this theory here in the limit of negligible drag between the two phases, whereby it becomes nearly equivalent to the Oster-Murray theory except for the cell migration and traction stress terms being anisotropic. In fact, the model (equations, initial conditions, boundary conditions and parameter values) analyzed here is identical in all respects to that proposed by Tranquillo and Murray (1992) except for these anisotropic terms (and a weighting of the total stress tensor for the cell/ECM composite by the local ECM volume fraction, which arises in the form of volume-averaging theory used – this is equivalent to making the elasticity and viscosity of the composite linear functions of the ECM volume fraction). A brief summary of the anisotropic terms follows.

Fiber alignment is defined in terms of the network deformation tensor through a fiber orientation tensor $\underline{\underline{\Omega}}_f(\underline{x}, t)$. This tensor is the integral of the weighted dyad product $\underline{n} \otimes \underline{n}\, P(\underline{n})$ where a fiber at the point $\underline{x}$ at time t has the probability $P(\underline{n})$ of having the orientation of $\underline{n}$:

$$\underline{\underline{\Omega}}_f = \oint \underline{n} \otimes \underline{n}\, P(\underline{n})\, d\underline{n}. \tag{1}$$

$P(\underline{n})$ is proportional to the differential area of the deformation ellipsoid defined at $\underline{x}$ based on the displacement $\underline{u}(\underline{x}, t)$, which has the advantage that no new parameters need to be used to relate tissue deformation to fiber orientation and has the interpretation that fibers tend to be aligned in the direction of maximum stretch or minimum compression.

A cell orientation tensor $\underline{\underline{\Omega}}_c$ is taken to be a monotonic function of $\underline{\underline{\Omega}}_f$,

$$\underline{\underline{\Omega}}_c = \left(\frac{3}{tr\underline{\underline{\Omega}}_f^\kappa}\right) \underline{\underline{\Omega}}_f^\kappa \qquad (2)$$

that is, cells align in the direction of aligned fibers with sensitivity κ (the bracketed term is a scaling factor to maintain $tr\underline{\underline{\Omega}}_c = 3$, which is also true for $tr\underline{\underline{\Omega}}_f$). The cell flux and traction stress terms are then asserted to be

$$\underline{J}_{cell} = -D_o \underline{\underline{\Omega}}_c \cdot \nabla c \qquad (3)$$

$$\underline{\underline{\sigma}}_{traction} = \tau_o \underline{\underline{\Omega}}_c \left(\frac{c\rho}{1+\gamma c^2}\right) \qquad (4)$$

in terms of cell (c) and network (ρ) concentrations, which are identical to the forms used in Tranquillo & Murray (1992) except that $\underline{\underline{\Omega}}_c$ appears instead of $\underline{\underline{I}}$. The cell flux expression for contact guidance in a spatially varying contact guidance field (Eqn. 3) can be derived from an underlying model of stochastic cell movement in a homogeneous contact guidance field (Dickinson II.6 this volume), under the conditions that cell speed is independent of direction (i.e. fiber alignment) and that persistence of movement direction is negligible compared to speed. While there are data supporting modeling fibroblast contact guidance in aligned fiber networks as an anisotropic diffusion (Dickinson et al. 1993), there are no data yet that bear on these conditions being satisfied. (There are data for neutrophils, however, indicating speed variation with orientation on planar substrata possessing structural anisotropy (Matthes & Gruler 1988), which motivated the heuristic modeling in Tranquillo et al. (1992) for the case of spatially-varying fiber alignment.)

The model of Tranquillo & Murray (1992) extended for contact guidance as indicated above was reanalyzed for the "base case" and "traction variation case." Briefly, the "base case" assumed for fibroblasts: random migration, passive convection (with the ECM), logistic growth; for ECM: passive convection; and for the cell/ECM composite: isotropic, linear viscoelastic solid rheology with distributed cell traction stress and linear elastic subdermal attachments. It assumed no role for inflammation. The "traction variation case" further modeled an enhancement of fibroblast traction due to some inflammation-derived factor at a pseudo-steady state during wound contraction by simply making the traction parameter increase toward the wound center (cf. Olsen et al. 1995).

3 Results

The inclusion of contact guidance does not affect the qualitative "base case" results reported for the original model, which were not consistent with wound contraction, that is, expansion of the wound followed by relaxation to the original dimension for both linear and circular wounds. (It remains to be seen whether the "base

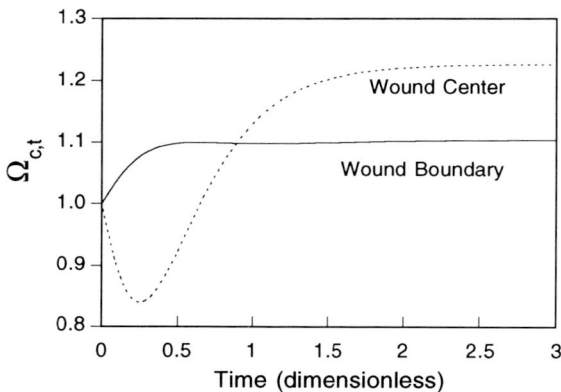

Figure 1: The degree of circumferential alignment versus time is plotted for locations at the (contracting) wound boundary and near the wound center. The range of $\Omega_{c,t}$ is 0-2, spanning complete normal (0) to complete tangential (2) alignment of the cells relative to the wound boundary.

case" yields predictions consistent with fibromatoses, where tissue contraction occurs without an inflammatory component, when relaxing the assumptions leading to Eqn. 3.) The same is true for the "traction variation case" results, which were consistent with wound contraction. Interestingly, however, both the predicted rate and extent of contraction are reduced (e.g. the wound dimension contracted by 8.8% and 7.3% for linear and circular wounds, respectively, without contact guidance ($\kappa = 0$), but only 3.7% and 6.3% with strong contact guidance described by $\kappa = 10$, wherein the cells align much more than the fibers).

The explanation for this lies in the assumption that contact guidance is manifested as anisotropic cell traction as well as migration: for both geometries, as wound contraction ensues due to the enhanced traction of fibroblasts populating the wound via migration and division, the strain field induces fibers to align tangential to the wound boundary. The fibroblasts therefore align and exert traction preferentially in the direction tangential to the wound boundary, reducing the driving force for wound contraction. This is seen in the plot of the component $\underline{\underline{\Omega}}_c$ tangential to the wound boundary ($\Omega_{c,t}$) at the position of the wound boundary in Fig. 1 (linear wound case), where $\Omega_{c,t} > 1$ indicates fiber alignment tangential to the wound boundary and $\Omega_{c,t} < 1$ normal to the wound boundary. This negative biomechanical feedback loop may be operative *in vitro* as well as *in vivo* in processes where cell-matrix mechanical interactions are significant (Barocas & Tranquillo 1996). Notice that at early times $\Omega_{c,t} < 1$ at the wound center, indicating that fibroblasts are initially aligned toward the wound center, which enhances their initial rate of populating the wound via migration. This effect is due to an initial wound expansion causing radial fiber alignment toward the wound. While the expansion arises in this model due to an imbalance of cell traction, rather than, more realistically, to a release of skin tension, it captures an initial fibroplasia-promoting role of contact guidance. Fiber alignment toward the wound has also been attributed to platelet-mediated fibrin clot contraction (Lackie 1986). This would be consistent with our view of strain-dependent fiber reorientation in (Eqn. 1), and our model would make similar predictions: initial contact guidance toward the wound, followed by contact guidance tangential to the wound boundary as wound contraction occurred.

4 Discussion

The migration and growth signals that determine the spatial and temporal distribution of fibroblastic cells in and around a wound, as well as the factors that govern the magnitude of force transmitted by the cells to local ECM fibers, are poorly understood. It is probable that cells are simultaneously responding by chemotaxis as well as by contact guidance. In order to account for this, Eqn. 3 will need to be extended to include a chemotactic contribution that is a function of $\underline{\underline{\Omega}}_c$, where $\underline{\underline{\Omega}}_c$ will depend on both the chemotactic factor concentration field and the fiber alignment field ($\underline{\underline{\Omega}}_f$) in a way that is as yet uncharacterized. Eqn. 3 should first be expanded in order to account for direction dependent speed and significant directional persistence, if data indicate such, and tested for the case of spatial variation of fiber alignment.

Two popular hypotheses for wound contraction are distinguished by the location of the cells responsible for the force driving wound contraction (Gross et al. 1995): in the "picture frame" hypothesis, it is ascribed to inward migration of fibroblastic cells located beneath a thin margin of dermis surrounding the wound; in the "central granulation" hypothesis, it is ascribed to fibroblastic cells located within the granulation tissue of the wound. All models of wound contraction to date are consistent with the latter, wherein the wound interior becomes a focus of contraction. It remains to develop a model that is consistent with the former; this will require an explicit relationship between cell migration and traction that has not previously been used. Whether the functional form of the traction stress should be different to reflect the "contraction force" associated with the hypothesized shortening of the entire cell body of smooth muscle cell-like myofibroblasts, which are prominent during wound contraction (Gabbiani 1994), as compared to the "traction force" associated with the continuous pseudopodal activity of stationary or migrating dermal fibroblasts (Ehrlich & Rajaratnam 1990, Harris 1982), and exactly what the functionality in the brackets of Eqn. 4 should be in either case, remains an open question.

Beyond just these issues pertaining to modeling the behavior of the fibroblastic cells, other important issues include the degree to which the chemical and physical fields in the cell's environment (e.g. growth factors and network stress) modulate extracellular matrix synthesis and secretion, division, and differentiation, as well as the origin and evolution of these fields. Moreover, there is the issue of accurate microstrusctural and rheological constitutive equations for the extracellular fiber matrix and how their defining parameters evolve with the evolving composition of the wound. Modeling these complexities involves a level of understanding about the biology and mechanics of wound healing that is presently tenuous at best but offers future challenges with potentially great fruition. Systematic testing of models by conducting well-defined in vitro assays that mimic elements of wound healing and contraction alone or in combination (Barocas & Tranquillo 1994) should be valuable for model development while this understanding advances. Continued investigations of fetal wound healing, where tissue regeneration rather than contraction and scarring results, should contribute to this advance (Adzick & Longaker 1992, Stocum 1995).

Acknowledgements The assistance of Roger Levy in the model and code development and the analysis of anisotropic diffusion by Mihir Wagle are gratefully acknowledged. R. L. was supported by a Minnesota Supercomputer Institute Internship. This work was supported by NIH GM-46052.

II.8

Wound Healing and Tumour Growth
– Relations and Differences –

A contribution to the discussion by Galina Solyanik (Kiev)

Inspite of the obvious formal difference between wound healing and tumour growth – the former is a normal physiological process while the latter is a pathological one – there are many similarities between them (Dvorak 1986). As tissue repair is a normal response to injury, the growth of solid tumours may be likened to an aberration of successful wound healing – an old concept which has been considerably developed in the past decade (Whalen 1990). Indeed, tumours are often referred to as "wounds that do not heal" (Dvorak 1986).

Regardless of the details of the signals which activate cells in wounded tissue, the result of normal healing is the renewal of cellular composition and spatial arrangement of tissue. Although the control mechanisms of this process are still unclear, tissue repair is believed to result from cell-host and cell-cell interactions, that are known also to play a significant role in tumour progression (Casciari et al. 1992).

Cell-host interactions are based on the ability of various host agents to induce or inhibit proliferation and motility of normal (and cancer) cells. It is known that the migratory response of cells depends on the concentration of attractants and includes both random and directed components of motion. In spite of numerous studies, the question of the relative contributions of the chemotactic and chemokinetic responses in wound healing and tumour growth is still an open question (Solyanik et al. 1995); see also Tranquillo & Barocas (II.7 this volume).

It should be noted that irrespective of the answers to the aforementioned questions, it is clear that regulation of cell proliferation and migration is of great importance in normal and pathological processes (Michelson & Leith 1994). For example, contact inhibition of motion and high density inhibition of cell proliferation are powerful inherent cellular regulators during the development, growth and repair of living organisms. Independently of the presence or lack of the external (host) signals which induce invasive and proliferative behaviour of normal cells within a wound, the process of wound healing can be stopped by these cell regulatory mechanisms, ensuring functional and morphological integrity of the host organism. A slackening in the cell-cell inhibitory influence (which is an attribute of tumour cells) may result in uncontrolled growth and emergence of some pathologies. In this connection, the role of cell-cell inhibitory mechanisms in abnormal wound healing and tumour growth is a problem of crucial importance to the understanding and treatment of fibrocontractive (Olsen et al. 1996c) and malignant diseases (Freitas & Baronzio 1994) and requires both experimental and theoretical investigation.

It appears that the behavioural repertoire of normal and cancer cells is genetically predetermined. The behavioural patterns realised, however, depend on cellular interactions and the dynamic microenvironment (Chaplain 1993), which in turn, depend on the spatial dimension of cellular arrangement. For example, it has been shown that the behaviour of cancer cells in a monolayer differs from that in a three-dimensional culture (Steeg et al. 1994). The open question here is: what can an additional spatial dimension contribute to the variety of cell behavioural patterns? The answer to this question is of extreme importance for a comprehension of wound healing and tumour growth as processes that are essentially three-dimensional. Moreover, studying this problem will make a valuable contribution to the general understanding of morphogenesis; see Chapter IV.

Discussion & Open Problems

The complexity of a cell is daunting, and the assertion that its function will ever be completely understood, even if its composition is completely defined, is doubtful. However, the ability to gainfully understand aspects of cell behavior is not hopeless, as evidenced by the state-of-art represented by the contributions of this Chapter. The diversity of the approaches used by these researchers underscores the highly interdisciplinary nature of cell behavior research. Progress will require the complementary use of advanced microscopy and imaging techniques (to quantify cell behavior and the disposition of key cellular components) and mathematical models (to formulate and test hypotheses about the cellular mechanisms underlying the interrelationships between adhesion, motility, traction, locomotion, and directed migration – collectively termed cell behavior here) in assays where the cell's environment is both well-defined and measurable. The paper by Dembo, Oliver & Ishihara (II.2) is an excellent example of this approach. More detailed remarks about future research of tissue cell behavior are now presented.

The driving force for the study of cell behavior has historically been understanding physiological processes such as development, wound healing and metastasis; see also Box II.8. Given that these generally involve a tissue environment, there must be a shifting focus to using assays that mimic the composition and structure of tissue rather than convenient substrata like plastic and glass surfaces. The dramatic differences in cell morphology and metabolism reported to occur when cells are cultured in tissue-like environments *vs.* on rigid surfaces makes any extrapolation of cell behavior on rigid surfaces rather tenuous. For example, the prerequisite of cell adhesion for traction and locomotion on surfaces may not be as strong for cells in fiber networks that are characteristic of tissues. Moreover, the nascent field of tissue engineering, wherein cells are entrapped directly into hydrated networks of entangled collagen fibrils in one approach to bioartificial soft tissue fabrication, provides another driving force for a shift in the paradigm of cell behavior studies to these so-called "tissue-equivalents".

A key feature of tissue-equivalents is the occurrence of aligned collagen fibrils and consequent cell contact guidance in instances where inhomogeneous compaction of the collagen network occurs, the compaction resulting from traction exerted by entrapped tissue cells. (The physiological analog of this *in vitro* phenomenon, dermal wound contraction, was the subject of the model by Tranquillo & Barocas (II.7).) Whereas the contact guidance cue on rigid substrata can be defined (and the mechanism of contact guidance understood in some cases), dissecting the contributions from chemical, mechanical and steric anisotropy in aligned fibril networks has proven difficult and thus the mechanism remains elusive. The inevitable reorganization of the fibrils by the cell as it exhibits contact guidance may prove to be the problematic analog of unsteady concentration gradients in the assay of chemotaxis (Vicker, II.3). There is little doubt, however, that the mechanism involves a spatially-integrated response by the cell to pseudopod activity around the cell periphery, as appears true for chemotaxis. Similarly, a successful mathematical model of contact guidance will require a "sensory" component inextricably connected to the cytomechanical component as has been proposed for chemotaxis (Tranquillo & Alt, II.4).

Another feature that opens new frontiers in the study of cell behavior is the effect of stress and/or strain (including fibril alignment) of the network on cell phenotype. It is known that DNA and protein synthesis are rapidly down-regulated when the tissue-equivalent is released from a mechanical constraint during compaction, concomitant with a rapid change in cell morphology from bipolar to stellate. The mechanisms by which cells transduce mechanical signals is being intensely studied but how the change of phenotype in "stressed" vs. "stress-free" tissue-equivalents affects cell behavior is largely unknown. The situation becomes even more complicated as the ramifications of "dynamic reciprocity" are considered, that is, the evolving interplay between the compositional and mechanical/structural state of the network, due to tractional structuring of network fibrils and secretion of collagen and other extracellular matrix components by the cells, and the cell metabolism and behavior, which depend on the network state.

As noted in the Introduction to this Chapter, understanding the interrelationships between adhesion, traction, locomotion, and directed migration, as well as their modulation by extracellular signals, remains a major challenge. Mathematical models of the underlying chemistry (e.g. receptor binding and signal transduction pathways) and physics (e.g. actin-based force generation in cortical and lamellar cytoplasm) are essential to understanding the logical consequences of defining mechanisms, since the interrelationships involve a complexity beyond the limits of intuition. However, even highly simplified mathematical models that attempt to capture just a subset of these processes, such as that proposed in the paper by Tranquillo & Alt (II.4), prove to be difficult to analyze and understand. This is complicated by the stochasticity manifest in cell behavior that likely has multiple origins. Model simulation and judicious time series/correlation analysis is not only invaluable, but will become increasingly feasible as computer technology advances. However, the power and utility of self-consistent "phenomenological" (diffusion approximation) forms of mechanistic models, exemplified in the paper by Dickinson (II.5), must not be lost.

Chapter III

Dynamics of Cell-Cell Interactions
– Collective Motion and Aggregation –

Coordinator: Philip K. Maini

> *The argument then is that not all the morphogenetic movements of* Dictyostelium *are controlled by external acrasin gradients, gradients in the outside environments, but this does not mean that there might not be internal gradients that we cannot see or measure. It is quite conceivable, although treacherously hypothetical, that there may be some sort of gradient within each cell and that these line up with respect to one another much as Rashevsky (1938) imagines might be possible in his purely formal analysis of morphogenetic movements of animals...*
>
> John Tyler Bonner (1952)

Introduction

Self-organization is a fundamental widespread process occurring in morphogenesis, wound healing and population dynamics. For example, in embryology, cell-cell interactions play a key role in tissue formation and cell aggregation prior to the formation of structures, such as skeletal elements in the limb, or skin organ formation, such as hair, teeth and feathers. In wound healing, cells respond to external guidance cues, as well as orientation cues from each other, to move into a wound and effect its closure. In populations, aggregation may be essential for shelter, reproduction, safety, etc, or to resist starvation conditions.

Recent advances in biotechnology have led to a great increase in experimental data on the biochemical and mechanical aspects of cell-cell interactions and mathematical modelling has a crucial role to play in providing a theoretical framework in which to interpret this data, as well as investigating the potential of hypothesized interactions to account for experimental observations. In this way, modelling can help elucidate the underlying processes involved in aggregation phenomena. In turn, there is the need to develop more sophisticated mathematical modelling, and analytical and computational techniques to account for the detailed observations at the cellular level. Phenomenological models, such as those of reaction-diffusion type, which have played an essential role in understanding processes at a gross level, have to be replaced by more detailed models to reflect the complexity of the underlying processes.

This Chapter contains four contributions which illustrate a number of different types of interactions that can lead to self-organization in cell populations. The interactions can be due purely to cell-cell contact cues, or can be mediated through substances secreted by cells which effect the motion of other cells. The papers also illustrate a number of different modelling strategies, ranging from pure continuum models, to hybrid models with discrete, stochastic, or cellular automaton components.

The paper by Mogilner, Deutsch & Cook (III.1) reviews a number of recently proposed models by the authors which try to capture different aspects of the general problem of spatio-angular self-organization of cells by mutual interaction and successive (gradual or abrupt) alignment. The authors briefly present their modelling approach and simulation/analysis results for a common class of stochastic diffusion/migration processes with interactions, each elucidating a different level of approximation by a nonlinear evolution system. An integro-partial-differential equation system is presented in which the independent variables are space and orientation angle (as well as time). Linear stability analysis reveals the possibility that model simulations can lead to simultaneous spatial aggregation of cells and alignment. This type of model can be analysed using the orientation tensor approach which approximates the full system by a system of reaction-diffusion-advection equations. It is shown that the model can exhibit total alignment. Finally, a stochastic model is considered using a lattice-gas cellular automaton approach.

The paper by Stevens & Schweitzer (III.2) models a very different type of cell-cell interaction, namely that of trail following wherein certain biological species, for example myxobacteria, can interact by laying down a substance that changes the substratum on which the bacteria are moving, thus influencing the motion of other bacteria. The model is based on a reinforced random walk description of cells moving on a two-dimensional lattice. The direction of motion is influenced by the concentration of the substance the cell experiences. Computer simulations of the model are presented, showing clustering, swarming and cell streaming. It is shown that this discrete interacting particle model can be approximated by a continuous model which can also exhibit the phenomena of blow-up and collapse.

A composite contribution follows (III.3), modelling cell streaming and aggregation in the slime mould *Dictyostelium discoideum*. Here, amoebae secrete the chemoattractant, cAMP, and aggregate under appropriate conditions. Three different approaches to modelling cell movement and chemical dynamics are presented together with simulations demonstrating that each model can account for the experimentally-observed phenomenon of cell streaming. At first, Dallon & Othmer describe a discrete-continuum model in which the chemoattractant concentration is modelled as a continuum field, while the amoebae are considered as discrete sources and sinks of the chemical. Model assumptions also include the adaptation of the cAMP signalling pathway. Simulations show spontaneous generation of spiral waves.

The model of Van Oss, Panfilov & Hogeweg is also of a hybrid discrete-continuum form but differs from that of Dallon & Othmer as it uses the model of Martiel & Goldbeter for cAMP relay and incorporates cAMP secretion and degradation via a discrete description of the amoebae. Analysis of the model shows the importance of the turnover rate of intracellular cAMP for cell streaming.

Höfer & Maini present a pure continuum model for *Dictyostelium* streaming and aggregation. This model shows behaviour consistent with two key experimental observations – first, during aggregation *in situ* the frequency of cAMP spiral waves increases while the speed of subsequent cAMP waves decreases; second, the appearance of a cell-free zone in the spiral core in the presence of caffeine which is thought to lower the excitability of the medium. This behaviour is also exhibited by the discrete-continuum hybrid models.

Finally, the paper by Savill & Hogeweg (III.4) considers a cellular automaton model in which cells are attached to each other by energy bonds which can be broken according to a certain probability function. By varying the key parameters, the model can simulate engulfment, cell dispersal and cell sorting. By coupling this model to a continuum model for chemical dynamics, it is shown that the composite model exhibits a number of key features. In particular, it can simulate a number of steps in the morphogenesis of *Dictyostelium discoideum*, including cell locomotion, mound formation, slug crawling, and cell sorting.

III.1

Models for Spatio-angular Self-organization in Cell Biology

Alex Mogilner (Davis)
Andreas Deutsch (Bonn)
Julian Cook (Los Angeles)

1 Introduction

In the past, models for population distributions have tended to dwell exclusively on spatial distribution and temporal dynamics. In nature, however, the relative orientations of individuals may have an important influence on their dynamics and interactions, increasing the level of complexity of corresponding biological phenomena. On the cellular level, a related *in vitro* phenomenon is the mutual contact and self-organization in mammalian cells such as fibroblasts on a surface (Elsdale & Bard 1972). When fibroblasts are grown in culture the following sequence of events is observed. As cells come into contact they assume a bipolar morphology. Cell density increases as a result of cell division and clusters of cells become aligned in parallel arrays (cells remain in a monolayer if collagenase has been added to the culture). The patchwork of arrays that form is relatively stable, though gradually the singularities in the alignment field annihilate each other so that finally all patches tend towards the same direction. The corresponding three-dimensional *in vivo* process plays an important role in morphogenesis.

Apart from the necessity of taking the angular distributions into consideration in the case of fibroblast populations, there is another feature of interaction that poses a challenging mathematical problem, namely, the non-local character of interactions between the cells. Stevens & Schweitzer (III.2 this volume) describe non-local models of aggregation depending on the exchange of chemical signals. Here, we focus on systems in which cells cannot sense substance concentrations but cell orientations. Non-local interactions call for integro-partial differential equation (integro-PDE) models. Such models (approximately) describe stochastic processes of cells correlating their orientations, processes in which any cell can meet and interact with a neighboring cell of any other relative orientation. Each such event is associated with some probability that one of the two cells will turn and/or shift and take on a new orientation and position. To account for all possible occurrences, one needs to sum up the probability of encounter, weighted by the likelihood of turning over all possible angles of contact. A theory of such stochastic processes is not treated here, rather we briefly describe three different types of approximations. This reasoning leads as a primary step to the formulation of an integral equation model; see also Civelekoglu & Geigant (I.12 this volume) for analysis of orientational modes in spatially homogeneous systems.

The clear disadvantage of integro-PDEs is that not all mathematical tools available to treat PDEs can be carried over to the analysis of integro-PDEs, see Levin & Segel (1985) for more discussion. However, linear stability analysis of integro-PDEs is not more difficult than that of PDE's. In the next section we discuss one of the integro-PDE models describing spatio-angular pattern formation and the results of its linear stability analysis.

In order to achieve results going beyond the linear stability analysis, certain simplifying approximations need to be made. A common approach to integro-PDEs is to assume that the solution changes slowly in space (spatial or angular coordinates). Then one expands the solution in Taylor series and keeps the first few terms. This technique reduces integro-PDEs to PDEs. Some aspects of partial differential equations are easier to analyse than individual-based models or integro-differential equations. Therefore, in the third section a reaction-diffusion-advection PDE model is analysed. This model can be either derived from a more general integro-PDE model, or independently formulated as another approximate description of the biological phenomenon.

PDEs as well as integro-PDE approaches are (macroscopic) mean-field descriptions of elementary (microscopic) stochastic processes. These can be formulated as biologically plausible rules of individual dispersal and interaction either continuous in space/time and/or orientation or as discrete approximation. As an example, a stochastic process is introduced, discrete in space, time and orientation serving as a cellular automaton model of cell alignment.

2   Integro-partial differential equation model

This section is based on the results of Mogilner & Edelstein-Keshet (1996). Populations of cells are represented using a continuum description. The density distribution C that represents the population of cells is a function of the space coordinate $r \in D$, orientational angle $\theta \in S$ of the cell's axis, and time t. The spatial domain, D, is either two-dimensional (e.g. a flat surface to which cells adhere in artificial *in vitro* growth conditions) or three-dimensional (e.g. cells *in vivo*). In the two-dimensional case, the angle describing the direction of a cell is $\theta \in [-\pi, \pi]$. In the three-dimensional case, the angle in spherical coordinates is $\psi \equiv (\phi, \vartheta)$ where $\phi \in [0, \pi]$ is the co-latitudinal angle and $\vartheta \in [0, 2\pi]$ is the longitudinal angle. The angular domain is denoted by S in both two and three dimensions. Here, we limit ourselves to the two-dimensional case. The generalization of the results to the three-dimensional situation is non-trivial, but straightforward. We consider only the cases in which the range of effective interaction between cells is at least a few orders of magnitude smaller than the size of the domain. This allows one to consider the spatial domain as infinite, and for the purposes of linear stability analysis, to ignore boundary effects. The model consists of a single equation:

$$\frac{\partial C}{\partial t} = \epsilon_1 \Delta_\theta C + \epsilon_2 \Delta_r C + C(K(C) * C). \tag{1}$$

Here the integral term is defined as:

$$K(C) * C(r, \theta) = \int_D \int_S L(C(r,\theta) - C(r',\theta'))G(r - r', \theta - \theta')C(r', \theta', t)d\theta' dr'. \tag{2}$$

The first term in Eqn. 1 describes random, density-independent turning of a cell. Parameter ϵ_1 is the corresponding rotational diffusion coefficient quantifying the turning rate. The second term accounts for spatial, density-independent diffusion characterized by the corresponding coefficient ϵ_2. Operators Δ_θ and Δ_r are Laplacians in angular and spatial variables, respectively. The third, non-linear, term is a rough approximation for the process of fast rotation of a small cluster of cells towards a more slowly moving big cluster and their final merging. The integral kernel is a product of two functions. We assume that the function L is odd, $L(-C) = -L(C)$, monotonic and bounded and $L'(0) > 0$. This density dependence of the function L reflects the tendency for a bigger cell cluster to grow at the expense of a smaller cluster. The function G reflects the rates of turning dependent on mutual position and orientation of the clusters as a result of spatio-angular interactions. This model has the property of conserving the total number of cells.

The detailed behaviour of the model depends in an interesting way on the symmetry properties of the kernel G. The latter are based on details of biological phenomena that are either observed through experiments or conjectured from some knowledge of the system. In the case of the fibroblasts, interactions causing alignment are known to be the weakest if the cells meet at 90°. Also, we simplistically consider "front" and "rear" ends of the cells as identical and approximate the kernel G as a doubly periodic function with maximum at 0° and minimum at 90° in the angular domain, and decreasing fast in spatial distance.

We performed linear stability analysis to determine when the solution that is homogeneous in the spatial and angular domains becomes unstable. The results of this analysis depend on the ratio of the amplitudes of the linear and non-linear terms in Eqn. 1. If ϵ_1 and ϵ_2 are large enough, the diffusion destroys order, and the homogeneous distribution is stable. If spatial diffusion is strong, while random turning is slow enough, then an angular pattern evolves while the spatial distribution remains homogeneous, i.e., the cells orient but they do not aggregate. If, on the other hand, cells turn randomly fast and their spatial diffusion is slow, then angularly disordered patterns with spatial inhomogeneities evolve. This means that the cells aggregate but do not align. One would then observe evolution of clumps of cells within which the cells are not directionally ordered (Fig. 1). Both of these cases do not resemble the experimental picture.

The most interesting case corresponds to the situation when both spatial and angular diffusion is slow. Then the linear stability analysis predicts simultaneous

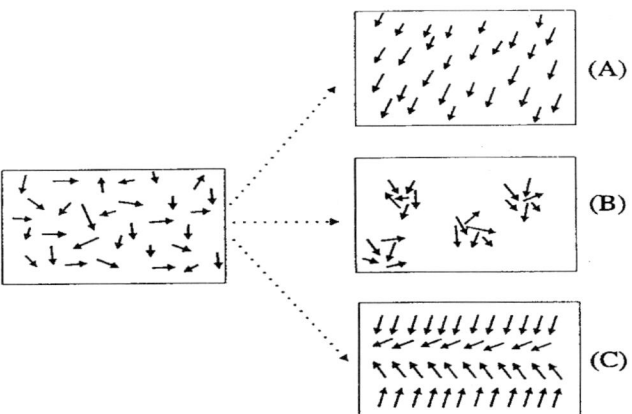

Figure 1: Depending on the growth protocol, any one of three possible bifurcation scenarios can take place. Starting from an initially disordered state (left), bifurcation may lead to (A) formation of angular order in a distribution that is spatially homogeneous, (B) aggregation and formation of regular clusters without common orientation, (C) development of "orientation waves" in patches of aligned objects.

aggregation of cells in patches and their alignment. Nevertheless, this analysis provides us only with establishing the fact of instabilities and with a hint on possible spatio-angular patterns.

3 Orientation tensors approach

This section is based on the results of Cook (1995). The integro-PDE model described above is sophisticated enough to account for various aspects of biological dynamics observed. At the same time it is rather complicated mathematically and sometimes tends to obscure the features of interest (the predominant direction of alignment, etc.). Our next step is, therefore, to approximate the orientation distribution using an apparatus of orientation tensors. This tool neatly describes the main features of the orientation distribution through space and time. Although elements of their derivation are formal and care must be taken to ensure that the equations are well-behaved, we maintain that the important properties of the system are preserved and that few artifacts are introduced. Thus, we have a method for constructing models for interacting, oriented populations which can be used to study the qualitative behaviour of such systems using equations about which much is known, reaction-diffusion-advection equations.

The idea is to write the Fourier series of the spatio-angular density distribution $C(r,\theta,t)$ in the form

$$C(r,\theta,t) = A_0(r,t) + A_2(r,t) \sin 2\theta + B_2(r,t) \cos 2\theta + \ldots \tag{3}$$

(Here A_i, B_i are the space dependent amplitudes of angular harmonics. All odd harmonics in the series are zero because of the symmetry constraints of the model: the front and rear ends of the cells are identical.) Then, assuming that the density distribution varies slowly in the angular variables, implying that certain linearization is appropriate, we can expand the truncated Fourier series into a Taylor series and express the result in terms of orientational tensors:

$$C(r,\theta,t) \simeq -(2\pi)^{-1} c(r,t) + (\pi/2)^{-1} \sum_{i,j} c_{ij}(r,t) \Theta_i \Theta_j. \tag{4}$$

Here

$$c(r,t) = \int C(r,\theta,t) d\theta \tag{5}$$

$$(c_{ij}(r,t))_{ij} = \left(\int C(r,\theta,t) \Theta_i \Theta_j d\theta \right)_{ij} = \begin{pmatrix} \frac{c(r,t)}{2} + Q(r,t) & R(r,t) \\ R(r,t) & \frac{c(r,t)}{2} - Q(r,t) \end{pmatrix},$$

where $\Theta_1 = \sin\theta$, $\Theta_2 = \cos\theta$ and R, Q are space dependent components of the first non-trivial orientational tensor (the one of the second order). The description of the systems in terms of these functions is informative, because they represent order parameters and the predominant directions of alignment. If we plug the series (4-5) into Eqn. 1 we obtain an infinite hierarchic system of linear PDEs for the orientational tensors. (In this section we consider the simplified model (Eqns. 1 & 2) where the spatial diffusion is neglected: $\epsilon_2 = 0$. Then $c(r,t) = c = \text{const}$). This system's equations for the functions Q and R depend on the higher third order tensors, which, in turn, are linked to the fourth order tensors and so on. Rigorous analysis of the complete system is impossible, and some closure approximation, making the system finite and self-consistent but non-linear, is needed. We used a simple linear closure approximation postulating that the third order orientational tensors are certain linear combinations of the lower order tensors, see Cook (1995) for details. This gives the following reaction-diffusion-advection system of equations for the components of the second order orientational tensors Q and R:

$$\frac{\partial Q}{\partial t} = Q(\beta^2 - (R^2 + Q^2)) + \alpha \nabla^2 Q \tag{6}$$
$$+ \gamma \left(6RQ_{xy} - 2QR_{xy} + Q(Q_{xx} - Q_{yy}) - R(R_{xx} - R_{yy}) \right)$$

$$\frac{\partial R}{\partial t} = R(\beta^2 - (R^2 + Q^2)) + \alpha \nabla^2 R \tag{7}$$
$$+ \gamma \left(-2QQ_{xy} + 2RR_{xy} - R(Q_{xx} - Q_{yy}) + 3Q(R_{xx} - R_{yy}) \right)$$

Here indices x, y stand for partial differentiation over the corresponding spatial variables. Parameters α, β, γ can be expressed in a cumbersome but straightforward way as functions of the diffusion constant ϵ_1 and of Taylor and Fourier transforms of the functions L and G from the definition of the integral kernel (Eqn. 2), respectively.

Eqns. 6 & 7 were solved numerically on a square domain with periodic boundary and random initial conditions. Results of the simulations reveal the picture of "waves of alignment" (Fig. 2): small clusters of aligned cells – whose (attraction) domains seem to build a "patchwork" with "orientation singularities" as corners – emerge from the angular disorder. There is no correlation between the orientation of the patches. The problem remains, if and how to determine the edges, by suitable tessellation, or just by connecting the "singularities". Finally, neighbouring patches merge leading to the formation of a small number of regions consisting each of totally aligned cells. Finally, one of the patches expands its fronts at the expense of the other patches, and at the end a stationary state of total alignment is achieved. This model scenario closely resembles the sequence of the events in fibroblast alignment experiments described above.

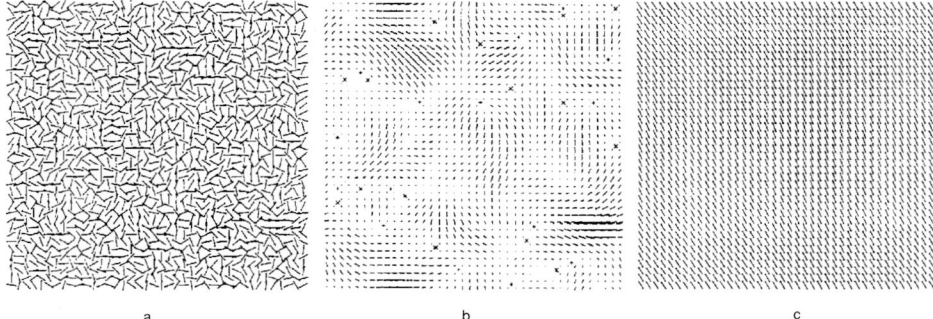

Figure 2: Simulations of the reaction-diffusion-advection equations for the spatial distribution of the orientational tensors. Eqns. 6 & 7 were solved numerically on a square domain with periodic boundary and random initial conditions. The initial random orientation (a) evolves into clusters of aligned cells (b). Crosses and plus signs mark the location of singularities. Neighbouring patches merge leading to total alignment (c).

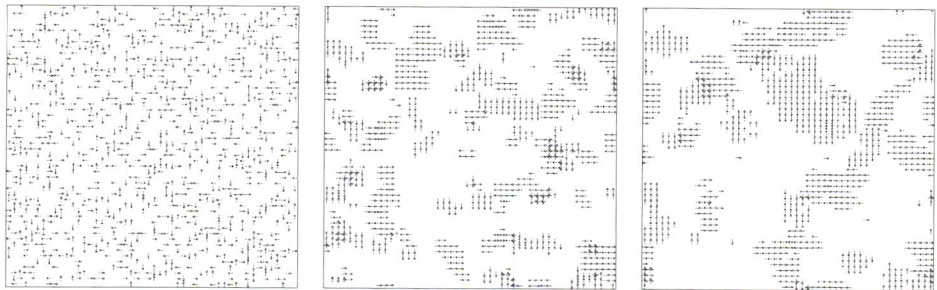

Figure 3: Temporal development of the lattice-gas cellular automaton. Distribution after $t = 0, 100$ and 1000 time steps (from left to right, $\beta = 2.5, n = 50$). Due to the transport properties of the model, formation of penetrating clumps (or patches) consisting of aligned cells can be observed.

4 Stochastic model of spatio-angular pattern formation

Here we turn our attention to an individual-based description of a cellular ensemble and introduce a stochastic process to tackle the dynamics of orientation-induced pattern formation. The model is based on a lattice-gas automaton approach motivated by Eqn. 1 (Frisch *et al.* 1986, Ermentrout & Edelstein-Keshet 1993, Deutsch 1995, 1996a) where temporal dynamics is formulated in terms of appropriately chosen automaton rules; see the box.

The model is discrete in space (a square lattice is assumed throughout), time and state domain. Cells are assumed to possess an intrinsic orientation and are able to actively move in the direction of their orientation (Eqn. 10). This inherent "transport behaviour" makes the model different from the spatio-angular models described in preceding subsections. The model is non-local – a finite sum ("local orientation field") mimics the integral in Eqn. 2 and counts cell orientations in a cell's local neighbourhood. Cells can change orientation interactively (compare kernel G in Eqn. 2) according to some intrinsic "behavioral pattern" which, for example, favors parallel alignment (Eqn. 9), like in the case of fibroblasts. The basic question is which kinds of structures result from such interplay.

In order to analyse the pattern-forming capability of the automaton model we introduce certain order parameters. In particular, the averaged mean velocity

$$\bar{\mu} = \frac{1}{N} \left| \sum_{r \in G} \sum_{i=1}^{b} c_i \eta_i(r) \right| \quad (0 \leq \bar{\mu} \leq 1)$$

is an efficient measure of "global directionality". Fig. 3 shows the outcome of one simulation. Clearly, formation of penetrating clumps of oriented cells and corresponding development of spatial correlations can be observed. Though there is analogy to the fibroblast alignment scenario in their initial period of migration when the cells are still dispersed, cells in these simulations keep on moving as swarms due to the transport property of the model.

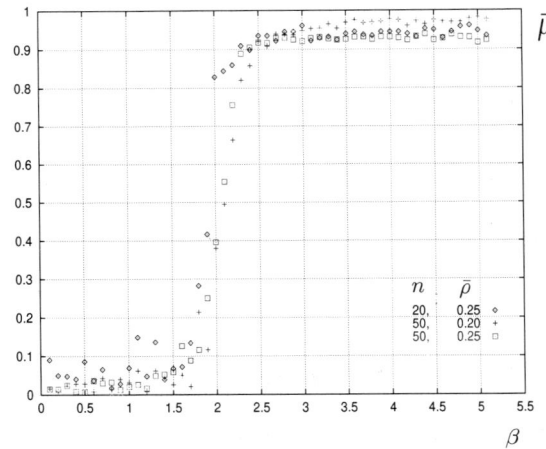

Figure 4: Mean velocity ($\bar{\mu}$) after 1000 time steps in dependence of the sensitivity (β) for various lattice sizes n (= 20, 50) and cell densities $\bar{\rho}$ (= 0.20, 0.25).

We systematically examined the influence of the sensitivity parameter β which determines the "strength of interaction" (Fig. 4). In the case $\beta = 0$ the local orientation field $O_{N(r)}$ has no influence on the interaction and a "pure diffusive" random walker model is obtained ($\bar{\mu} \approx 0$). This means that a homogeneous distribution is stable, as already pointed out for the continuous model. For increased β the actual value becomes important and the outcome of an interaction is chosen amongst those states for which the functional $O_{N(r)} \otimes s$ is maximal. In the $\bar{\mu}$-plot over β (Fig. 4) there is indication for a dynamic phase transition at $\beta \approx 2$ which is also visible in the density distribution (Deutsch 1996b). We currently investigate the character of this transition. Phase transitions were already observed in related models (Vicsek *et al.* 1995). Ongoing work is dedicated to characterize the dynamic instabilities of the model, in particular by means of suitable approximations of the underlying distributions and subsequent linearization. The phase transition could already be reproduced in a mean-field approximation by numerically solving the nonlinear Boltzmann equation associated with the cellular automaton (Bussemaker *et al.* 1997).

Definition of the cellular automaton

A. Geometry: square $(n \times n)$ lattice $D \subseteq \mathbb{Z}^2$ with nodes r. Periodic boundary conditions are assumed throughout.

B. State space: $S = \{s = (s_i)_{i=1}^4, s_i \in \{0,1\}\}$, $\eta(r) = (\eta_i(r))_{i=1}^4 \in S$ denotes a configuration at node r. $s_i = 1$ (0) means that there is one (no) particle at node r with orientation i, where i indicates one of the four directions in the square lattice (i=1: right, 2: top, 3: left, 4: down). Node density $\rho(r) = \sum_{i=1}^4 \eta_i(r)$ is thus limited by 4 (exclusion principle).

C. Local orientation field: $N(r) = \{r + c_1, r + c_2, r + c_3, r + c_4\}$ contains the next neighbours of node r ($c_1 = (1,0), c_2 = (0,1), c_3 = (-1,0), c_4 = (0,-1)$). Then, the orientation field $O_{N(r)} = \sum_{p=1}^4 \eta(r + c_p) \in \{0, 1, \ldots, 4\}^4$ gives the summed orientations of all next neighbour configurations of node r.

D. Temporal dynamics: Initially ($t = 0$) N particles are randomly distributed on the lattice leading to an average node density $\bar{\rho} = \frac{N}{n \times n \times 4}$. Temporal development $t \longrightarrow t + 1$ is performed by means of subsequent application of interaction and transport steps to every node in the lattice, simultaneously.

Interaction: $\eta(r) \longrightarrow \eta^I(r)$ according to the transition probability (depending solely on the distribution in a local neighbourhood)

$$A_{s,s'}(O_{N(r)}) = P(\eta^I(r) = s' / \eta(r) = s, \eta|_{N(r)}) \tag{8}$$

$$= \begin{cases} \frac{1}{Z(s)} e^{\beta \cdot (O_{N(r)} \otimes s')} & \text{if } \rho(s') = \rho(s) \\ 0 & \text{else} \end{cases}.$$

Here, $Z(s) = \sum_{s' \in S, \rho(s')=\rho(s)} e^{\beta \cdot (O_{N(r)} \otimes s')}$ is a normalization constant and

$$O_{N(r)} \otimes s : \mathbb{N}_0^5 \times S \longrightarrow \mathbb{N}_0$$

is a bilinear functional. Different interaction rules $O \otimes s$ can mimic various biological hypotheses (Deutsch 1996a). Favoring parallel orientations is incorporated by means of the simple scalar product:

$$O \otimes s := <O, s>. \tag{9}$$

The interpretation is as follows: If, for example, $s = (1, 0, 1, 0)$ and $O = (1, 1, 2, 2)$ are pre-interaction state and local orientation field, respectively, then, with high probability a post-interaction state $s' = (0, 0, 1, 1)$ would result since $\rho(s) = \rho(s') = 2$ and according to the definition $O \otimes s$ becomes maximal if $s' = (0, 0, 1, 1)$. Furthermore, $\beta \in \mathbb{R}_0$ is a sensitivity parameter determining the 'strength of interaction'.

Transport:
$$\eta_i(r) \longrightarrow \eta_i^T(r) := \eta_i(r - c_i) \tag{10}$$

5 Conclusions

Similar to the models presented by Civelekoglu & Geigant (I.12 this volume) dedicated to the formation of angular structure in oriented F-actin skeletons the models introduced here which in addition take care of spatial heterogeneity (and transport in the case of the cellular automaton) share common qualitative features. In particular, they are mass-invariant and the kernels (in the continuous models) as well as orientational functionals (in the discrete stochastic process) possess equal symmetry properties. Depending on the ratio of diffusion constants (orientational/spatial in the PDE models) and "interaction strength" (in the discrete model) various scenarios are observed ranging from no order over scattered clusters to homogeneously distributed patterns. Simulation results yield the following picture: neglecting transport (as in the orientation tensors approach) implies that a random initial distribution might develop into a pattern appearing either as a tessellation of more or less aligned smaller patches or as a huge cluster of global alignment. Aggregation is only possible if transport, either passive (diffusion, integro-PDE) or active (orientation tensors approach), is incorporated into the model. Then, cells cannot just reorient without changing local position but reach other patches of "alignment activity". As a result, oriented cell rafts keep on crawling and "feeding" on smaller oriented rafts by incorporating smaller into bigger clusters if parameters are chosen appropriately. In the cellular automaton model no hard core repulsion is assumed implying that cells can move upon each other leading to the picture of penetrating clusters. In addition, patch boundaries are much more complicated.

We stress again that the crux of our approach is the particular way interaction leads to communication. In the three models presented, cells communicate directly via sensing of orientations of neighbouring cells while in (chemo-) tactic models (see Stevens & Schweitzer III.2, Dallon *et al.* III.3, Savill & Hogeweg III.4 this volume) communication is achieved indirectly by means of sensing concentrations of chemoattractant substances. The question is how cells can measure orientations of their neighbours. One possibility is by means of differential stimulation of mechano-receptors or external organelles. For example, myxobacteria possess pili, extracellular appendages, which seem to play an important role in the cooperative motion of these microorganisms (Dworkin & Kaiser 1993).

Nevertheless, in most biological systems various mechanisms of communication and interaction are prevalent. Therefore, hybrid models should be analysed in the future.

Acknowledgements We wish to thank W. Alt, E. Geigant and L. Edelstein-Keshet for valuable discussions and comments. A. Deutsch's work is supported by the Deutsche Forschungsgemeinschaft "DFG" (SFB 256 "Nonlinear Partial Differential Equations" at the University of Bonn).

# III.2

Aggregation Induced by Diffusing and Nondiffusing Media

Angela Stevens (Heidelberg)
Frank Schweitzer (Berlin)

1 Introduction

Gathering of individuals is a widespread phenomenon in biology. The reasons are, for instance, to give each other shelter, to reproduce, to explore new regions, to feed or to endure starvation conditions. In the case of myxobacteria (Dworkin & Kaiser 1993) it is known that they glide cooperatively and aggregate under starvation conditions. During gliding they prefer to use paths which were laid down by themselves. When the final aggregation takes place, they glide in streams towards developing mounds which later grow to form so-called fruiting bodies. Other examples of collective aggregation are known from larvae, e.g. of the bark beetle *Dendroctonus micans* (Deneubourg et al. 1990) which clumps to feeding groups. Group feeding often improves individual survival, or allows better exploitation of food resources (Tsubaki 1981, Tsubaki & Shiotsu 1982).

In many cases, the aggregation process occurs via exchange of chemical signals between the entities. Chemotaxis is one of the major communication mechanisms and has been found, for instance, in the aggregation of cells as human leukocytes (Gruler & Boisfleury-Chevance 1994, Tranquillo & Alt 1994) and in the slime mold amoebae (Keller & Segel 1970).

In order to discuss the dynamic process of aggregation, we first introduce a discrete model suitable for lattice-based computer simulations, where particles interact by locally changing the surface they move on. This model can be applied either to the formation of trails or to the formation of aggregates, as found for myxobacteria and insect larvae. The discrete model is approximated by a PDE-system, a so-called chemotaxis system. We discuss critical parameters of the aggregation process, such as initial population density, production rate of the chemotactic substance, the presence or lack of diffusion of this substance, and dependence on the chemotactic sensitivity.

2 The discrete interacting particle model

Our model is based on particles moving on a two-dimensional plane which are able to produce and lay down a substance that changes the surface state locally. The particles are sensitive to this substance which may change their further movement, thus resulting in a non-linear feedback between the particles and the secreted substance. This model has been applied to interactive structure formation processes, for instance, in Stevens (1992), Schweitzer & Schimansky-Geier (1994), Schweitzer et al. (1997), Schimansky-Geier et al. (1996) and Othmer & Stevens (1997). Here the particles should represent a biological species. We will discuss first the case of myxobacteria which produce so-called slime trails. Each particle sensing a slime trail will prefer to encounter it instead of gliding into an untrailed area. Second we consider the case of larvae that aggregate to feed. The aggregation occurs due to a chemotactic response to a diffusing chemical substance produced by the larvae. This might also be the case for myxobacteria.

Both cases can be discussed within a unified approach. In a discrete model, the movement of the particles is described as a random walk on a two-dimensional lattice, where A^2 denotes the lattice size. For the simulations, two different kinds of lattice are used: (i) square lattice, (ii) triangular lattice, which has the advantage of spatial isotropy. In each case, we assume periodic boundary conditions for the lattice and define N_x to be the set of nearest neighbours of $x \in A^2$.

To keep close to biological motion, movement of the particles is characterized by the directional persistence q, indicating that the particles tend to move in the direction of the last step rather than choosing a completely random direction. Let $h_i(t)$ be the orientation of the ith particle at time t and location x_0, that is, the direction in which it has been moving in the previous time step. Then the preference for the direction of orientation is described by

$$d(x_0, y, h_i(t)) = \begin{cases} q \geq 1.0, & \text{if } y \text{ is a neighbour in direction } h_i(t) \text{ of the orientation of the particle which is located at } x_0 \\ 1.0, & \text{else.} \end{cases}$$

Additionally, the movement of the particles is affected by the substance to which they are sensitive. This can be the slime, as in the case of myxobacteria, or a chemotactic substance, as in the case of larvae and myxobacteria. Let $S : \mathbb{N} \times A^2 \to \mathbb{R}_+$ describe the concentration of the substance in time and space, where $S(0, x) = 1.0$. Then, the probability that the ith particle, located at x_0 at time t, moves to point $x \in N_{x_0}$ is

$$P_{i,x_0}(t+1, x) = \frac{S(t,x) \cdot d(x_0, x, h_i(t))}{\sum_{y \in N_{x_0}} S(t,y) \cdot d(x_o, y, h_i(t))} . \tag{1}$$

Let $I_x(t)$ denote the number of particles covering the point $x \in A^2$ at time t. Then $S(t, x)$ can be changed during the next time step by decay via decomposition due to the loss rate λ, by production via $Q(S(t,x), I_x(t))$ and by diffusion, with a diffusion constant D_S. Hence,

$$S(t+1, x) = (1-\lambda)S(t,x) + Q(S(t,x), I_x(t)) - D_S \left(S(t,x) - \tfrac{1}{N_x} \sum_{y \in N_x} S(t,y) \right) .$$

We assume that the particles produce "$\alpha \times$ the existing concentration" at the point where they are located and, additionally, a fixed amount β, where $\alpha, \beta \geq 0$. So

$$Q(S(t,x), I_x(t)) = \left(\alpha^{I_x(t)} - 1\right) S(t,x) + \beta \frac{\alpha^{I_x(t)} - 1}{\alpha - 1}.$$

For $\alpha = 1$ a linear production term results: $Q(S(t,x), I_x(t)) = \beta I_x(t)$ since $\frac{\alpha^{I_x(t)}-1}{\alpha-1} = I(x)$ if $\alpha \to 1$.

When dealing with the model for the chemotactic response of larvae we use $\alpha = 1$. For the case of myxobacteria we discuss both linear ($\alpha = 1$) and superlinear ($\alpha > 1$) production of the slime, which is similar to Davis (1990) who investigated a one-dimensional reinforced random walk without diffusion of the substance. He found that a particle localizes at a random place if the substance is produced superlinearly and does not localize if it is produced only linearly.

Our model describes a reinforced random walk which may result in an aggregation of the particles due to the non-linear coupling between the concentration of substance and the movement of the particles. A more realistic model of myxobacterial aggregation is given in Stevens (1992) where both slime trail following and response to a diffusing chemoattractant are needed to get stable centres of aggregation. Here, we restrict ourselves to the simpler model of only one substance in order to get a better control of the parameters and to check their relevance.

3 Computer simulations

In our model of interacting particles, there is an interplay between the parameters describing the performance of the particles themselves, like the persistence, q, or the production of substance, α, β, and the parameters which describe the evolution of the chemical substance, such as decay, λ, or diffusion, D_S.

In our computer simulations we consider a square lattice A^2 of 70×70 grid points with periodic boundary conditions. This gridsize is chosen to avoid excessively strong boundary effects and at the same time to guarantee a clear output. Initially, $N = 1000$ particles are randomly distributed on the inner square lattice A_1^2 of 30×30 grid points (see Fig. 1a). At each time step, the particles move to one of their four nearest neighbours and interact with the surface as described above.

No diffusion, no decay, linear production, no persistence: For $D_S = 0.0, \lambda = 0.0, \alpha = 1.0, \beta = 0.1$, with $S(0,x) = 1.0$, and $q = 1.0$, the simulations show swarming of the particles if the chemical substance is laid down and measured inbetween the grid points (see Fig. 1b, 200 time steps), as described by Davis (1990) for a single particle.

If the substance is laid down and measured directly on the grid points as described in the model equations, a stronger taxis effect results; compare Fig. 2a and Othmer & Stevens (1997). If the initial particle density, N/A_1^2, exceeds a critical value, the swarm remains more local and the particles form small clusters.

Figure 1: a) (left) Initial conditions. b) Reinforced random walk of 1000 particles, where the substance is laid down and measured in between the grid points. Grey dots mark the paths the particles have used. Black squares mark a single particle, the squares of different grey levels mark 2 to 9 particles and white squares mark 10 and more particles.

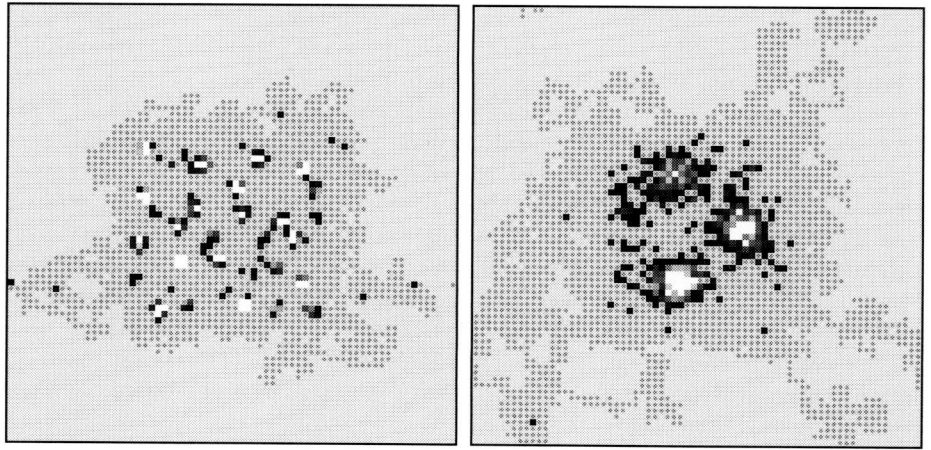

Figure 2: a) (left) Reinforced random walk of 1000 particles, where a non-diffusing substance is laid down on the grid points. b) Same situation as in a) but with diffusion of the substance.

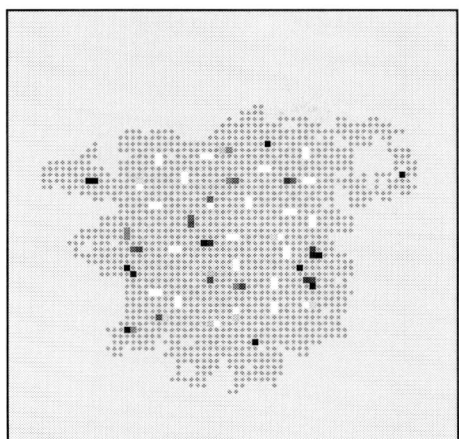 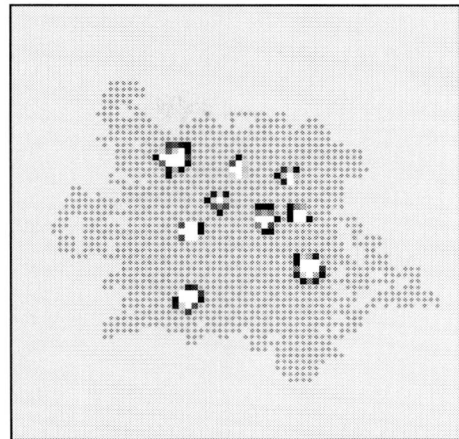

Figure 3: (a) (left) Superlinear production of the substance. (b) Same situation with diffusion of the substance.

Diffusion, no decay, linear production, no persistence: Fewer clusters appear if the substance diffuses with $D_S = 0.05$, but more particles are trapped, compared to the situation without diffusion; see Fig. 2b, 1000 time steps. In this case the initial conditions play a less important role since diffusion of the attractive information effaces initial clusters and guides particles from regions with only few particles towards regions with developing aggregation centres. For increasing D_S, the aggregation centres interfere even more than shown in Fig. 2b and no separated clusters are formed. Hence, in this case, the existing clusters can only be stabilized when the production of substance is increased and a higher initial concentration of the particles is given.

No diffusion, no decay, superlinear production, no persistence: For a superlinear production ($\alpha = 1.01$) and no diffusion of the chemical substance, well separated aggregates are formed very quickly; see Fig. 3a, 200 time steps. This effect is amplified if the particles have a high initial density. However, if the particles are more distant initially, they are trapped in many very small clusters.

Diffusion, no decay, superlinear production, no persistence: Adding diffusion ($D_S = 0.05$), the streaming towards the clusters becomes stronger and the clusters get bigger and become well separated (see Fig. 3b).

Decay of the chemical substance amplifies the effect of aggregation; however, particles far away from the centres have a tendency to jump back and forth between two grid points. This effect will be smoothed out by increasing D_S, which again results in streaming towards the clusters.

Diffusion, decay, superlinear production, persistence: To keep close to the myxobacterial behaviour, the simulations are now carried out with directional persistence, $q = 3.0$. The particles swarm out and return to the aggregates more easily once they have chosen the correct direction. This behaviour is close to reality and can be supported by choosing $D_S \neq 0$ (Fig. 4 shows 5000 particles after 200 time steps). Here diffusion, $D_S = 0.05$, and decay, $\lambda = 0.05$, stabilize the aggregation centres. In a more complex model for myxobacterial aggregation described by Stevens (1992), superlinear slime production does not affect aggregation in the way discussed here. Further research will be done to understand this.

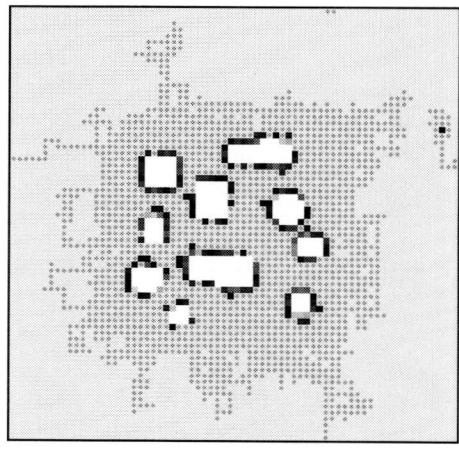

Figure 4: Persistence of the particles and decay of the substance.

In our simple model a fine tuning of the parameters accounts for swarming of the particles, aggregation and stabilization of the aggregation centres. A further explanation of the computer simulations is given in the following section based on a continuity approximation.

4 Continuity approximations for the discrete particle model

The interacting particle model introduced in the previous section can be approximated by a continuous model. First a diffusion approximation can be carried out for the simplest version of the four nearest neighbour reinforced random walk of *one* particle in two dimensions, that is Eqn. 1, where $q = 1.0$. This results in the following chemotaxis equation for the probability of the particle to be located at point x at time t (Othmer & Stevens 1997):

$$\partial_t p = D \nabla \left(\nabla p - \chi \frac{p}{s} \nabla s \right). \tag{2}$$

Here $\chi = 2$ and $s(t,x)$ denotes the density of the chemical substance, which satisfies the reaction diffusion equation $\partial_t s = D_s \Delta s + b(s,p) - \lambda s$, where D_s and $\lambda \geq 0$, and b is a suitable functional for the growth of the chemical substance.

In the following we consider $b(s,p) = \beta \cdot p$. For $D_s = 0, \lambda = 0$ and with an initial peak for $p(0,x)$, only a high production rate of s accounts for blowup of p in finite time. For a low production rate of s an initial peak of p breaks down. This is closely related to Davis' (1990) results on the reinforced random walk of a single particle, where the approximation yields Eqn. 2 with $\chi = 1$. Then blowup for p occurs only for superlinear growth of s, which reflects the localization result. If the decision of the particles is gradient based, i.e. the transition rates equal $a_1 + a_2(s(t,x) - s(t,x'))$, where $x' \in N_x$ and $a_1, a_2 \in \mathbb{R}$, one obtains

$$\partial_t p = D' \nabla (a_1 \nabla p - 2 a_2 p \nabla s)$$

which has the form

$$\partial_t p = D \Delta p - \nabla p (\chi \nabla s), \tag{3}$$

where $D = D' a_1$ and $\chi = 2 D a_2 / a_1$ (Othmer & Stevens 1997).

Now Eqn. 3 can be derived as a limiting equation for the behaviour of *many* interacting particles. In our model the chemical substance determines the motion of the particles. They search for local maxima of $s(t,x)$, so the dynamics of the ith Brownian particle in the N-particle system is described by the following Langevin equations:

$$\frac{dx_i}{dt} = v_i, \qquad \frac{dv_i}{dt} = -\gamma v_i + \nabla s(t, x_i) + \sqrt{2\epsilon\gamma}\, \xi_i(t), \tag{4}$$

where γ is the friction coefficient and $\xi_i(t)$ is Gaussian white noise with intensity ϵ. The probability of finding N particles in the vicinity of $x_1, \ldots, x_N$ on a surface A at time t can be formulated in terms of the canonical N-particle distribution function $P(t, x_1, \ldots, x_N)$. In the limit of strong damping, $\gamma \to \infty$, it reads:

$$\frac{\partial}{\partial t} P(t, x_1, \ldots, x_N) = -\sum_{i=1}^{N} [\nabla_i (\chi \nabla_i s) P - D \Delta_i P] . \tag{5}$$

Here $\chi = 1/\gamma$ denotes the mobility of the particles and $D = \epsilon/\gamma$ the spatial diffusion coefficient for the density of the Brownian particles. In the mean field limit, we obtain the particle density $p(t,x)$ from:

$$p(t,x) = \int dx_1 \ldots dx_{N-1}\, P(t, x_1, \ldots, x_{N-1}, x) . \tag{6}$$

Finally, Eqn. 3 results from integrating Eqn. 5 due to Eqn. 6.

From a bifurcation analysis of the mean field equation one finds the following condition for the instability of the homogeneous state p_0, s_0:

$$\beta\, p_0 / \gamma > \epsilon (\lambda + \kappa^2 D_s),$$

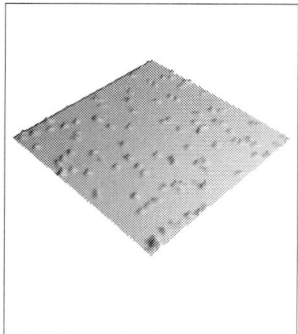 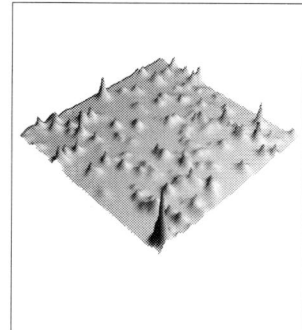 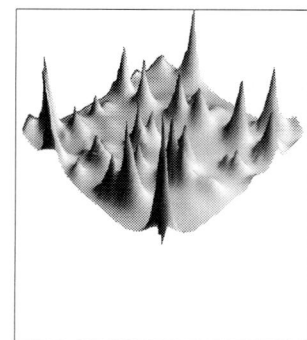

Figure 5: Evolution of $s(t,x)$ generated by $N = 100$ particles during the growth regime. Time in simulation steps: a) (left) $t = 10$, b) $t = 100$, c) $t = 1000$ (lattice size: $A = 100 \times 100$).

 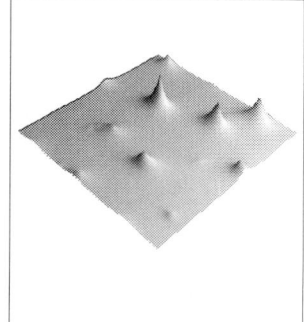 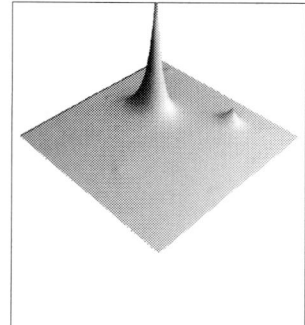

Figure 6: Evolution of $s(t,x)$ generated by $N = 100$ particles during the competition regime. Time in simulation steps: a) (left) $t = 1000$, b) $t = 5000$, c) $t = 50000$. The density scale is $0.1 \times$ scale of Fig. 5. Hence, the left part of Fig. 6 is the same as the right part of Fig. 5.

where $p_0 = N/A$, $s_0 = p_0 \beta / \lambda$, A is the surface area, and κ the wave number of a fluctuation. If β is large, or if ϵ is small, the particles quickly form several clusters, which result in a local growth of $s(t,x)$, as shown in the simulations below.

Initially the particles are distributed on a triangular lattice and s is produced linearly. First the spikes of the chemical substance grow independently as presented in Fig. 5. If the production of the substance becomes stationary, i.e. the decay compensates for the production, a transition into a second regime occurs. Here, the different spikes, which have bound the Brownian particles, compete, leading to a decrease in the number of spikes (see Fig. 6).

The selection among the different spikes can be described in terms of a selection equation of the EIGEN-FISHER type (Schweitzer & Schimansky-Geier 1994), which is

$$\partial_t s(t,x) = \frac{\gamma \, s_0}{\langle \exp[(\chi/\epsilon)\, s(t,x)]\rangle_A} s(t,x)$$
$$\times \{\frac{\exp[(\chi/\epsilon)\, s(t,x)]}{s(t,x)} - \frac{\langle \exp[(\chi/\epsilon)\, s(t,x)]\rangle_A}{s_0}\} + D_s \Delta s(t,x) \,,$$

where $\langle \exp[(\chi/\epsilon)\, s(t,x)]\rangle_A = \frac{1}{A}\int_A \exp[(\chi/\epsilon)\, s(t,x')] \, dx'$ is similar to the mean "fitness", representing the global selection pressure. Further growth of a spike occurs only as long as $\exp[(\chi/\epsilon)\, s(t,x)] \, s_0 > \langle \exp[(\chi/\epsilon)\, s(t,x)]\rangle_A \, s(t,x)$ holds. Otherwise, the spike will decay again due to the competition process. Provided a suitable neighbourhood, eventually the largest spike will survive, as indicated by the simulations.

A rigorous approach to derive density equations from an interacting many particle system can be found in Stevens (1992). The position of the ith particle $x_N^i(t)$ in a N-particle system is given by $\frac{dx_N^i(t)}{dt} = \chi \nabla s_N(t, x_N^i(t)) + \sqrt{2D}\,\xi_i(t)$. Here $s_N(t, x_N^i(t))$ describes the amount of the chemical substance not only at the point $x_N^i(t)$ but also in its neighbourhood, due to a weight depending on N. Hence the interaction is not local. It is chosen to be moderate, which means that the main interaction range of each individual particle shrinks for $N \to \infty$, but the number of other particles in this range tends to ∞. Under these conditions the many particle system and the continuous model are a good approximation of each other.

5 Other results on the limiting equations

Several results are known about chemotaxis equations. The stability analysis for a quite general situation was done by Schaaf (1985). Blowup results were given by Jäger & Luckhaus (1992), Nagai (1996), Herrero & Velasquez (1995, 1996a, 1996b) and Biler (1995). The qualitative behaviour in generally nonsmooth domains was considered by Gajewski & Zacharias (1996). In all cases the substance diffuses, sometimes with $D_s \gg 1$. For chemotaxis equations with $D_s = 0$ qualitative results were given by Rascle & Ziti (1995) (blowup), Othmer & Stevens (1997) and Levine & Sleeman (1997). In the last two papers blowup of p in finite time, finite stable peaks and collapse of developing peaks are disussed.

6 Discussion

We have simulated the aggregation of interacting particles due to a substance produced by the particles themselves. Different parameters have been discussed with respect to their effect on aggregation. We note that a large production rate of the slime, combined with a supercritical initial concentration of the particles, results in the formation of aggregation centres but, on the other hand, prevents swarming. Diffusion of the slime enhances the aggregation effect but effaces the centres a little bit. Decay of the slime stabilizes the aggregation centres but increases the chance that single particles are trapped in certain regions. The persistence of particle movement does not change this qualitative behaviour; however, it makes the simulations more realistic. Further research should be done to compare the qualitative behaviour of the interacting particle model and its continuous approximation. From a numerical point of view, it would be interesting to use the particle model with moderate interaction to simulate the chemotaxis system.

III.3

Models of *Dictyostelium discoideum* Aggregation

John C. Dallon, Hans G. Othmer (Salt Lake City)
Catelijne Van Oss, Alexandre Panfilov, Paulien Hogeweg (Utrecht)
Thomas Höfer, Philip K. Maini (Oxford)

1 Introduction
Philip Maini & Thomas Höfer

Since its discovery in the 1940's, the life cycle of the cellular slime mould *Dictyostelium discoideum* has attracted the interest of developmental biologists. It involves a relatively simple transition from unicellular to multicellular organization. Briefly, amoebae feed on bacteria in the soil and divide. Exhaustion of the food supply triggers a developmental sequence which leads, via cell aggregation, to the formation of a migrating slug-like "organism". The slug eventually culminates into a fruiting body, aiding the dispersal of spores from which, under favourable conditions, new amoebae develop. To date a variety of species in different taxonomic groups are known whose life cycles follow a similar pattern (Margulis & Schwartz 1988). Over the past fifty years, many of the molecular and cellular mechanisms which are involved in cell aggregation, collective movement and differentiation have been identified, and much work is devoted to the understanding of the interaction of these mechanisms in shaping *Dictyostelium* development. Mathematical modelling has proved a useful tool with which to study these interactions on a quantitative basis.

A typical aggregation sequence in *Dictyostelium discoideum* begins by the formation of concentric and spiral concentration waves of the extracellular messenger cyclic 3'5'-adenosine monophosphate (cAMP) which induce cell chemotaxis in periodic steps towards the aggregation centre (Alcantara & Monk 1974, Tomchik & Devreotes 1981). The onset of multicellularity is marked by the establishment of a branching pattern of cell streams in which direct cell-cell contacts are established.

The first attempt to develop a quantitative theory of slime mould aggregation was made by Keller & Segel (1970). They proposed a system of two coupled partial differential equations for the dynamics of the cell density, incorporating an advective chemotaxis term as suggested earlier by Patlak (1953), and for the change of cAMP. The emergence of aggregation centres is linked to a chemotaxis-driven instability in the model which leads to cell clustering. While this description appears to be valid for *Dictyostelium* species without periodic chemoattractant waves, such as *D. minutum* and *D. lacteum*, the situation for *Dictyostelium discoideum* is now known to be more complex. Cohen & Robertson (1971a) developed a rule-based model of cAMP signal relay which accounted for its pulsatile wave-like character, and they considered cell movement in a subsequent paper (Cohen & Robertson, 1971b). Nanjundiah (1973) carried out a stability analysis of the Keller-Segel system on a circular domain with a signalling centre and found unstable "azimuthal" modes which he linked to the occurrence of cell streaming. cAMP signalling and cell movement was combined in the rule-based computer simulations of *Dictyostelium* aggregation by Parnas & Segel (1977, 1978) and MacKay (1978). In particular, the two-dimensional simulations by MacKay appear to be the first demonstration of cell streaming in an aggregation model.

While these earlier rule-based models largely relied on phenomenological observations on the aggregation dynamics, the elucidation of the molecular mechanisms of the cAMP signalling dynamics since the 1980's (e.g. Devreotes 1989 and references therein) was accompanied by the development of mechanistic models of cAMP signalling. The models incorporate the detailed biochemical dynamics of both the activation and desensitization of the adenylate cyclase pathway by binding of extracellular cAMP to its surface receptors, and concentrate in detail on the temporal aspects of signalling as revealed by experiments on stirred cell suspensions. The models essentially differ with respect to the presumed mechanisms of cAMP-induced desensitization. These include receptor phosphorylation (Martiel & Goldbeter 1987), desensitization through inhibition of adenylate cyclase via external calcium influx (Rapp *et al.* 1985, Othmer *et al.* 1985, Monk & Othmer 1989) and G-protein mediated desensitization (Tang & Othmer 1994, Goldbeter 1996). Although these and other mechanisms have been implicated by experiments, the question of which of these play the dominant role *in situ* remains a source of controversy. It is not unlikely that multiple mechanisms operate concomitantly (Van Haastert *et al.* 1992). Current evidence suggests that the G-protein mechanism is the primary one.

Incorporation of diffusion of extracellular cAMP into these models yields a description of the signalling dynamics in a stationary cell layer, which turns out to be a valid approximation for the situation at the beginning of aggregation (Tyson *et al.* 1989, Monk & Othmer 1990, Tang & Othmer 1995). With the help of these reaction-diffusion models, the experimentally observed cAMP waves have been characterized as a particular case of chemical wave patterns in so-called excitable media (Tyson & Murray 1989). However, the signalling models neglect cell movement and thus can not describe the actual aggregation process and in particular cell streaming.

This contribution consists of three models that couple cell movement with the chemical dynamics leading to cell streaming. The models of Dallon & Othmer (Sect. 2) and Van Oss *et al.* (Sect. 3) consider cells as discrete entities responding to a continuum field of chemoattractant concentration. Höfer & Maini (Sect. 4) model the cell distribution as a continuous density, coupled with the chemical dynamics.

2 A discrete cell model with adaptive signalling

John C. Dallon and Hans G. Othmer

In this section we describe a model in which *Dictyostelium discoideum* cells are treated as discrete points that detect and respond to the continuum field of the chemoattractant. This model, details of which can be found in Dallon & Othmer (1996) (I hereafter), is comprised of two main parts: (i) the mechanism for signal transduction and cAMP relay response, and (ii) the cell movement rules. The transduction of the extracellular cAMP signal into the intracellular signal is based on the G-protein model developed in Tang & Othmer (1994, 1995), and the reader is referred to those papers for details. The equations for the intracellular dynamics of the ith cell can be written as a system of the form

$$\frac{d\mathbf{w}^i}{d\tau} = \mathbf{G}^i(\mathbf{w}^i, w_5), \qquad (1)$$

where $w_5(\mathbf{x}, \tau)$ is proportional to extracellular cAMP, the components w_j^i, $j = 1,\cdots,4$, represent intracellular quantities in the signal transduction and cAMP production steps, and $\mathbf{w}^i$ is a vector of these four internal variables for the ith cell. When the cells are treated as discrete points the evolution of extracellular cAMP is governed by the partial differential equation

$$\frac{\partial w_5(\mathbf{x},\tau)}{\partial \tau} = \Delta_1 \nabla^2 w_5(\mathbf{x},\tau) - \hat{\gamma}_9 \frac{w_5(\mathbf{x},\tau)}{w_5(\mathbf{x},\tau)+\gamma_8}$$

$$+ \sum_{i=1}^N \frac{V_c}{V_o}\delta(\mathbf{x}-\mathbf{x}_i)\left(sr(w_4^i) - \gamma_7 \frac{w_5(\mathbf{x},\tau)}{w_5(\mathbf{x},\tau)+\gamma_6}\right). \qquad (2)$$

Here $\mathbf{x}_i$ denotes the position of the ith cell, the first term represents diffusion of cAMP, the second represents the degradation of cAMP by extracellular phosphodiesterase, and the summation represents the localized sources and sinks of cAMP at the cells. The precise definitions of the variables and the parameter values can be found in Tang & Othmer (1995).

The second part of the model involves the cell movement rules, the following two of which are common to most of the simulations. They are (i) the cell moves in the direction of the gradient of cAMP when the motion is started; (ii) the cell moves at a speed of 30 microns per minute (Alcantara & Monk 1974). Various rules for initiating movement and determining its duration were explored.

As we show in I, formal rules based on a fixed duration of movement can produce aggregation. However, if the duration is too short aggregation does not occur, but by adding other mechanisms such as directional persistence the problem can be corrected. For example, when the duration is set at 20 seconds, which is the experimentally observed turning time (Futrelle *et al.* 1982), the cells do not aggregate successfully. By adding cell polarization (Varnum-Finney *et al.* 1987) or a memory of recently encountered gradients, the aggregation patterns are restored.

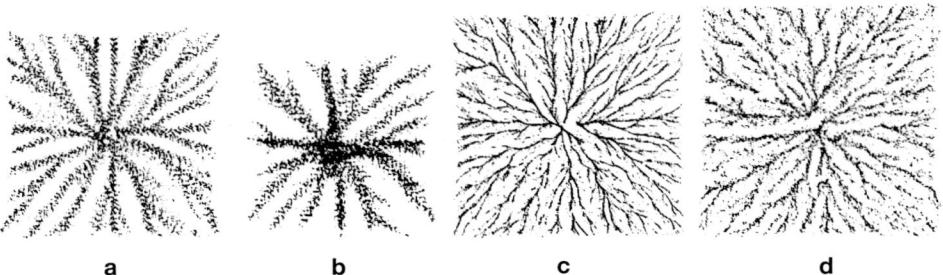

a b c d

Figure 1: Aggregation patterns which formed with a pacemaking region at the centre and different rules for the duration of movement. Cells move for 100 seconds (a), 500 seconds (b), and according to an internal variable (c). In (d) the direction of movement is randomly perturbed by up to 90 degrees from the direction of the local gradient. All simulations are shown at 150 minutes. Eqn. 2 is solved using an Alternating Direction Implicit method with the intercellular variables lagged in time. The domain is 1 cm by 1 cm with 200 grid points in each direction. The number of cells used was 10089, each weighted by 16, which corresponds to a volumetric density of about 0.2.

Because the cAMP signal a cell sees is very rough, the cell may move away from the aggregation centre, and our simulations indicate that the cell must commit to a direction for a sufficient length of time to aggregate successfully.

These formal rules based on a fixed duration of movement ignore some essential biological facts. For example, if the profile of the cAMP wave is altered the 100 second rule used in Fig. 1a will certainly not be applicable. A detailed model of how a cell chooses the direction of motion and the length of a "run" is not available, nor would it be feasible to use such a model at present. However we have developed more realistic rules based on internal variables as follows. It is known (see I) that cAMP activates the cGMP pathway via G-proteins in addition to activating the cAMP production pathway (Newell *et al.* 1990). It is also known that cGMP is near the beginning of the chemotactic response pathway and that cGMP production adapts to the cAMP stimulus on a time scale of about 30 seconds. If cGMP adapts then downstream components will also adapt except in unusual circumstances, perhaps on a longer time scale. Thus we assume that there is a downstream "motion controller", the identity of which is not known. However, it must be used in such a way that the cell moves only when cAMP is increasing, for it is known that cells only move in the rising phase of the cAMP wave. In the absence of detailed information about the controller dynamics, we used as a stand-in a quantity in the cAMP pathway that has the appropriate time course. This mechanism is biologically more realistic than the *ad hoc* rules and it gives results which match very well with experimental results (cf. Fig. 1c). This rule shows how a cell can respond to temporally-increasing cAMP levels by predicating motion on a threshold of an intracellular variable. Our simulations agree with the conclusion reached by Soll *et al.* (1993), that cells seem to orient during the beginning of the wave of cAMP and then move in a relatively blind fashion. We also show in I that aggregation is very robust under the combination of our signal transduction mechanism and the movement rule is based on an internal variable. For example, Fig. 1d shows that successful aggregation can occur as long as the cells choose their direction within the correct half space determined by the gradient and a line orthogonal to it. Our simulation indicates that many strategies can lead to successful aggregation.

This model also gives insight into the occurrence of target pattern waves *vs.* spiral waves. In particular we have found that spiral waves can arise spontaneously at higher densities when cells are initially distributed randomly, and that they may coexist with target patterns. This is in agreement with the results of recent laboratory experiments (Lee *et al.* 1996).

Finally, the simulations give a compelling argument that the mechanisms which are relevant in stream formation are finite amplitude instabilities. There are many factors involved in stream formation including random density variations, random cell parameter variations and variations in cell speeds. Each of these have a host of consequences which contribute to stream formation.

Acknowledgements This research was partially supported by NIH Grant #29123 and a grant of computer time from the Utah Center for High Performance Computing.

3 Streams and spirals in a discrete cell model of *Dictyostelium* aggregation

Catelijne Van Oss, Alexandre Panfilov, Paulien Hogeweg

We have modelled the aggregation of *Dictyostelium discoideum* (*Dd*) considering the *Dd* cells as discrete points that move in a continuous cAMP field (Van Oss *et al.* 1996). In the model, the process of cAMP production is described by the Martiel-Goldbeter (MG) equations for cAMP relay (Martiel & Goldbeter 1987, Tyson *et al.* 1989). Discrete cells are added to the MG equations in the following way. The position of cell i in space is given by $\vec{R}_i(t) = (x_i, y_i)$. By using the delta function $\delta(\vec{r} - \vec{R}_i)$, where $\vec{r} = (x, y)$, we can write down the equations governing cAMP production as follows:

$$\frac{\partial \rho}{\partial t} = [-f_1(\gamma)\rho + f_2(\gamma)(1-\rho)] \sum_{i=0}^{N} \delta(\vec{r}-\vec{R}_i)$$

$$\varepsilon_i \frac{\partial \beta}{\partial t} = [s_1 \Phi(\rho,\gamma) - \beta] \sum_{i=0}^{N} \delta(\vec{r}-\vec{R}_i) \qquad (3)$$

$$\frac{\partial \gamma}{\partial t} = D\nabla^2 \gamma + \frac{1}{\varepsilon_e}[s_2 \beta - \gamma] \sum_{i=0}^{N} \delta(\vec{r}-\vec{R}_i),$$

where $f_1(\gamma) = \frac{1+\kappa\gamma}{1+\gamma}$; $f_2(\gamma) = \frac{L_1 + \kappa L_2 c\gamma}{1+c\gamma}$; $\Phi(\rho,\gamma) = \frac{\lambda_1 + Y^2}{\lambda_2 + Y^2}$; $Y = \frac{\rho\gamma}{1+\gamma}$, $\rho(\vec{r},t)$ is the fraction of active cAMP receptors, $\beta(\vec{r},t)$ is intracellular cAMP, $\gamma(\vec{r},t)$ is extracellular cAMP and N is the number of cells. Chemotaxis can be written as:

$$\frac{d\vec{R}_i}{dt} = \mu \vec{\nabla} \gamma(\vec{r},t)\Big|_{\vec{R}_i}. \qquad (4)$$

The parameter $\mu=0$ if one of the following conditions is satisfied:

1. $\rho < 0.7$. In other words, to explain the fact that amoebae do not respond to cAMP gradients in the back of the cAMP wave, we assume that the chemotactic apparatus becomes desensitized as a result of prolonged cAMP$_{[ex]}$ stimulation.
2. $\vec{\nabla}\gamma(\vec{r},t) < \theta$, where θ is a threshold value preventing motion towards very small gradients.
3. $\vec{R}_i(t+\Delta t) = \vec{R}_j(t+\Delta t)$, $i=1,\ldots,N; i \neq j$, where Δt is the time step of the simulations. Two cells cannot be at the same position at the same time.

Otherwise, $\mu=1$.

The results show that the parameter ε_i, which is inversely proportional to the turnover rate of internal cAMP, is important for the spatial pattern that arises. Cell streams do not form at the experimentally determined parameter setting used by Martiel & Goldbeter (1987) and Tyson et al. (1989) (Fig. 2a). However, if ε_i is decreased, streams do form (Fig. 2b). The smaller the value of ε_i, the faster the streams form and the more pronounced they are, culminating in the case where ε_i is so small that β can be assumed to be at quasi steady state and (1) reduces to two equations for ρ and γ ("the two-variable model"). In the simulations in which no streams were formed (ε_i is high), the speed of the cAMP wave does not depend on the cell density. Interestingly, the decrease in ε_i leads, in addition to stream formation, to a dependence of the cAMP wave speed on cell density: wave speed is high at high density and *vice versa*. Our hypothesis is that this dependence of wave speed on cell density is the underlying mechanism for stream formation (for further details see Van Oss et al. 1996). This view is supported by Vasiev et al. (1994), who showed that cell streaming in their (much more simplified) model

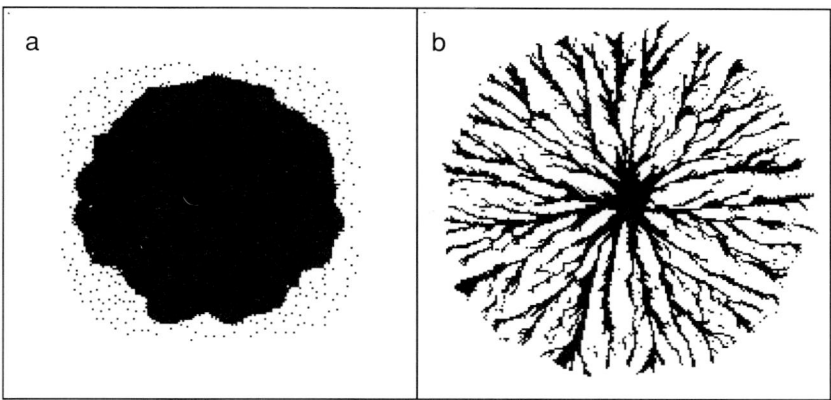

Figure 2: Aggregation in the model defined by Eqns. 3 & 4, (a) $\varepsilon_i=0.019$ and (b) $\varepsilon_i=0.0038$. Time $\approx$ 200 min. Cells are shown in black. Initially, cells are randomly distributed on a circular domain, stimulation occurs by periodically raising γ in the centre of the field, period 8.3 min. To integrate the explicit Euler method is used, space step 0.01 and time step 0.001 (space scale 4.5 mm, time scale 8.3 min), field size is 200×200 meshpoints $\sim 1 \times 1$ cm^2, boundary conditions are zero Dirichlet. The number of cells is $1.8 \cdot 10^4$, $c=10$, $s_1=950$, $s_2=0.05$, $\lambda_1=10^{-3}$, $\lambda_2=2.4$, $\kappa=18.5$, $\varepsilon_e=0.002$, $L_1=10$, $L_2=5 \cdot 10^{-3}$, $D=0.01$, $\theta=1$.

of Dd aggregation is due to the density-dependent wave speed. Experiments of Siegert & Weijer (reported in Van Oss et al. 1996) on the aggregation phase of Dd show that the speed of the cAMP wave at high cell density is higher than at low density, indicating that wave speed is indeed dependent on cell density. The stability analysis of the cell distribution carried out by Höfer & Maini (see Sect. 4) is derived from a two-variable caricature of Eqn. 3 which does not contain the parameter ε_i. An interesting open question concerns the role of ε_i in the stability criterion if a similar analysis was made of Eqn. 3.

Besides aggregation due to concentric waves, we also studied aggregation due to a spiral wave. Simulations (using the two-variable model) show that the spiral wave behaviour depends strongly on the initial cell distribution and shows a great amount of variability. The spiral wanders or anchors, sometimes breaks up and forms several spirals or a double-armed spiral, or an empty (no cells present) core is formed around which the spiral rotates. This diverse behaviour of the spiral wave, which is also observed experimentally (Durston 1973, 1974) is due to the continuously changing excitable medium, which is caused by chemotaxis. During aggregation, the increasing cell density (and thus increasing excitability) in the aggregation centre leads to the experimentally often observed (Gross et al. 1976, Siegert & Weijer 1989) decrease in spiral wave period.

4 A continuum model of slime mould aggregation

Thomas Höfer and Philip K. Maini

It is intuitively clear that the dynamics of the cell distribution and of cAMP signalling are closely coupled: cell movement is induced by the cAMP waves, while cells themselves act as sources for extracellular cAMP and also for its degrading enzyme, phosphodiesterase; hence they also act as cAMP sinks. Thus a model of the aggregation process must 1. include a description of (chemotactic) cell movement and the resulting dynamics of the cell distribution, and 2. extend the model of the cAMP dynamics to the case of spatially and temporally varying cAMP sources. Recently, this problem has been tackled in two different ways. The first approach consists in modelling discrete cells equipped with cAMP-dependent movement rules and coupled to a finite-difference approximation for the continuous cAMP dynamics (Dallon & Othmer 1996, Van Oss et al. 1996, Kessler & Levine 1993). In a second approach, the cell distribution is approximated by a continuous density, resulting in a system of coupled partial differential equations for the cell density and the cAMP dynamics. (Vasiev et al. 1994, Höfer et al. 1995a,b). Höfer et al. (1995a) propose the following continuum model of the aggregation process

$$\frac{\partial n}{\partial t} = \nabla \cdot (\mu \nabla n - \chi(v) n \nabla u) \qquad (5)$$

$$\frac{\partial u}{\partial t} = \lambda[\phi(n) f_1(u, v) - (\phi(n) + \delta) f_2(u)] + \nabla^2 u \qquad (6)$$

$$\frac{\partial v}{\partial t} = -g_1(u)v + g_2(u)(1 - v), \qquad (7)$$

where n, u and v denote cell density, extracellular cAMP concentration and fraction of active cAMP receptors, respectively. The cell density dynamics (Eqn. 5) include random cell movement with a cell diffusion coefficient μ, and chemotactic drift in gradients of cAMP. The magnitude of the chemotactic response is assumed to depend on the cellular sensitivity towards cAMP, measured by the fraction of active cAMP receptors per cell. Accordingly, the chemotactic coefficient is taken to be of the form $\chi(v) = \chi_0 v^m / (N^m + v^m)$, $m > 1$ (Höfer et al. 1994). The following functional forms are used for the rates of cAMP synthesis and degradation per cell: $f_1(u, v) = (bv + v^2)(a + u^2)/(1 + u^2)$, and $f_2(u) = du$. These are somewhat simplified versions of the kinetic terms derived by Martiel & Goldbeter (1987). The cell density dependence of the local rates of synthesis and degradation is reflected in the factor $\phi(n) = n/(1 - \rho n/(K + n))$. Similarly, the rate functions of receptor desensitization and resensitization are simplified expressions of the corresponding terms in the Martiel-Goldbeter model, $g_1(u) = k_1 u$, and $g_2(u) = k_2$.

This relatively simple continuum model yields a good description of the key features of the aggregation process. A typical aggregation sequence is shown in Fig. 3. Linear stability analysis predicts the break-up of the initially homogeneous cell density distribution perpendicular to the direction of wave propagation on

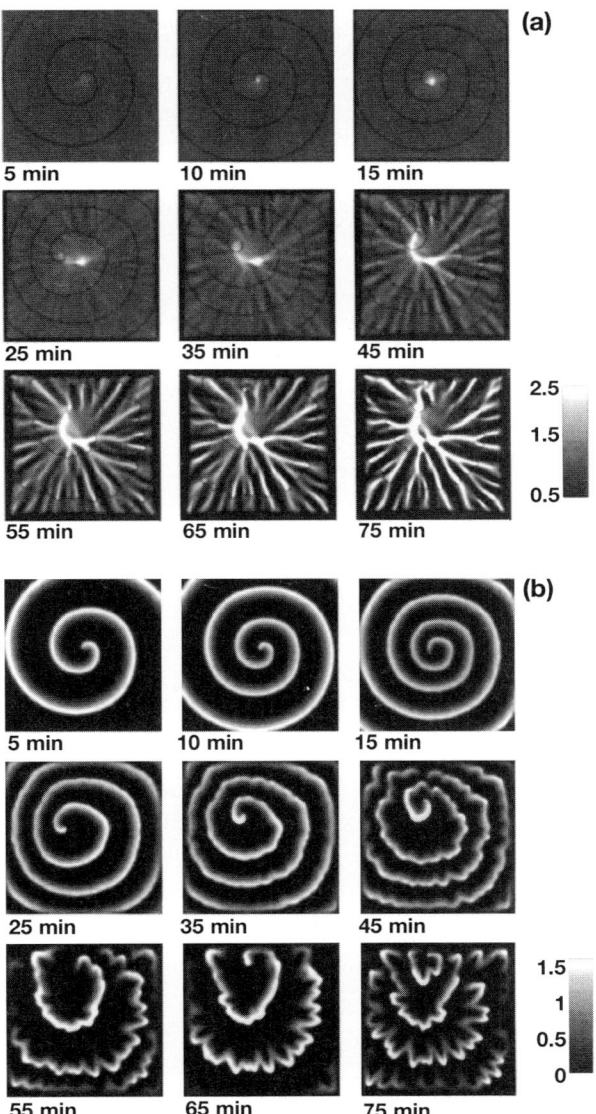

Figure 3: Spatio-temporal evolution of (a) cell density, and (b) cAMP concentration in a numerical simulation of system (5)–(7). The dimensional domain size is 6.5 mm × 6.5 mm, and snapshots are taken at the times indicated. Initial conditions were chosen to be a plane wavefront with a free end at the centre of the domain and homogeneous cell density (1.0) with a random perturbation between −0.075 and 0.075 added at every mesh point. Boundary conditions are zero-flux. Parameter values: $\lambda = 70.0$, $a = 0.014$, $b = 0.2$, $\rho = 0.7$, $K = 8.0$, $d = 0.0234$, $\delta = 0.11$, $k_1 = k_2 = 2.5$, $\mu = 0.01$, $N = 1.2$, $\chi_0 = 0.5$, $A = 0.72$, and $m = 10.0$.

a characteristic length scale, causing the formation of a branching cell stream pattern, cf. Höfer et al. (1995b). Detailed computational studies (unpublished) confirm the prediction of the growth of a small-amplitude pattern in cell density in a quantitative fashion. The linear analysis of a model caricature gives an explicit instability criterion which is typical of a chemotaxis-driven instability.

The evolving cell density pattern feeds back into the cAMP wave dynamics, as wave propagation speed depends on cell density. This can explain two apparently unconnected experimental observations (Höfer et al. 1995a,b). First, the model reproduces the experimentally observed decrease in cAMP wave propagation speed and the concomitant increase in wave frequency as a cAMP spiral evolves with time. Second, the model predicts the formation of closed cell loops in the centre of a spiral wave pattern at low excitability of the medium. The formation of central cell loops has indeed been induced by the application of caffeine, which lowers excitability by interfering with the adenylate cyclase pathway (e.g. Steinbock & Müller 1995).

Acknowledgements T. H. acknowledges support from the Boehringer Ingelheim Fonds, the Engineering and Physical Sciences Research Council of Great Britain, and a Jowett Senior Scholarship at Balliol College, Oxford.

III.4

A Cellular Automata Approach to the Modelling of Cell-Cell Interactions

Nicholas J. Savill and Paulien Hogeweg (Utrecht)

1 Introduction

Cellular automata representations of biological cells are typically modelled as one automaton per cell and their interactions are between the nearest and maybe next nearest neighbours. This has several disadvantages, for example:

- the cells have a fixed size and shape,
- the dynamics can be influenced by grid effects,
- all cells interact with a fixed number of neighbours,
- time and space are highly discretised.

Not all of the above points will be important or affect a model's dynamics but it is possible that they could play a significant role in the observed behaviour. A recent paper (Agarwal 1995) introduces what is called a "cell programming language". A "cell", in this language, is represented by a collection of connected squares. The cells can perform global (to the cell) functions, for example grow, divide, roundup, etc. However, there is an inelegant aspect of this model: Before a cell can perform any function it must recompute its neighbourhood to take account of any changes that have taken place since it was last updated.

A more elegant model is that of Glazier & Graner (1993) who use subcellular processes. They only considered cell sorting through differential cellular adhesion. We show that their model can be easily and elegantly extended to include such things as growth, division, etc. also on the subcellular level. It then becomes a useful tool in modelling cell-cell interactions and possible self-organized complex behaviours.

2 The original model

The model of Glazier & Graner (1993) is based on two-dimensional cellular automata (CA). The state of an automaton, σ, represents the unique identification number of a cell (state zero being empty). Cells can be of different types $\tau(\sigma)$ and there exist dimensionless energy bonds acting across automaton boundaries.

These bonds connect i) different cell types with energy J_{τ_1,τ_2}, ii) a cell and the medium with energy $J_{\tau,M}$ and iii) automata internal to cells with energy 0. The total energy of a cell is given by: $H = \sum \frac{J_{\text{cell,cell}}}{2} + \sum J_{\text{cell,medium}} + \lambda(a - A)^2$, where a is the area of a cell (number of automata), A is the target area of a cell and λ is the elasticity of a cell. The final term ensures that the area of a cell remains close to A. The CA is updated asynchronously: An automaton is chosen at random and the state of one of its neighbours is copied into it with a probability P given by

$$P = \begin{cases} 1 & \Delta H < 0 \\ e^{-\frac{\Delta H}{kT}} & \Delta H >= 0, \end{cases}$$

where ΔH is the change in energy *if* the copying were to occur and T is "temperature".

By simply varying the values of the energy bonds many different types of behaviour can be simulated, including engulfment, cell dispersal and cell sorting (Fig. 1).

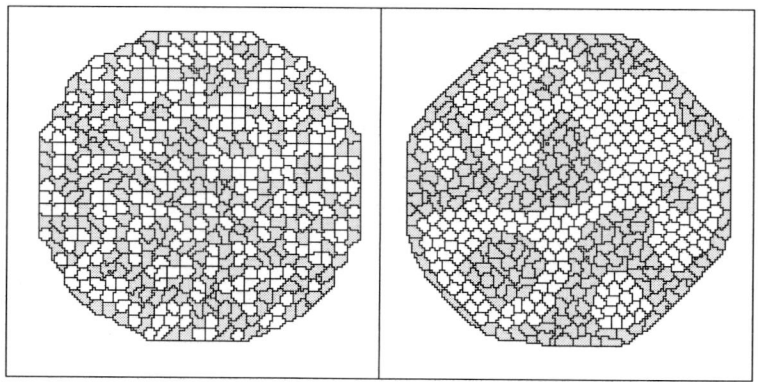

Figure 1: Cell sorting from a random initial distribution of two cell types.

3 Extensions of the original model

Cell growth can be modelled by slowly increasing the target area, A, of a cell (for example when a cell consumes some nutrients). Because a remains close to A the cell's area will slowly increase. Cell division occurs when a cell's area reaches a maximum, say $a = a_{\max}$. Then the states of the automata in one half of the cell are changed to an, as yet, unused identification number. The daughter cells can inherit the parents' properties with or without mutation. Extension to three dimensions is simple and gravity can be modelled as a positive bias in the downward copying probability for example.

Chemotaxis can be modelled by coupling the CA with a partial differential equation for the diffusion and production of a chemical (e.g. cAMP produced by D. discoideum). The basic idea is to modify the change in energy, ΔH, if a copying were to occur, so as to make the cell migrate more favourably in a given direction. In order to couple the PDE to the CA one node in the PDE solution grid should correspond to one automaton in the CA. The modification is: $\Delta H' = \Delta H + (c' - c)$, where $\Delta H'$ is the new change in energy if the copying were to occur, c' is the chemical concentration of the automaton whose state is being copied into the automaton with concentration c. If $c > c'$ then $\Delta H' < \Delta H$ and hence the copying probability increases, if $c < c'$ then $\Delta H' > \Delta H$ and hence the copying probability decreases and if $c = c'$ there is no change in the copying probability. So over time the cell copies more states into the automata with higher concentrations than with lower concentrations and hence the cell migrates up the chemical gradient. This can obviously be changed so that a cell migrates down a chemical gradient.

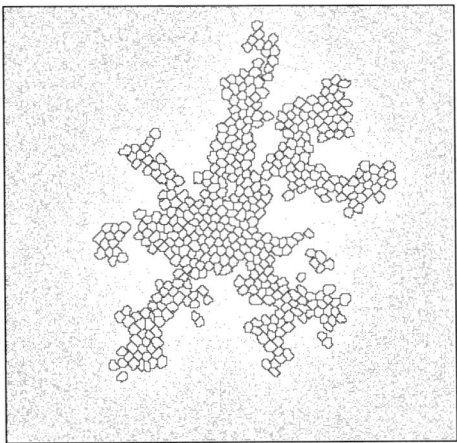

Figure 2: Diffusion limited growth of cells. Black dots represent diffusing nutrients.

4 Some applications of the new model

4.1 Diffusion limited growth

As an example of the growth and division rules we looked at a simple two-dimensional system of diffusion limited growth of bacterial cells. There is only one cell type that adheres to itself and does not chemotactically move. Starting with only one bacterium and a sparse distribution of nutrients represented as an automaton with a state of 1 in a second layer of the CA, the bacterium target area A increases by a certain amount each time a bacterium consumes a nutrient particle. When the bacterial area a reaches 50 the bacteria divide in half either horizontally or vertically. For low growth per nutrient rates we find fractal-like structures (Fig. 2), for higher rates we find more compact structures. Other extensions could include random walks and chemotaxis (Ben-Jacob et al. 1992) and mutations for selection of growth strategies (see section 4.3) (Ben-Jacob et al. 1994).

4.2 Migration velocity

An interesting result to come out of this model is that groups of adhering cells migrate faster along a chemical gradient than individual cells. Fig. 3 shows an example of this on a gradient increasing from top to bottom. This behaviour is due to cells at the rear of the group "pushing" cells in front of them forward.

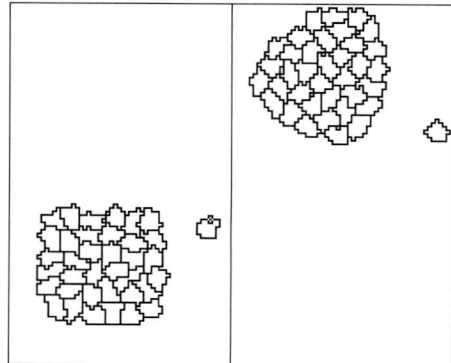

Figure 3: A group of adhering cells migrate faster than individual cells; a) (left) snapshot at t=0 (initial condition), b) snapshot after a certain number of time steps (t=100).

4.3 Selection for dispersal

If we couple the growth, division and death of cells with chemotaxis to randomly appearing large patches of food sources which secrete a diffusible chemoattractant, we can examine which cell strategies are selected for in this environment. That is to say, we evolve the strength of the adherence/repulsion between cell species. It appears that selection favours those cell types that repel themselves from other cell types and overrides the advantage of faster cell migration endowed upon groups of adhering cells (see section 4.2). By dispersing, a small population of repelling cells cover more area than an adhering population and hence are more likely to encounter chemoattractant to other nutrients (Fig. 4). Compare with *D. discoideum* that grows and divides as single cells but then aggregates at the onset of starvation.

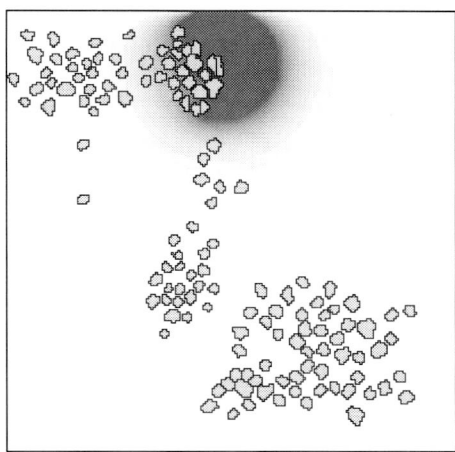

Figure 4: Cells migrate towards the circular food source secreting chemoattractant. Cells that repel each other out-compete cells that adhere over time.

4.4 Morphogenesis of *Dictyostelium discoideum*

We have extensively studied the morphogenesis of *D. discoideum* from the aggregation to the slug stages (Savill & Hogeweg 1996). The amoebae are modelled in three dimensions and can have one of three cell types; prespore, prestalk and autocycling prestalk. The bond energies are chosen so that the amoebae adhere and sort into homogeneous groups. The autocycling cells periodically produce cAMP. Cyclic AMP is modelled as an excitable medium (Tyson *et al.* 1989) using modified Fitzhugh Nagumo equations (Panfilov & Winfree 1985).

We start with a random distribution of 500 cells in the ground layer (Fig. 5A) with five autocycling amoebae in the centre. After several waves the mound and streams start to form (Fig. 5B). When most of the cells are in the mound (Fig. 5C) it falls over under the influence of gravity and then crawls as a slug (Fig. 5D). Cell sorting is occuring continuously due to the differential adhesion. The prestalk cells are sorting into the top/anterior of the mound/slug respectively.

We have demonstrated that the basic morphogenesis of *D. discoideum* can be entirely explained by three processes: the production of and chemotaxis to cAMP and cellular adhesion, all working on a spatial scale smaller than the diameter of the amoebae.

The model also makes some experimentally testable predictions: Cells in aggregates move faster than when alone. This observation suggests an evolutionary explanation for stream formation, which occurs if and only if there is fast intracellular cAMP production (Van Oss *et al.* 1996). The sorting of cell types can be partially or fully due to differential adhesion between the cell types. Siegert & Weijer (1992) were able to observe, with high resolution digital image processing,

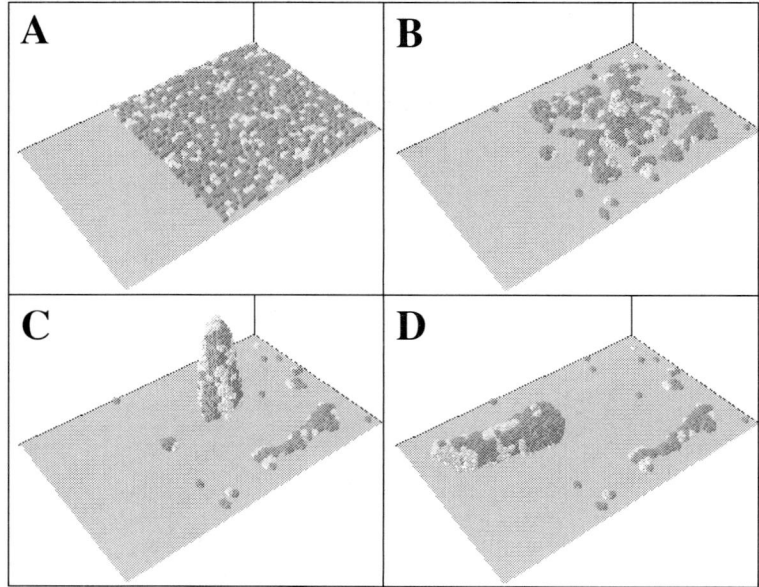

Figure 5: Developmental time sequence of *D. discoideum* showing mound and stream formation and the crawling slug. Light prestalk cells and dark prespore cells sort via differential adhesion into the anterior and posterior of the slug, respectively.

a circular motion of individual amoebae in the anterior of slugs that produce scroll waves of cAMP. We predict that for slugs of strains not producing spiral and scroll waves all amoebae will be seen to move with the same average speed and direction compared to that of the whole slug.

5 Conclusion

We have extended the model of Glazier & Graner (1993) to incorporate other cell functions and have given three examples of the application of this model. The elegance of the model is due to the localized process of minute adjustments to a cell's membrane in order to minimize its free energy. We believe that this modelling approach may be useful in examining and explaining behaviour in many cellular systems where cell-cell adhesion plays a critical role in the formation of patterns.

Acknowledgements N. Savill is supported by the Priority Program Nonlinear Systems of the Netherlands Organization for Scientific Research. The Dynamics Systems Lab provided the computer facilities.

Discussion & Open Problems

Although there are a number of models focussing on different aspects of the biochemical dynamics and signal transduction in *Dictyostelium discoideum*, there is little known about force transduction and the mechanisms of cell motion. All of the papers presented in this Chapter which address pattern formation in *Dictyostelium discoideum* model chemotactic cell movement in a phenomenological way. Ongoing work by Othmer and collaborators aims to couple the internal states of cells with receptor dynamics and external cell-substrate interactions with the goal of deriving a more biochemically realistic form of cell and tissue motion (see also Chapters I & II).

The analysis of spatial interactions and aggregation also has ecological applications in which individuals (organisms) move due to various internal and external cues (Grünbaum 1996).

In the models presented in this Chapter, cell-cell interaction occurs due to a single cue, for example, either directly through the sensing of orientations of neighbouring cells, or indirectly through the sensing of chemical concentrations. Other cues include, for example, haptotaxis, the response to gradients in adhesivity, galvanotaxis, motion due to ionic gradients, contact guidance, wherein cells orient the fibres on the substratum on which they move, influencing the direction of motion of other cells (see also Chapter II). In any one system, more than one form of cell-cell communication is probably occurring. As yet, there has been no detailed analysis of how these differing cues interact and if new phenomena can arise which can not be deduced from one form of cell-cell communication, alone.

Discrete-continuum hybrid models of cell-chemical interactions are more accurate, physically, than pure continuum models which "blur" the scales by averaging both chemical and cell distributions. The latter type of model is more amenable to mathematical analysis whereas the properties of the former can usually only be determined by extensive numerical simulation. Very few studies so far address the relation between biologically relevant interacting particle systems, microscopic interactions and appropriate macroscopic continuum descriptions (notable exceptions being Stevens 1992, Othmer & Stevens 1996). A balance needs to be drawn between biologically realistic models and analytically tractable systems.

The non-local interaction models presented in this Chapter lead to systems of integro-partial differential equations. Linear stability analysis of these systems is straightforward but a detailed nonlinear analysis is, as yet, elusive, although in some cases, the orientation tensor approach can be used (as illustrated by Mogilner, Deutsch & Cook, III.1 this volume).

The problems brought up in this Chapter are closely linked with wound healing insofar as they concern the movement of cells in response to chemical cues which, in turn, are modified by the cells themselves (see Chapter I). In wound healing and also cancer growth there is the further complication of additional mechanical cues (see Chapter II). For example, in wound healing, stress- and flux-induced alignment of cells plays a crucial role in the quality of the healed tissue (Olsen *et al.* 1995, 1996a,b).

Chapter IV

Dynamics within Tissues
– Morphogenesis and Plant Movements –

Coordinators: Sharon R. Lubkin and Lev V. Beloussov

> *Das Lebensgeschehen, die Lebensregungen spielen sich nicht nur* innerhalb *der Zellen, sondern auch im Raume* zwischen *denselben ab. Ein* physiologisches Feld *drängt sich uns mit der gleichen zwingenden Gewalt, wie das embryonale Feld auf. Stellt man sich auf diesen Standpunkt, so ergeben sich mannigfache Konsequenzen auch bezüglich des Zellgetriebes.*
>
> *Alexander Gurwitsch* (1923)

Introduction

This Chapter contains papers on the general subject of tissue stresses in the context of both plants and animals. The contributions by Hejnowicz & Sievers (IV.4), Green (IV.5), Rennich & Green (IV.6), Stein, Rutz & Zieschang (IV.7) and Nakielski (IV.8) deal explicitly with phenomena in plants. The papers by Beloussov (IV.2) and Lubkin (IV.3) specifically examine animal development, while those of Beloussov, Bereiter-Hahn & Green (IV.1) and Rivier, Dubertret & Schliecker (IV.9) discuss issues of cell geometry and development in a more unified context.

This diversity of topics reflects the underlying biological similarities and differences between plant and animal tissues. The first and most obvious difference between plants and animals is not, as the layperson assumes, that animals move, whereas plants do not, for of course plant movements are ubiquitous (for an entertaining yet comprehensive review, see Simons 1992). Root movements are modeled in a number of different ways in the paper by Stein, Rutz & Zieschang (IV.7) and movements of plant stems are discussed in the contribution of Hejnowicz & Sievers (IV.4). The major difference between plants and animals, for our purposes, is that plants (and fungi) possess a cell wall, whereas animals do not. This simple fact informs almost all discussion of stresses in the tissues of plants and animals. The second major – but not universal – difference is that, with a few exceptions in the context of reproduction, plant cells remain with their neighbors, whereas this is not at all given in animal tissues, where cell migration may play an important role (or be the subject of study itself, cf. Chapter II). Thus often an important distinction in modeling is drawn between tissues composed of contiguous stationary cells and those composed of potentially migratory cells embedded in an extracellular matrix. The latter models may often have more in common with models of aggregation (cf. Chapter III) than with models of plant or non-migratory animal tissues.

Though the major distinction between plant and animal tissues is in the presence or absence of a cell wall, the extracellular matrix (ECM) of animal tissues shares many of the properties of the extracellular matrix of plant tissues. Most crucially, the ECM in both cases governs the material properties, confering most of the stiffness and rigidity of the tissue. Both types of ECM undergo extensive remodelling over time, changing their stiffness, (an)isotropy, etc. and, importantly, their residual stresses.

Stresses in plants are primarily generated by gradients in stiffness and turgor pressure, regulated by ion pumps and intermediates, as extensively discussed in the paper by Hejnowicz & Sievers (IV.4). Animal tissues have this mechanism at their disposal, as mentioned in the article by Beloussov (IV.2), however, in animal tissues, stresses are primarily generated not by osmotic gradients but by molecular motors, also regulated by ion channels and intermediates (cf. Chapter I).

It is these similarities between such clearly different systems as plant and animal tissues that makes modeling a valuable exercise. For example, the model, presented by Green (IV.5) and specified by Rennich & Green (IV.6), for generation of periodic structures in plants (e.g. leaves, flowers), is equally applicable to the generation of periodic structures in animal tissues (e.g. hairs, scales, teeth). There is nothing in the model which is intrinsically plant-related and inapplicable to the formation of periodic structures elsewhere.

The paper by Beloussov, Bereiter-Hahn & Green "Morphogenetical dynamics in cellular tissues: expectations of developmental and cell biologists" (IV.1) contains a historical review of concepts of control in morphogenesis, and a critique of the accepted concept of positional information.

The next contribution by Beloussov "Mechanical stresses in animal development: patterns and morphogenetical role " (IV.2) reviews the origin and effects of mechanical stresses in animal embryos, and their methods of measurement.

The paper by Lubkin "Mechanisms for branching morphogenesis of the lung" (IV.3) discusses issues of the formation of branched, glandular structures in terms of morphogenesis in epithelia, in mesenchyme, and in the intact organs, where mechanical and chemical linkages join the two tissue types.

Hejnowicz & Sievers, in "Tissue stresses in plant organs: their origin and importance for movements" (IV.4), discuss the origin of many tissue stresses in plants in the interaction between variable turgor and variable elasticity of different tissue layers.

Next, Green presents his mechanical model for "Self-organization and the formation of patterns in plants" (IV.5) as a two-dimensional time-varying buckling problem with an elastic foundation, which, with appropriate assumptions, yields propagating patterns of a characteristic wavelength.

The mathematical basics of this mechanical model is then presented in "The mathematics of plate bending" by Rennich & Green (IV.6).

Stein, Rutz & Zieschang, in "Mechanical forces and signal transduction in growth and bending of plant roots" (IV.7), present three models of motion and growth in plant roots, one based on feedback between turgor and cell wall elongation, another based on a delayed response to gravitational stimulus, after Israelsson & Johnsson (1967), and the third taking into account mechanical interactions with the soil.

In the contribution by Nakielski "Growth field and cell displacements within the root apex" (IV.8) we see a model of cell division and elongation leading to cell files of different geometry.

Rivier, Dubertret & Schliecker, discussing "The stationary state of epithelial tissues" (IV.9), consider epithelial (plant or animal) tissues to behave like foams, and cells to be equivalent to bubble-elements in the foam. This represents an abstract treatment of cell division in terms of entropy maximization.

IV.1

Morphogenetic Dynamics in Tissues: Expectations of Developmental and Cell Biologists

Lev V. Beloussov (Moskva)
Jürgen Bereiter-Hahn (Frankfurt)
Paul B. Green (Stanford)

The idea of the participation of mechanical forces and stresses (balanced forces) in the development of organisms has more than a century of history (His 1874, Thompson 1942). Until quite recently it has resided, with few exceptions, at the very periphery of developmental biology. Although almost nobody denied, in principle, the role of mechanical components in such developmental processes as gastrulation, neurulation, etc. (to say nothing about plant development), these have been considered as trivial epiphenomena, having nothing in common with the more fundamental problems of embryonic determination (speaking in terms of classical embryology) or of the regulation of gene activity (in more modern terms).

It is our belief meanwhile that many substantial gaps and contradictions in problems of developmental biology can be largely overcome just by considering a developing embryo to be an "active solid body". Mechanical stresses are one of its main dynamical components. In order to substantiate such a suggestion, let us trace briefly the main ideas and the unsolved problems in the theory of development.

1 Driesch's rules and morphogenetic field theories

As the point of departure for modern developmental biology one can take the conclusions made more than a century ago by the German embryologist Hans Driesch (1921). By tracing the development of isolated blastomeres of sea-urchin eggs, he made two main statements. Here we give the first one in a somewhat modified fashion while the second one is in its original formulation:

1. A "delocalization rule": For any morphological structure, to appear at a certain advanced stage of embryonic development, one can always point to a certain earlier stage when this structure still lacks any individual spatially restricted precursor (it has no definite prelocalization).

2. A "position-dependence rule": The fate of a part of an embryo (that is, of a morphological structure, or of a cell type which it will produce) is determined by

its position either within an entire embryo (at the early stages) or within certain regions (at more advanced stages).

An enormous amount of empirical data obtained since Driesch's time supports these generalizations. More than that: the idea of an initial delocalization (a lack of one-to-one imprinting) which bore in Driesch's times an almost unavoidably vitalistic odor, looks nowadays to be completely in line with general self-organization theories equally applicable to living and non-living matter. In no way, however, can that fact remove the enormous difficulties facing anybody who attempts to build upon Driesch's generalizations a kind of a coherent and heuristically valuable theory of development.

From their very beginning, such theories have been associated with the idea of a morphogenetic field, directly deduced from Driesch's statements. In its broadest sense, a morphogenetic field means nothing more than an area where a certain set of potencies is delocalized and positionally dependent. During the first half of this century, the main contributions to morphogenetic field theory were made by Alexander Gurwitsch (1922, 1930), Paul Weiss (1939), Conrad Waddington (1940) and a few others. Among them, Gurwitsch was the first to realize that both of the Driesch statements can be applied not only to some unique experimental situations and to final structures or "fates" to be realized, but also to each successive step of normal development. In other words, he advanced the idea that the field should act continuously as a motive force of development and should be expressed as an invariable vectorial equation. Weiss was the first to point out that the field vectors can be material, e.g. oriented submicrostructures. Waddington, in his "epigenetical landscape" allegory introduced into morphogenetic field theory the ideas of structural and dynamical stability/instability, largely promoting their further integration into the general self-organization framework.

2 "Positional information"

Unfortunately, for a number of reasons, these prophetic insights have not yet been brought together into a unified theory. It so happened that for most developmental biologists it was the concept of "positional information" (PI) (Wolpert 1969), which appeared to be an adequate phrasing of what was rational in Driesch's ideas, particularly his position-dependence law. A more detailed analysis shows that this is not the case. One of the main distinctions between the two concepts is that in Wolpert's construction, contrary to Driesch's, there is the presence of privileged material elements or locations to which PI is referred. Usually these elements are defined as the "sources" and "sinks" of some morphogenetically active substances. For a non-symmetrical two-dimensional body, two such points are required. For a three-dimensional body, three points are required. On the other hand, Driesch's statement about the fate-determining "position within a whole" implies that, instead of few privileged reference points, there is a global coordinate set, somehow linked to an upper level entity, a "whole". It is easy to demonstrate that it is the

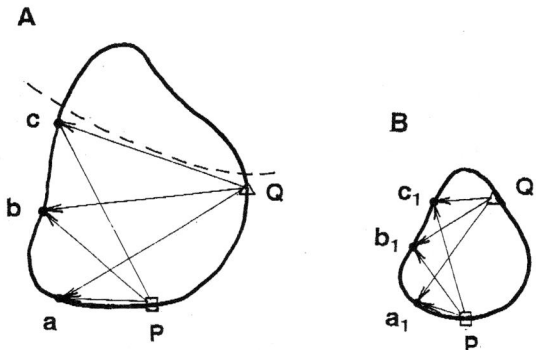

Figure 1: The concept of "privileged elements" (P and Q) as the sources of positional information (PI) is incompatible with embryonic regulations. A: suppose that the fates of an intact embryo elements a, b and c are determined by the ratio of their distances from P and Q: Pa/Qa for an element a, Pb/Qb for an element b, etc. A curved dotted line in A shows the direction of an experimental dissection of an embryo to obtain a partial embryo B (from the material remaining below the dissection line) and to trace its regulatorial capacities. B: Even if suggesting that both PI sources P and Q are included in the partial embryo (PQ distance, as measured along embryo surface, remaining unchanged), their mutual positions, due to wound closure, will be geometrically non-homologous to those of an intact embryo. As a result, for the elements a_1, b_1 and c_1, their positions being roughly homologous to those of a, b and c, the distance ratios from the sources will considerably differ from those in the intact embryo. Thus, they will get an abnormal PI which is incompatible with embryonic regulations.

initial Driesch concept, rather than the PI one, which is more compatible with the phenomenon of embryonic regulation. As a regulation proceeds in a partial embryo, any one previously selected material element takes a position geometrically non-homologous to that occupied by the same element within an intact embryo. Therefore, the PI referred to any individual element should be greatly distorted as compared with normal development. This makes embryonic regulation virtually impossible (Fig. 1). On the other hand, the shape of a whole remains largely the same in an intact embryo, and in partial embryos after wound closure, providing relatively preserved patterns of holistic position-dependence.

Another principal shortcoming of the PI concept is that it ignores completely the differential, step-by-step approach, forwarded by Gurwitsch. At any stage of development, PI is assumed to be directly projected onto the final state. That makes it impossible, within the PI framework, to reconstruct a realistic course of embryonic development, particularly that part related to shape change.

Many of the PI concept's shortcomings are due to the idea that, in the view of many investigators, PI is firmly bound with the diffusion gradients of certain substances considered as sole PI bearers. One cannot certainly doubt that chemical substances affect developmental processes; this is given. But in no way does that mean that the substances create precise "prepatterns", co-localized with the future morphological structures. In any case, according to Driesch's first rule (see above), there should always be an early enough stage when these pre-patterns are definitely absent. Even in the most perfect experiments designed to create chemical gradients in embryonic tissues (Gurdon *et al.* 1994) the cell differentiation patterns assumed to be produced by these gradients appear to be much less spatially precise than those of normal development. The main role of the chemical mediators of development (from activin in amphibian embryos up to auxins in plants) may be just in generating mechanical stresses or in making possible the adequate morphogenetical responses to these, rather than directly establishing morphological patterns. Using the terminology of a self-organization theory, the chemical mediators should be attributed to the parameters, rather than to dynamical variables of the developing organisms. A self-organizing potential of the mechanical models is, at least no less (if not greater) than that of the chemo-diffusional ones: they are stable to considerable perturbations of the initial conditions (Belintzev *et al.* 1987) and can produce definite patterns even if starting from a completely homogeneous state (Harris *et al.* 1984, Green, IV.5 this volume).

3 Fields of mechanical stresses

Thus, most of the shortcomings of the PI concept are eliminated, and new advantages are gained, if instead we associate morphogenetical fields with *fields of mechanical stresses* (FMS). The main point here is that the FMS, in order to be established, do not require, in general, any privileged sites within an embryo (although if they can produce such sites, the stress singularities, as a consequence of their action). On the other hand, these fields depend very much upon holistic geometry (symmetry, topology) of a stressed body, transforming its initial simple geometry to a definitely new one. Such a transformation is, as a rule, associated with a reduction in symmetry, that is, with the production of new, non-prelocalized, structures. In other words, both the initial delocalization of embryonic structures and their dependence upon holistic geometry fit quite well with the FMS concept. Also, mechanical stresses can spread very rapidly (in principle, at the rate of an elastic wave) and in a directed manner throughout entire embryos; the agents required for their transmission (the elements of a cytoskeleton) are in any case less specific than those needed for providing the reception of any chemical signals. That gives universality to FMS-based morphogenetical concepts. Returning to Driesch's terminology, his "vis vitalis" can be actually a "vis mechanica", or, better to say, a set of the active cytomechanical reactions within a mechanically closed system of a developing embryo.

In applying FMS to the interpretation of morphogenetical processes, one can roughly distinguish the following two steps. In the first, the formation of morphological structures can be simply considered as a movement towards a minimal energy state, i.e., as a relaxation of some pre-existing, more or less homogeneously distributed, stresses. Already this approximation is enough to provide some important and quite non-trivial results (see Green IV.5 this volume). One can go, however, a step further (this seems to be indispensable) by suggesting that the mechanical stresses not only are passively relaxed, but also affect in some way some active mechanochemical devices which produce further stresses. [This has been shown in a mechanical experiment on sunflower (Hernandez & Green 1993)]. In such a way an effective feedback loop between passive and active stresses can be established, leading to the progressive development of shape.

There is an increased amount of evidence that FMS affect not only the mechanochemical processes proper, but also the whole chain of intracellular processes, regulating gene expression (for references, see Ingber et al. 1994, Opas 1994). A new trend in molecular biology, appreciating the solid properties of signal transducers (e.g. Agutter 1994), considerably reinforces such a view. In general, this opens the way to link morphogenesis and cell shape dynamics to the refined processes of intracellular differentiation.

IV.2

Mechanical Stresses in Animal Development: Patterns and Morphogenetical Role

Lev V. Beloussov (Moskva)

While the role of *mechanical stresses* (MS) in plant development is now widely accepted, only few researchers are interested in studying their functions in animal morphogenesis. This is largely due to the fact that, as compared with plants, MS in animal tissues are much more various in their origin and space arrangement and less available for direct measurements and manipulations. Also, the relations between the "passive" stresses and the active stress-generating devices in animal embryos are as a rule more complicated than in plants. Nevertheless, a considerable bulk of data indicates not only the presence and a regular patterning of MS at any stage of animal development, but, in some cases at least, their crucial role for morphogenesis and cytodifferentiation.

1 Generation and transmission of MS in animal embryonic tissues

Animal embryos use several ways for producing MS. One of them, the osmotically driven turgor pressure, is the same as in plants. For the morphogenesis of multicellular rudiments, not so much turgor pressure within the cytoplasm as that in vacuoles and embryonic cavities (blastocoel, gastrocoel, subgerminal cavity of Amniotes, intraocular cavity) is of morphogenetical importance. Hydra budding is arrested if the turgor pressure in its gastral cavity is reduced (Wanek *et al.* 1980) and amphibian development is greatly disturbed while doing the same with blastocoel turgor pressure (Beloussov & Luchinskaia 1995a and in preparation). Eye development in chicken embryo is also blocked by releasing the hydrostatic pressure in the eye cavity (Coulombre 1956). In all these cases the normal turgor pressure is maintained by a directed apico-basal ionic transport via epithelial cells. Its directionality is believed to be provided by a predominant localization of ionic channels in the apical domain of cell membranes and ionic pumps in their basolateral domains (Stern 1984). In such a way sodium (accompanied by chloride) is pumped from the external environment into subcellular cavities. The created turgor pressure within cavities maintains the covering epithelial layers in a stretched state.

In a number of cases the osmotic pressure within the vacuoles and the cavities exhibits pulsations with periods of a few minutes. The most pronounced example is provided by hydroid polypes which grow and change their shape due to the finely regulated pulsations of the vacuoles swelling and the associated changes in tangential pressure within a cell layer (Beloussov et al. 1989, 1993).

In most of animal embryos, meanwhile, it is the microfilamental contraction, reinforced and stabilized by resorption of the apical membranes of the contracted cells (and balanced by osmotically driven turgor pressure) to be the main MS-producing engine. Again in an obvious contrast with plants, microtubules are generally of less importance, although during so-called rotation of fertilization there are just microtubule-mediated translocations which seem to be crucial for initiating embryonic development (Houliston & Elinson 1991).

Also worth mentioning, the intercellular generated MS are very much modified and reinforced on a supracellular level due mostly to cooperative polarization (columnarization) and movements of cells. Schematically, two main kinds of stress-producing cooperative cell activities can be distinguished: (1) involution of a part of a cell layer or immigration of a cell group – this generates tangential stretching in the adjacent parts of a layer; (2) intercalation (mutual insertion) of cells (Keller 1987), which exerts a pressure stress, extending a given tissue piece perpendicularly to the direction of cell intercalation. At least in the vertebrate embryos, practically all the morphogenetic processes can be considered, to a fairly good approximation, as a combination of these two cell activities and the associated MS.

MS generated by individual cells and cell groups can reach substantial values (Bereiter-Hahn 1987, see Tables 1, 2). In most cases they are propagated with an elastic wave rate, which is estimated for actin networks as about 1200 m/s (Forgacs 1995) and which is able to propagate throughout distances comparable with the entire embryo's dimensions (up to millimeters) using as a substrate most probably intermediate filaments, actin microfilaments and extracellular matrix structures. Due to such a fast MS propagation, we may consider the stresses to be properly counterbalanced at any given material point of embryonic tissue. In the other words, we can neglect accelerations and use the condition of a mechanical equilibrium (CME) for any material point.

2 Methods for measuring and localizing MS

For measuring the absolute stress values, sucking methods as well as those hampering morphogenetic movements and tissue deformations by a known force value are used (review: Bereiter-Hahn 1987). However, they are not so good for providing information about the detailed spatial patterns of stresses. For the latter purposes, we used two methods: (1) tracing the "immediate" (occurring within fractions of second) and at the same time passive (proceeding, for example, under low temperatures) deformations of embryonic tissues after accurately positioned dissections; by making a series of such dissections in various locations, one can create a detailed "map" of the stresses within a given area of an embryo. Balancing the deformations by a given weight value, one can also roughly estimate the elastic energy stored within a tissue (Beloussov et al. 1975). (2) Estimating the *relative* stretch values of the given structures (fibers, cell walls, files of stretched cells etc.) by measuring their intersection angles. This classical method is based upon Laplace's law and CME. It may be used, for example, to reveal the tension gradients in tissues (Beloussov & Lakirev 1988). Laplace's law can be also used for roughly estimating the tensions on eggs or embryo's surfaces as the function of their surface curvatures. A combination of the mentioned methods increases the reliability of the results obtained.

3 Some typical patterns of MS in developing organisms

A developing organism is mechanically stressed from its earliest stages and MS patterns are reorganized during development in quite a regular and "morphogenetically reasonable" way. Even the starting step of development, namely egg maturation is already associated with a drastic increase of tension within a subcortical region, due to tangential rearrangement of the microfilaments (Ryabova 1995). Cytoskeletal patterns within the mature and fertilized eggs indicate a field singularity at the animal pole (Fig. 1A). Fully unexplored are MS in insect eggs; taking into consideration however their regular asymmetry and regarding, according to Laplace's law, the tensions on their surface to be inversely proportional to the local curvatures, one should expect the existence of both antero-posterior (ap) and dorso-ventral (dv) tension gradients, descending from p to a and from d to v (Fig. 1B). So far as the expression stripes of the pair-rule genes in a normal egg coincide with a direction of a zero ap-gradient and in the *bicoid* mutants the egg shapes and hence the tension gradients become ap-symmetrical (see, e.g. Nusslein-Volhard et al. 1987) one cannot exclude that the tensile patterns may be one of the intermediate links in regulating gene expression.

If ascribing some tension values to the yolk/cytoplasmic borders within ooplasm, redistribution of amphibian and fish egg components following fertilization should lead to further complication of MS patterns (Fig. 1C). It follows from CME, that in the nodule d, corresponding to the presumptive dorsal side, the tension on the surface of the animal hemisphere is greater than that of the vegetal surface ($t_{da} \gg t_{dweg}$), whereas for the ventral nodule v the reverse is true ($t_{va} < t_{vveg}$). So far as an animal surface of an egg can be considered as a part of a closed mechanical contour, the tensile lines should converge in the vd direction (Fig. 1D). In fish eggs just this tensile pattern has been detected (Cherdantzeva & Cherdantzev 1985). The subsequent movements of the blastoderm cells obey this pattern. Remarkably, a chicken embryo blastoderm is self-stretched at the very onset of incubation due to contractile activity of the *area opaca* cells (Kucera & Monnet-Tschudi 1987).

At the very beginning of cleavage it is the just initiated first furrow which becomes a focus of the cortical and subcortical tensions (Fig. 1E), attracting a flow of a subcortical actin (White 1990). In the spirally cleaved mollusk eggs, ad-

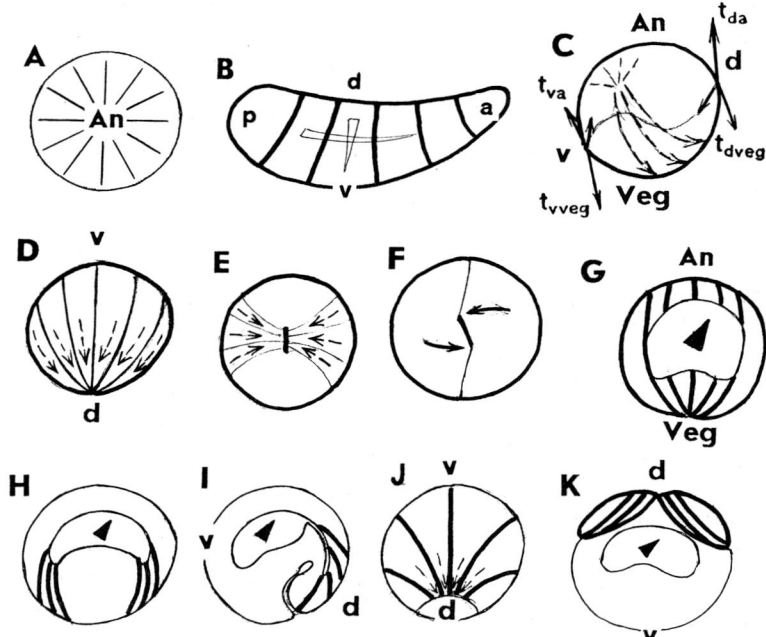

Figure 1: Patterns of mechanical stresses (MS) at different stages of animal development. In all the pictures heavy lines indicate the direction of the main tensile stresses. A: patterns of microtubules around the animal pole (An) of a sea-urchin egg, indicating an earliest developmental singularity of a stress field. B: MS pattern of an insect egg, as estimated theoretically by Laplace's law. Empty triangles indicate the suggested tension gradients and dense arches the equitensed lines. The latter fit the orientation of the genes expression stripes. Antero-posterior (ap) and dorso-ventral (dv) egg polarity is indicated. dv means the same in other frames as well. C: interfacial tensions in the amphibian egg established by the rotation of fertilization. A qualitative estimation according to CME (see text). Curved arrows indicate the direction of the microtubules extension. D: convergence of stress lines and flow of cell material (dotted arrows) in fish egg blastoderm. E: stress pattern in holoblastic eggs during initiation of a first cleavage furrow (heavy vertical line). Dotted arrows indicate a direction of an actin flow. F: arising of shear stresses in a mollusk *(Lymnaea)* egg due to disymmetrical shifts (arrows) of a surface material. Oblique heavy line is the distorted central part of the first cleavage furrow. G: MS pattern at the pregastrula stage of a sea-urchin embryo. In this and the next frames the filled oblique triangles indicate the turgor pressure. H-K: MS patterns at the successive developmental stages of amphibian embryos. H: blastula, cross-section. I: early gastrula, saggital section. J: gastrula, dorsal view. K: early neurula, cross-section. MS patterns in G-K have been detected by the local incisions technique. A from Isaeva & Presnov (1990). E composed according to White (1990) and F according to Mescheryakov (1991). Other frames are original.

ditional shear stresses emerge within the subcortical cytoskeleton during each next anaphase stage (Mescheryakov 1991). Those arising within the second cleavage division distort a central part of the first division furrow (Fig. 1F) which later on bears this tension. This part corresponds, as a rule, to the sagittal plane of a future embryo.

From the beginning of a gastrulation either the initial MS pattern is reinforced (Figs. 1G&J, cf. 1D) or a new one (as during embryonic regulation) arises, always being characterized by the convergence of stress lines towards the vegetal pole (Figs. 1G&H) and also, in vertebrate embryos, to the dorsal side (Figs. 1I&J). A subsequent development of a vertebrate embryo is characterized by a series of relatively fast transformations of MS patterns, alternated by more prolonged periods of their topological invariability. The latter periods perfectly coincide with those known in classical embryology as gastrulation, neurulation and (for amphibians) tail-bud formation. The most crucial transformation of MS pattern takes place at the beginning of neurulation and consists in formation of a powerful bundle of stretched cell files converging towards the dorsal midline of the embryo (Fig. 1K). Pronounced and morphogenetically important MS in the advanced neural rudiments have been described by Saveliev with co-workers (see Beloussov *et al.* 1994) and "residual stresses" in the walls of large arteries by Vaishnav & Vossoughi (1987). In both cases (and this seems to be true for all of the epithelial tubes) MS patterns are characterized by substantial stresses of the internal pressure within the walls and the tension along the outer surfaces. Considerable tensile stresses in the extracellular matrix of embryonic and adult connective tissues are permanently maintained by fibroblasts (Harris 1994a). Very complicated stress patterns in the different organ rudiments and body parts of human embryos were detected by Blechschmidt & Gasser (1978).

4 Effects of MS modulations upon embryonic development and tissue dynamics

As mentioned above, the decrease of turgor pressure within cavities leads to a complete arrest of development. This effect seems to be universal. A hypertonically shrunk zygote also stops cleaving, resuming cleavage immediately after being placed in normal medium (Harris 1994b).

Considerable stage-dependent morphogenetic effects can be produced by *relaxations* of tissue tensions in amphibian embryos (which however cannot be long-term, since the relaxed tissues actively restore tensions). These imply a rapid reduction of cell-cell contact areas, extensive columnarization of cells and, as a postponed result, various morphological abnormalities, mostly concerning the structure of the axial rudiments (Beloussov *et al.* 1990, Beloussov & Luchinskaia 1995a). On the other hand, 1.5-2 fold artificial tissue *stretching* rapidly (within several minutes) promotes the formation of microfilaments bundles (Kolega 1986, Beloussov *et al.* 1988), increase in cell-cell contact areas and, as a postponed result of no less than

30 min stretching, the reorientation of the embryonic rudiments in the stretching direction (Beloussov et al. 1988). Generally, tissue stretching triggers cell intercalation movements and orients them perpendicularly to the stretching direction (Beloussov & Luchinskaia 1995b); due to such stretch-induced intercalation an initially stretched rudiment continues to actively elongate itself (and modify its shape) long after the end of the forced extension.

A crucial role of tensions in imposing morphological organization onto cultures of mesenchymal cells has been shown by Harris et al. (1984). In these experiments tensions were produced by the fibroblasts themselves and extended over large distances along elastic substrates. This demonstrates, that the mechanical forces produced by cells themselves are necessary and sufficient for creating a definite macromorphology even if starting from a completely homogeneous state.

The mentioned experiments argue for a crucial role of mechanical factors during at least some important developmental periods. One can hardly imagine, on the other hand, how the described stress remodellings (in no way associated with a disruption of embryonic tissues) could have, as their primary target, a disorganization or reorientation of any presumed chemical prepatterns.

5 An interplay of the passive stresses and the active stress responses in animal morphogenesis

For a student of morphogenesis, the main interest in MS patterns is in considering them as links in the feedback loops which might drive forth the developmental processes. The main idea of such an approach is quite simple: a given pattern of mechanical stresses affects somehow the active stress-generating devices of the embryonic cells so that a new stress pattern is produced, and so on. The first paper employing such an approach was by Odell et al. (1981), who suggested that the tangential tension of a cell layer switches-on the active contractile mechanisms. The model of Belintsev et al. (1987) was based upon the idea that cooperative columnarization of cells, a starting event in epithelial morphogenesis, is stimulated by the relaxation of tangential stresses and inhibited by increase in tangential stretching of a layer. Each of these assumptions proved to be true under certain conditions. While the model of Odell et al. reproduces a wide-spread stretching-contraction response, it is the advantage of the model of Belintsev et al. that it is dynamically stable to considerable initial perturbations and is scale invariant (imitating thus Driesch's regulations).

Is it possible to unify both models under a wider, although up to now purely phenomenological framework? As a tentative generalization of such a kind, we forwarded an idea of a hyperrestorative response of a cell or a tissue piece to stress changes. By this idea (Beloussov & Mittenthal 1992), any changes in MS, exerted either by an artificial external force or caused by another region of the same embryo will generate an active mechanochemical reaction directed towards a restoration of an initial stress value but as a rule overshooting it to the opposite side. Thus, a relaxed tissue piece tends to restore and even to overwhelm its initial tension, while a stretched tissue piece tends to reduce stretching up to developing the active pressure force within. In amphibian embryonic tissues it is quite easy to find experimental evidence of such reactions, the above described stretch-dependent cell intercalation being one of these (see Beloussov 1994, Beloussov *et al.* 1994 for more details). On a qualitative level at least, the normal development of an amphibian embryo from the blastula to late neurula stage can be considered as a succession of such reactions.

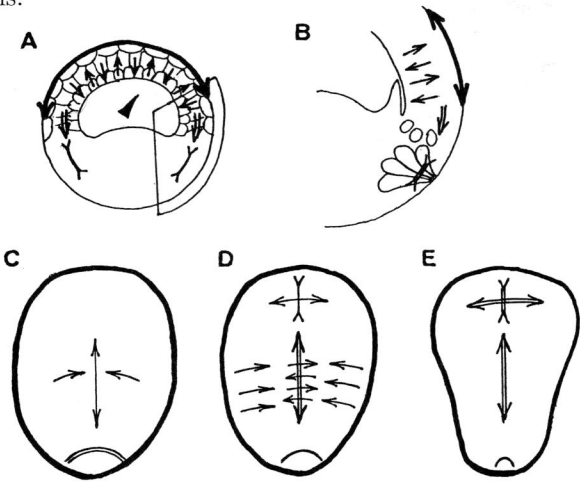

Figure 2: A sketch of the passive and active MS which may drive the development of an amphibian embryo from blastula up to a tail-bud stage. Passive stresses are denoted by single-lines bidirectional arrows while the active ones by double-lines arrows. Small arrows in A-D, pointing to the opposite directions, indicate cell intercalation movements. A: blastula stage. B: a marginal zone of an early gastrula, framed in A. Shown are the apically contracting bottle cells and the rounded contact-free cells of a deep layer (both apical contraction and loss of cell-cell contacts is a typical reaction to a relaxation, see Beloussov *et al.* 1988, 1990 for more details). C-E: MS patterns at the successive stages of a neurulation, dorsal view, anterior pole to the top. In a first approximation, the MS patterns in the ectoderm and the underlying chordomesoderm are roughly the same.

Briefly, the scenario might be as follows. We start from a blastula stage, the blastocoel roof being stretched by turgor pressure within the blastocoel (Fig. 2A). The radial cell intercalation within the roof (small vertical arrows) may be considered as a hyperrestorative reaction, transforming passive stretching into active internal pressure, which expands the roof and promotes epiboly. Such an expansion would lead at least to the relaxation (if not compression) of the marginal regions; a hyperrestorative reaction within these regions will express itself in apical cell contraction and, possibly, immigration of some cells (Fig. 2B). This, in turn, will promote further stretching and hence intercalations in the blastocoel roof, creating thus a kind of a positive stretching-contraction feedback loop. A dissymmetry (a preferential axis of intercalation, contraction and immigration) is imposed here by the initial axial differences in MS patterns established already immediately after fertilization (see Fig. 1C). As a result, we get anisotropic (longitudinal) expansion of the dorsal part of the embryo, accompanied by a transversal shrinkage. Both processes are actively reinforced by the same rules (Figs. 2C&D). A resulting segregation of the dorsal ectoderm into an extensively polarized domain of neuroepithelial cells and the flattened surrounding areas of ectodermal cells fits the predictions of the model of Belintsev *et al*. In longitudinal direction, the active elongation of the dorsal part will produce an area of a relaxation-compression in front of it (Fig. 2D); its active reinforcement, fulfilling the same rules, will produce the transversely extended forebrain (Fig. 2E). As detected by local incision methods, MS patterns in the dorsal ectoderm and underlaid chordamesoderm are roughly the same.

We see no difficulties in interpreting from the same point of view many other morphogenetic processes, including those associated with curvature changes in epithelial layers. In all these cases temporal delays in the active responses of the tissues to the imposed stresses should be of great importance. There are some reasons to suggest that, in the course of metazoan evolution, these delay times have been considerably increased, reaching in vertebrate embryos several dozens of minutes as compared to only few minutes in hydroid polyps. Such a retardation could provide less monotonous, more generalized and spatially extended tissue responses, contributing thus to the establishment of a global macroscopic order.

6   Concluding remarks

In spite of numerous gaps in our present-day knowledge of MS in animal embryos, some simple things are obvious: all of the developing animal species, studied in this respect, are mechanically stressed, the stress patterns are quite regular and "morphogenetically reasonable" and their modulations lead to severe and in many cases predictable morphogenetic abnormalities. Considering in addition the increased evidences of MS participation in cell signalling (Forgacs 1995) and regulation of gene expression patterns (Beloussov *et al*., IV.1 this volume) one may expect "cyto"- and "histomechanics" to soon become hot points in developmental and cell biology not only of plants, but also of animals.

Acknowledgements   The investigations summarized in this paper have been supported by the International Science Foundation, grants NAZ300 and NAZ000 and by the Russian Fund for Basic Researches, grant 93-04-77-15.

IV.3

Mechanisms for Branching Morphogenesis of the Lung

Sharon R. Lubkin (Seattle)

1 Introduction

The human lung is a dichotomously branched, hollow structure, which forms in a similar manner to other glandular organs such as the kidney, pancreas, mammary and salivary glands. A hollow finger of epithelium grows into surrounding mesenchyme, and branches repeatedly, over some 23 generations. In turn, the surrounding mesenchyme condenses around the epithelium. Although a great deal is known about branching morphogenesis, primarily from studies of the mouse submandibular gland, branching is not a fully understood process.

In particular, the differences in the development of the various glandular organs, which exhibit quite different morphologies (Fig. 1), are not fully understood. The lung ultimately has a highly adapted branching structure, which appears to be optimized for efficient fluid transport (Weibel 1963, 1991). How the same branching process can yield such different morphologies as the lung, kidney, mammary and salivary glands is not yet clear, and it would be of great interest to determine the range of morphologies possible with each proposed mechanism. It is entirely possible that multiple mechanisms of glandular morphogenesis exist, and that some organs use one – or more – of the available mechanisms. For example, after the embryonic period of lung development (and possibly during that period) there are fetal breathing movements, which have been shown to be essential for the proper development of the lung (Crystal *et al.* 1991). We might hypothesize that the initial branching of a lung is due to the same mechanism as that in the salivary gland, and that the subsequent adaptation of the lung's form is due to a dynamic remodeling of the existing structure under the influence of fetal breathing. The latter is an entire line of research itself, and shall not be addressed here.

In this review, we shall discuss epithelial, mesenchymal, and epithelio-mesenchymal models in turn, and the implications and use of each type of model.

2 Epithelial models

Although the evidence is very good for an epithelial basis to branching morphogenesis in the general case (Nogawa & Takahashi 1991, Takahashi & Nogawa

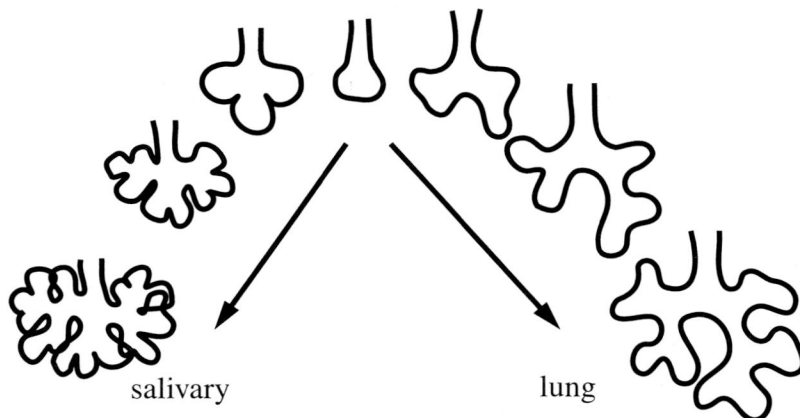

Figure 1: Salivary gland and lung are similar examples of branching morphogenesis, but the morphology of each is substantially different. The lung rudiment branches dichotomously for many generations, whereas the salivary gland can form multiple clefts in each lobe, and exhibits often elongated stalks. This indicates different physical or chemical parameters, or a somewhat different morphogenetic process (Figure after Spooner (1973)).

1991, Nogawa & Ito 1995), branching morphogenesis has not been addressed in an epithelially-based model.

Epithelial models have been used to address other issues, such as gastrulation and neurulation (Odell et al. 1981, Weliky et al. 1991), arthropod leg morphogenesis (Mittenthal & Mazo 1983), formation of microvilli (Oster et al.. 1985), and epiboly (Weliky & Oster 1990). It is known from studies of submandibular glands that clefts form in a branching epithelium in the absence of mitosis (Nakanishi et al.. 1987), and that recently-formed, but not established, clefts can be reversed by treatment with cytochalasin B (Spooner & Wessells 1972). Although individual clefts can form in the absence of mitosis, a glandular structure cannot form if mitosis (growth) is halted, because new tissue is needed for each new generation of clefting/branching. A complete epithelial model of branching morphogenesis must therefore take into separate account (a) formation of clefts, and (b) growth, both of which are necessary for the formation of a branched epithelial structure, but each of which is clearly governed by its own rules. Few epithelial models have included growth.

Odell et al. (1981) examined the implications of the purse-string mechanism, and hypothesized that apical microfilament bundles exhibit a strain-dependent excitability. Cells were modeled as collections of linked springs, subject to mechanical stretch. A subthreshold stretch of any one spring would return to equilibrium length, but a superthreshold stretch of an apical microfilament bundle (spring) would induce a contraction of that bundle, leading to a new, shorter rest length for that spring. Simulations with spherical or cylindrical symmetry and an initial superthreshold stretch of a single element strongly resembled *in vivo* gastrulation, ventral furrow formation, and neurulation respectively.

Mittenthal & Mazo (1983) modeled epithelial deformations leading to arthropod limb morphogenesis from the point of view of differential adhesion gradients. The model investigated energy minimization and the family of strain sequences and adhesive gradients which would lead to the observed (*in vivo*) morphologies, using an energy functional which had components of differential-adhesion energy and of strain (bending) energy.

Murray & Oster (1984) formulated a model of epithelial pattern formation based on calcium-regulated contraction of a sheet of cytoplasm. Oster *et al.* (1985) used this epithelial model to propose a mechanism for the formation of microvilli on a single cell, using essentially the same contractile cytogel mechanism as proposed for epithelial sheets, but on a substantially smaller length scale. In both cases, the proposed contractile mechanism leads to the formation of hexagonal pattern on a sheet, which is a necessary condition for branching morphogenesis. It remains to be seen if such models, on the appropriate geometry and with quantitatively realistic parameters (such as the appropriate length scale), would exhibit the tip splitting characteristic of branching morphogenesis, and if, with the addition of the continued growth essential to continued branching, the models would exhibit the appropriate cascade of dichotomous branching. Work in progress indicates that this is so.

Weliky & Oster (1990) and Weliky *et al.* (1991) modeled epithelial morphogenesis in *Fundulus* epiboly and in *Xenopus* notochord formation, respectively, by hypothesizing particular behaviors of the constituent cells. In the former study, they accounted for circumferenial elastic forces and hydrostatic and osmotic pressure, and found that isotropic contractility of interior cells, and protrusive activity of marginal cells, with simple stress relaxation, was sufficient to account for known behavior in *Fundulus* epiboly. In the latter study, they found that, among all the mechanisms they investigated, only the specific combination of polarized protrusive activity, contact inhibition, and refractory tissue boundaries gave the known behavior of *Xenopus* notochord formation. In these studies, the epithelial margin was the major morphogenetic player, but in branching morphogenesis, there are not generally any epithelial edges, or if there are, they are not important.

3 Mesenchymal model

Our mesenchymal model (Lubkin & Murray 1995), summarized in this section, is based on *in vitro* evidence of mechanical stimulation of mitosis in embryonic lung tissue (Liu *et al.* 1992, Bishop *et al.* 1993). The studies indicate that in various different culture systems on flexible substrates such as latex sheets or 3D sponges, lung fibroblasts or mixed lung tissues have a basal mitotic rate which is dramatically enhanced by cyclic mechanical deformation of the substrate. We interpret this as evidence of a mechanical stress-sensitivity of embryonic lung tissues *in vivo*, and construct a model of branching morphogenesis in the lung on this premise. It is not known whether other glandular tissues exhibit the same stress-mitosis response to mechanical stimulation, but the phenomenon is widespread in a variety of tissues (Desmond & Jacobson 1977, Curtis & Seehar 1978, Folkman & Moscona 1978).

The simplest interpretation of the stress-growth rate data of the *in vitro* studies is that the mitotic rate, or equivalently the strain rate, of embryonic lung mesenchyme, is proportional to the time-averaged local mechanical stress. This leads to an immediate mathematical description of mesenchymal deformations as those of a viscous fluid, and therefore described by the Navier-Stokes equations. The luminal fluid is modeled with the same equations. As we estimate the Reynolds number of the flow to be approximately 10^{-21}, we trivially justify the use of the Stokes equation

$$\nabla p = \mu \nabla^2 u \qquad (1)$$

as an approximation of the Navier-Stokes equations. In the basic model, we consider the divergence of the mesenchyme $\nabla \cdot u = c$ to be a nonzero constant (incompressible fluid, but with a source term), though in simulations, we remove this computational difficulty. The epithelium is treated as essentially passive for the purposes of this mesenchymal model. It is assumed that lung epithelium responds to tangential mechanical stresses by growing/relaxing to a certain residual tangential stress (Takeuchi 1979), though it is known, as discussed in the previous section, to be substantially more active than this simplifying assumption, when isolated from its mesenchyme *in vitro*. It therefore behaves equivalent to a surface tension on a fluid, and can be described by the Laplace-Young condition,

$$p_2 - p_1 = -\tau(\kappa_1 + \kappa_2) \qquad (2)$$

where τ is the surface tension and κ_i are the two principal curvatures.

For the purposes of simulation, it was necessary to make a reduction of an essentially 3D model to two dimensions, since the difficulty and cost of 3D simulations is clearly prohibitive and impractical for an initial model. We therefore made the choice of a Hele-Shaw approximation to 2D flow: the flow is assumed to take place between parallel plates a set distance b apart. This reduces the 3D Stokes equation to the 2D Darcy equation

$$\nabla p = -\frac{12\mu}{b^2} u \qquad (3)$$

which with the assumption of a constant divergence reduces to Laplace's equation $\nabla^2 p = \tilde{c}$. The free boundary at the epithelium, which has a nonlinear (curvature) term in it, presents a computational difficulty, however. We implemented our simulations using a Monte Carlo algorithm developed by Kadanoff (1985) and Liang (1986) which simultaneously solves Laplace's equation in the mesenchyme and computes the curvature-dependent motion of the free boundary. Boundary

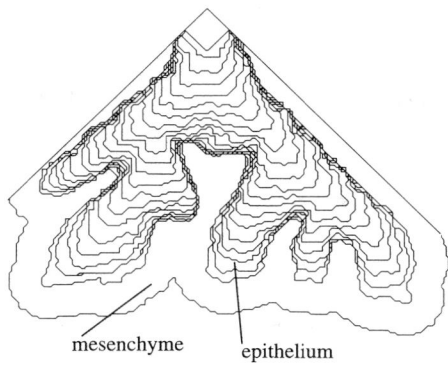

Figure 2: Representative simulation of mesenchymal model of initial lung branching. Contours represent the shape of the epithelial interface over time, starting with the arbitrary initial condition at top; see Lubkin & Murray (1995) for details.

conditions were taken to consist of a uniformly expanding boundary far out in the mesenchyme, corresponding to gross growth of the embryo, or to a positive lumenal pressure. The simulations give excellent qualitative agreement with the form of the branching lung for a certain range of the single nondimensional control parameter (Fig. 2). It remains to be seen, in a much more costly simulation, if the same branching appears in a 3-D model. A 3-D model should also reproduce the known alternation of direction of the lung's dichotomous branching (Weibel, pers. comm.).

Linear analysis of the model with both two and three fluids gives an expression for the dominant wavenumber of pattern formation in terms of the physical parameters of the system. Using our best estimates of the physical parameters of the actual developing lung, we find that even the highly simplified model gives excellent *quantitative* agreement with *in vivo* development (Lubkin & Murray 1995). This is unusual in mathematical models of morphogenesis, where often the parameters cannot even be estimated. We further conclude from the analysis of an extended 3-fluid model, where the mesenchyme is considered to consist of an inner and an outer tissue-fluid of different "viscosities", that the outer, less dense, mesenchyme must have a greater proliferative response to stress, that is a lower "tissue viscosity", than the inner, condensed, mesenchyme.

4 Tissue interaction model

Our mesenchymal model of the previous section has demonstrated that it may be possible for branching to occur solely due to mechanical direction by the mesenchyme, with passive epithelium. However, it is known from experiments on the submandibular gland (Nogawa & Takahashi 1991, Takahashi & Nogawa 1991) and lung (Nogawa & Ito 1995) that epithelia mechanically separated from inductive mesenchyme can still grow and branch if embedded in a gel containing appropriate growth factors and substrate components, in particular those of the basal lamina. Therefore, mesenchymal "viscosity", as in our model, while we believe it to be a sufficient condition for branching morphogenesis, is not a necessary condition. However, it is known that different mesenchymes induce different branching patterns of epithelium, and that mechanically isolated epithelium does not branch precisely as it would *in vivo* (Mori *et al.* 1994). Therefore an important unanswered question is the relationship between epithelial branching and mesenchymal induction and mechanical modulation of that branching. Full exploration of that question requires an investigation both of epithelial models and of tissue-interaction models.

Although we have shown that stress-responsive mesenchyme with a passive epithelium can form a hollow, branched structure, it is also known that lung and salivary epithelium mechanically isolated from mesenchyme grow and branch to form a hollow, branched structure. Can we conclude therefore that branching morphogenesis is mechanically under the control of the epithelium alone? *In vivo*, the epithelium is surrounded by mesenchyme. It is not possible, quantitatively, for the stresses produced in a mechanically isolated epithelium *in vitro* to deform to the same extent the mesenchyme in which they would be imbedded *in vivo*, just as a hand which can make a fist in air cannot make the same fist embedded in clay, without generating substantially more force.

In order to fully understand the different roles of epithelium and mesenchyme in branching morphogenesis, we need to integrate models of epithelial and mesenchymal behavior, and the mechanical interaction between them *in vivo* and in *in vitro* configurations where there exists a mechanical linkage, as well as in the *in vitro* transfilter recombination configurations and mesenchyme-free cultures which do or do not exhibit branching (Mori *et al.* 1994). Such models have not been hitherto attempted, because of the considerable computational difficulties that they present, namely those associated with (a) constitutively different active viscoelastic fluids in two and three dimensions (epithelium and mesenchyme respectively), and (b) free boundaries.

In particular, it is known that *in vivo*, branching morphogenesis is mediated primarily through the linkage of the epithelium and mesenchyme across a basal lamina, and that turnover of laminar components provides some mechanical direction and control of the morphogenesis (Bernfield & Banerjee 1982, Bernfield *et al.* 1984). A complete model of epithelial-mesenchymal interactions in lung development would include known behavior of epithelium in isolation, known behavior of mesenchyme in isolation, and the known linkage between the two via basal lamina turnover. Such a model, our ultimate goal in this line of research, would unite what is known about epithelial and mesenchymal morphogenesis, and lead to an enhanced understanding of the generation of glandular organs, and ultimately to an enhanced understanding of morphogenesis in general.

IV.4

Tissue Stresses in Plant Organs: Their Origin and Importance for Movements

Zygmunt Hejnowicz (Katowice)
Andreas Sievers (Bonn)

1 Introduction

To understand what the term *tissue stresses* means, imagine a herbaceous stem composed of turgescent tissues. The walls of all cells are under tensile stress caused by turgor, which is the hydrostatic pressure in cells. However, in addition to this, some tissues may be under tensile stress while others are under compressive stress. On isolation from the organ, the former will become shorter, the others longer than in the intact organ. These additional stresses acting on the tissues, superimposed on the tensile stress caused directly by turgor, are called tissue stresses. Tissue stress (TS) is defined as the stress which acts on a layer or a strand of tissue in an organ in excess of the turgor-induced tensile stress in cells. So, the singular form, TS, refers to a particular tissue in the organ. In static equilibrium (no movement in the organ), the sum of the forces which generate TSs in all tissues, as well as the sum of their moments (torques), must be zero in each direction. This means that, for instance, in a cylindrical organ, there are tensile and compressive longitudinal forces involved in the generation of longitudinal TSs, and the forces are equal in magnitude. The plural form (TSs) must therefore be used when considering an organ. Though the tensile and compressive forces have equal magnitude, the corresponding TSs are different if there is a difference in the cross-sectional areas over which the forces are distributed.

We have determined TSs, longitudinal and transverse, in the hypocotyl of the sunflower (Hejnowicz & Sievers 1995b). The tensile TS in the epidermis of sunflower hypocotyl is 4 times as high as turgor pressure. Certainly, such a high TS must be important for the various structural and physiological properties of the tissue and also for the growth of the whole hypocotyl.

The finding that there are considerable TSs in sunflower hypocotyls prompted the question: how are the stresses generated? It is commonly assumed by plant physiologists that the TSs originate from a tendency of various layers within an organ to grow at different rates (differential growth) in the organ (e.g. Thimann & Schneider 1938; Burström et al. 1967), though it is generally accepted by all the plant biomechanicists we know that TSs are due to turgor pressure as Vincent &

Jeronimides (1991) state. The TSs occur both in elongating and non-elongating organs. Our own unpublished work on TSs in the sunflower hypocotyl indicates that the longitudinal TSs: (i) do not change in any obvious way when the growth rate is changed by temperature; (ii) disappear when the organ is immersed in a solution which causes incipient plasmolysis in the cells but recover fully after transferring the organ into water; (iii) disappear reversibly when the organ wilts. The last two observations indicate that turgor pressure is involved in the origin of TSs, i.e. besides the well-known direct effect of turgor pressure – the walls of all living cells in a plant organ are under tensile stress – there is also an indirect effect manifested by TSs in an organ x – the parenchyma as a whole is usually longitudinally compressed while the more rigid tissues such as the epidermis or the collenchyma are stretched. Working with a one-dimensional model, we have shown that longitudinal TSs arise due to turgor pressure in cells of a cylindrical organ composed of turgid tissues which differ with respect to their longitudinal modulus of elasticity (Hejnowicz & Sievers 1996a). In this study we wish to present a three-dimensional model of the origin of TSs in different directions. We shall show that the TSs originate in the symplastic, turgor-induced elastic extension (strain) of tissues which differ in moduli of elasticity. Such an origin does not exclude the possibility that the stresses can be modified by a transient differential growth of the tissues.

2 The model

Our model is a cylinder composed of two tissues, a core of the inner tissue (IT), and a layer of the outer tissue (OT). The two tissues differ in moduli of elasticity and Poisson ratios. The moduli refer to the cross-sectional areas through the tissue (not for the cell walls only). When there is turgor in the cells, the tissues show elastic strains (ε_d) in different directions. We assume that a strain in a particular direction may have a component caused by the TS in this direction, besides the obvious component caused directly by turgor pressure (P) which is the same in all directions. The TS in a direction d is caused by the force (F_d) acting in this direction on the tissue and distributed over the cross-sectional area (A_d) through the tissue perpendicularly to d.

It is commonly accepted that plant organs are characterized by symplastic growth. This means an integrated growth of neighboring cells in which the adjacent cells do not alter their position relative to each other and no new areas of contact are formed, in contrast to animal organs where integrated growth is accompanied by movements of cells which thereby change their contacts. Physically, this means that a plant organ is pieced together. It is obvious that symplasticity concerns not only the growth, which applies to plastic extensions of cells (not considered here), but also concerns the elastic extensions which we consider in this study. In our model, the symplasticity gives rise to the following relationships:

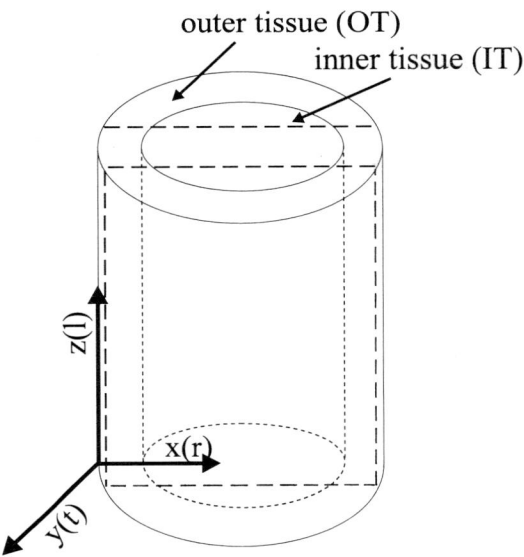

Figure 1: Scheme of a cylinder composed of two tissues, a core of the inner tissue (IT), and a layer of the outer tissue (OT). Cartesian coordinate (x, y, z) applied to an imaginary flat layer indicated by dashes, where x is in the radial (r), y in the tangential-transverse (t), and z in the longitudinal direction (l).

(i) the longitudinal (l) strain is the same in both tissues, i.e.

$$\varepsilon_l^i = \varepsilon_l^o \tag{1}$$

(the superscripts refer to the tissues, i for IT, o for OT);
(ii) the tangential-transverse (t) strain is the same in both tissues at the border between the IT and OT, i.e.

$$\varepsilon_t^i = \varepsilon_t^o \quad \text{(at the border)}. \tag{2}$$

Though cylindrical coordinates are naturally suited to our model we will avoid them for simplicity. We shall apply Cartesian coordinates (x, y, z) to the description of the mechanical state in an imaginary flat layer cut through the organ so that the layer contains the axis (Fig. 1); x is in the radial direction (r) of the layer, y is in the tangential-transverse direction (t), and z is in the longitudinal direction (l). The cylindrical geometry of the organ can then be taken into account by calling on Eqns. 1, 8, 9 & 10. Instead of (x, y, z) we shall use (r, t, l) to facilitate the identification of the directions. We make three additional simplifying assumptions: the cylinder obeys Hooke's law; it does not twist on increasing stress so that the principal axes of mechanical symmetry coincide with the direction of coordinates; and shear stresses do not cause pure extension or contraction along the mechanical axes. With these assumptions the simplified stress-strain relationship in the layer can be described by the equations

$$\varepsilon_r = S_{rr} \cdot T_r + S_{rt} \cdot T_t + S_{rl} \cdot T_l \tag{3a}$$

$$\varepsilon_t = S_{tr} \cdot T_r + S_{tt} \cdot T_t + S_{tl} \cdot T_l \tag{3b}$$

$$\varepsilon_l = S_{lr} \cdot T_r + S_{lt} \cdot T_t + S_{ll} \cdot T_l. \tag{3c}$$

ε_j ($j = r, t, l$) are strain components in a direction j. S_{ik} (i or $k = r, t, l$) are the elastic coefficients (compliances) determining strain components resulting from a given stress T_k ($k = r, t, l$) which is composed of two parts: turgor pressure P which is a multiaxial stress (the same in all directions), and a uniaxial stress F_k/A_k caused by the force F_k acting on the area A_k which is the cross-sectional area through the tissue perpendicular to the direction k. F_k/A_k represents the presumed TS in this direction, hence

$$T_r = P + F_r/A_r, \qquad T_t = P + F_t/A_t, \qquad T_l = P + F_l/A_l. \tag{4}$$

The general properties of S_{ik} may be summed up as follows:

$$S_{ik} = S_{ki}. \tag{5}$$

This is an important result of consideration of elastic energy in a deformed body obeying Hooke's law. It reduces the number of independent coefficients. A coefficient with two identical subscripts, S_{ii}, relates the strain to stress in the same direction; $S_{ii} = 1/E_i$, where E_i is the elastic modulus for the direction i. We thus have: $S_{ll} = 1/E_l$, $S_{rr} = 1/E_r$, $S_{tt} = 1/E_t$. A coefficient with different subscripts, S_{ik} ($i \neq k$), relates the strain in the direction i to the stress in the perpendicular direction k. It is related to the Poisson ratio (Hejnowicz & Sievers 1995a) by:

$$\nu_{ik} = -S_{ik}/S_{ii} = -S_{ik} \cdot E_i. \tag{6}$$

Eqn. 6 is equivalent to three equations in our model: $S_{lt} = -\nu_{lt}/E_l$, $S_{lr} = -\nu_{lr}/E_l$, $S_{rt} = -\nu_{rt}/E_r$. By combining Eqns. 5 & 6 one gets:

$$\nu_{ik}/E_i = \nu_{ki}/E_k. \tag{7}$$

This equation reduces the number of independent E_i and ν_{ik}. It should be noted that the relationship given by Eqn. 7 is fulfilled not only by ideally elastic materials but also by a material which does not perfectly obey Hooke's law, namely by wood (see Hearmon 1948). One may, however, doubt whether Eqn. 7 is also valid in the case of a turgescent tissue. We have experimentally verified this equation for the anisotropic OT of tulip stems and sunflower hypocotyls (unpublished) which proves the general applicability of Eqn. 7 to the analysis of elastic properties of the tissues. Probine (1963) and Sellen (1983) used a system of equations analogous to Eqns. 3-7 to interpret the mechanical properties of characean cells.

As already indicated, we assume that we know the material coefficients, S_{ik}. We assume that turgor pressure P is also known and, for simplicity, is the same in the two tissues. Thus, six unknown quantities (three strains and three forces) remain for each tissue, i.e. in general there are 12 unknown quantities. We have six equations (Eqns. 3 a-c), for both tissues. We also have two equations resulting from symplasticity (Eqns. 1 & 2). We need four more equations to calculate the forces. Three equations can be obtained from static equilibrium – zero sum of forces. In the case of a cylinder, the equilibrium requires that:

$$F_l^o = -F_l^i, \qquad (8)$$

$$\frac{F_r^i}{A_r^i} = \frac{F_t^i}{A_t^i}, \qquad (9)$$

and $$\frac{F_r^i}{A_r^i} = -\frac{F_t^o \cdot T}{A_t^o \cdot R}, \qquad (10)$$

where T is the thickness of OT and R is the radius of IT. We assume that a force (and the resulting strain) is positive when it causes an extension (as would be the case after increasing turgor), i.e. positive F_j results in a tensile stress.

The second condition of static equilibrium – the vector sum of torques exerted by all forces (F_j) acting on the organ must be zero – does not need any formulation, because the assumed radial symmetry of the model assures that the condition is fulfilled. Namely, in a cylinder the sum of torques exerted by F_j is zero because the torques exerted by the two equal forces acting in the opposite angular sectors of the cylinder cancel mutually. However, if the radial symmetry is lost, but ΣF_j is still equal to 0, which may occur during a tropic response of a stem, the resulting non-zero sum of the torques on opposite sides will tend to bend the organ.

The last required equation results from the fact that the OT is on the surface, therefore

$$F_r^o = 0. \qquad (11)$$

Our task is to check whether TSs, i.e. the forces F_j, do appear in the model when the cells are turgid. From Eqns. 1, 3c, 8, 9 & 11 we obtain

$$F_l^o \cdot B = P \cdot C + \frac{F_t^o}{A_l^o} \cdot D \qquad (12)$$

and from Eqns. 2, 3b, 8, 9 & 10

$$\frac{F_t^o}{A_t^o} \cdot E = P \cdot F + F_l^o \cdot G, \qquad (13)$$

where

$$B = \frac{1}{E_l^i \cdot A_l^i} + \frac{1}{E_l^o \cdot A_l^o}, \qquad C = \frac{1 - \nu_{lt}^i - \nu_{lr}^i}{E_l^i} - \frac{1 - \nu_{lt}^o - \nu_{lr}^o}{E_l^o},$$

$$D = \frac{\nu_{lt}^o}{E_l^o} + \frac{T}{R \cdot E_l^i} \cdot (\nu_{lr}^i + \nu_{lt}^i), \qquad E = \frac{1}{E_t^o} + \frac{T}{R \cdot E_t^i} \cdot (1 - \nu_{tr}^i),$$

$$F = \frac{1 - \nu_{tl}^i - \nu_{tr}^i}{E_t^i} - \frac{1 - \nu_{tr}^o - \nu_{tl}^o}{E_t^o}, \qquad G = \frac{\nu_{tl}^o}{E_t^o \cdot A_l^o} + \frac{\nu_{tl}^i}{E_t^i \cdot A_l^i}.$$

Solving Eqns. 12 & 13 for F_l we obtain

$$F_l^o = P \cdot (C \cdot E + F \cdot D) / (E \cdot B - G \cdot D) \qquad (14)$$

and

$$\frac{F_t^o}{A_t^o} = P \cdot \left(\frac{F}{E} + \frac{C \cdot E + F \cdot D}{E \cdot B - G \cdot D} \right). \qquad (15)$$

The model predicts that longitudinal TSs occur in an turgid organ ($P > 0$) as soon as the longitudinal moduli and Poisson ratios of various tissues differ. Namely, if $E_l^o \gg E_l^i$ and the IT is parenchymatic so that $\nu_{jk}^i < 0.5$ (ν_{lt}^o may be > 0.5 while ν_{tl}^o is then $\ll 0.5$), then $B, C, D, E, F > 0$, $E \cdot B > G \cdot D$. This means that F_l^o is positive, i.e. the longitudinal force acting on the OT is tensile and that in the IT is negative. A tensile longitudinal TS in the OT is accompanied by a tensile tangential TS in this tissue. The IT is then under transverse (tangential and radial) compressive TSs. The TSs disappear when turgor is lost.

If Eqns. 12 & 13 are correct, they should lead to a 1:2 ratio of the longitudinal and transverse stresses in the wall of a cylinder filled with liquid under pressure. To "change" the IT into liquid we assume that $E^i \to 0$ (weaker and weaker cell walls) and Poisson ratios for the IT (ν_{jk}^i) $\to 0.5$ (weak cell walls filled with incompressible liquid). We obtain $(F_t^o \cdot T)/(A_l^o \cdot R) = F_l^o/A_l^i$. Introducing $A_l^i = \pi \cdot R^2$ and $A_l^o = 2 \cdot \pi \cdot R \cdot T$ we obtain $F_t^o/A_t^o = 2 \cdot F_l^o/A_l^o$ which means that the expected ratio is indeed realized by the model. Similarly, the assumption of liquid IT leads to the expected tensile longitudinal force acting on the OT: $F_l^o = P \cdot A_l^i$. These results can be considered as a proof of the consistency of the applied theory.

We know that transverse stresses are small in comparison with P (Hejnowicz & Sievers 1995b) so at the first approximation we can omit these stresses in Eqn. 12, thus:

$$\frac{F_l^o}{A_l^o} \simeq \frac{A_l^i \cdot E_l^i \cdot E_l^o}{A_r^i \cdot E_l^i + A_l^o \cdot E_l^o} \cdot \left(\frac{1 - \nu_{lt}^i - \nu_{lr}^i}{E_l^i} - \frac{1 - \nu_{lt}^o - \nu_{lr}^o}{E_l^o} \right) \cdot P. \qquad (16)$$

Eqn. 16 fits that obtained from the one-dimensional model of the cylindrical organ (Hejnowicz & Sievers 1996a). When making the comparison it should be taken into account that the modulus of elasticity in the longitudinal direction for multi-axial stress ($E_{m,l}$) is related to that for uniaxial stress in this direction ($E_{n,l}$) in the following way: $E_{m,l} = E_{n,l}/(1 - \nu_{rl} - \nu_{tl})$ when transverse Poisson ratios are different, and $E_{m,l} = E_{n,l}/(1 - 2 \cdot \nu_{rl})$ when the ratios are the same. Similarly, it can be shown that the term with P in Eqn. 13 is smaller than that with F_l^o. Thus to obtain the first approximation for F_t^o/A_t^o the term with P can be omitted. Moreover T/R is a small number so we can assume that $1/E_l^o \gg T/(R \cdot E_l^o)$. Also $A_l^o \gg A_l^i$. Hence, the term with T/R, and also that with A_l^i can be omitted. We obtain $F_t^o/A_t^o \cong F_l^o \cdot \nu_{tl}^o/A_l^o$ which is the equation used for the OT of *Helianthus* (Hejnowicz & Sievers 1995b).

3 Applications

Now that we have the equations for the TSs in different directions we could use them to calculate the TSs for the hypocotyl of sunflower, for which the moduli of elasticity and Poisson ratios are known (Hejnowicz & Sievers 1995a, 1995b, 1996a). However, the obtained equations refer to the case when the moduli and the ratios are constant (i.e. do not depend on P and TSs). We know that this is not the case (Hejnowicz & Sievers 1995a, 1996a). Therefore, the Eqns. 3, 4, 8, 9 & 10 should be rewritten for small increments of strains and stresses to obtain the Eqns. 14 & 15 in the differential form; e.g. $dF_l^o = dP(C \cdot E + F \cdot D)/(E \cdot E - G \cdot D)$. Knowing the moduli and Poisson ratios as functions of P and F_l, such equations should be integrated similarly as we did in the case of one-dimensional model (Hejnowicz & Sievers 1996a). Nevertheless, we can obtain the first approximation for the F_l^o and F_t^o from the Eqns. 14 & 15 in their original form, using the values determined for the hypocotyl at full turgor (Hejnowicz & Sievers 1995a, 1995b, 1996a); $P = 0.51$ MPa; $E_l^o \cdot A_l^o = 15.2$ N; $E_l^i \cdot A_l^i = 7.94$ N; $E_l^o = 117 \cdot 10^6$ N·m^{-2}; $E_l^i = 3.28 \cdot 10^6$ N·m^{-2}; $\nu_{lt}^o = 1.4$; $\nu_{tl}^o = 0.3$; $\nu_{lr}^i = 0.32$; $T = 0.02$ mm; $R = 1$ mm; $E_t = 25 \cdot 10^6$ N·m^{-2} (E_t is an experimentally determined, unpublished value). We assume that the IT is isotropic ($\nu_{lt}^i = \nu_{lr}^i = \nu_{rl}^i$, $E_l^i = E_r^i$) and that the two ratios for the transverse stress in the OT are similar ($\nu_{tl}^o = \nu_{rl}^o$). F_l^o calculated in this way is tensile, and amounts to 0.56 N for the whole circumference of the hypocotyl, while F_t^o also tensile, amounts to 0.02 N for the segment of OT 1 mm long. The obtained values are approximately 50% higher than those determined experimentally (Hejnowicz & Sievers 1995b).

Though the origin of TSs presented above refers to a model of a very simple geometry, we may extrapolate the basic conclusion – TSs arise in a turgid organ when it is composed of tissues which differ in moduli of elasticity – to organs of other geometries. Such an extrapolation can easily be verified for a dome-shaped organ like a shoot apex. If the surface layer of the tunica is characterized by a higher modulus in the tangential direction, it will be under tensile tangential TS while the deeper-located tissue will be under radial and tangential compressive stress.

High-tensile TS occurs in the peripheral tissue of pulvini of *Mimosa pudica*, known for rapid seismonastic bending movements of leaflets and leaves (see Haupt

1977), as can be demonstrated by considerable shrinkage of this tissue on isolation. The same can be said about the traps of *Dionaea* (Hodick & Sievers 1989). Before the stimulation, there is a static equilibrium with respect to the forces involved in TSs in the pulvini and traps. This means that the vectorial sum of the forces, and also the vectorial sum of the moments of these forces are equal zero in the steady state. If the stimulation changes the TSs, and this almost certainly occurs, then there is a high probability that a net non-zero moment of the forces will arise which will act as a bending moment for the movements. The fast movements in plants have not been studied as yet from this point of view.

A study of the gravitropic bending of stems of *Reynoutria* (Hejnowicz & Sievers 1996b) has shown that the stem remains straight as long as the TSs are distributed symmetrically around the stem axis. If the tensile stress undergoes relaxation on one side, a net bending moment appears, and the organ must bend. Indeed, we have found that the first mechanical event occurring during the gravitropic response of the stem is the relaxation of the tensile TS on the lower side (Hejnowicz & Sievers 1996b).

It is known that expression of some genes is extremely sensitive to mechanical stimuli (Braam & Davis 1990). Tissue stresses may provide such stimuli to allow a feed-back between the differentation of tissues with respect to the mechanical properties of the cell wall and the expression of genes in the tissues. Since TSs result from differences in mechanical properties of cell walls, they will signal the differences to the system regulating the expression of genes. If, for instance, the cell walls which are on the surface of the organ become thicker, e.g. according to an actual topographical situation, the modulus of elasticity of the outer layer increases, a tensile TS appears in this layer, but simultaneously a compressive TS appears in the more deeply located tissue. This may be a factor in the feedback between the existing structure and the genetical control of the further development.

4 Summary

Tissue stress (TS) is defined as the stress which acts on a tissue in an organ in excess of the turgor-induced tensile stress, and which would also act in the tissue after isolation from the organ. It is shown by means of a three-dimensional model of a cylindrical organ, composed of a layer of outer tissue and a core of inner tissue which differ in moduli of elasticity, that turgor pressure brings about TSs in longitudinal and transverse directions. The outer tissue with the higher modulus is under tensile TSs, the core is under compressive TSs in longitudinal and transverse directions. The validity of the model is reinforced by the fact that if the core is considered to behave like a liquid, i.e. its modulus of elasticity is made to equal zero and its Poisson ratio to equal 0.5, the ratio of transverse stress to longitudinal stress in the outer tissue becomes 2:1, as expected for a thin-walled cylinder filled with liquid. It is proposed that the bending moments of the forces involved in generating TSs play a role in movements of plant organs.

IV.5

Self-organization and the Formation of Patterns in Plants

Paul B. Green (Stanford)

Beloussov *et al.*, the introductory article (IV.1 this Chapter), pointed out contrasting assumptions in positional information theory (PI) and in theory based on more epigenetic principles as expounded by Driesch. One key feature of PI theory, in addition to the mentioned use of privileged sites to establish a geometrical frame of reference for development, is that the relation between a control and the response to it need to be defined only by correlation. At any moment this gives PI theory enormous flexibility with regard to cause and effect. That is, a given positional value may be coupled to virtually any kind of activity. This feature can be considered a virtue because of the obvious versatility. This can also be considered a shortcoming because it places, at least temporarily, a black box between cause and effect. Control elements ideally have an explicit connection to the responding system. Also, what happens at one time needs to lead into what happens next. This normal behavior is hard to account for explicitly with a black box present in the causal chain. The theses in IV.1 thus makes plea for a rational explanation of the self-evolving and self-correcting properties of developing systems.

1 Pattern formation mechanisms

In light of these issues, we present here a summary of our work on patterns in plants. It happens to involve: development initially without privileged sites, explicit connections between control and response, and interacting self-organization features. All these aspects can be derived from an established pair of differential equations that pertain to the folding or buckling of plates (epithelia). The readiness of the transfer from physics to biology relates, almost certainly, to the fact that in plants the patterns are radial, repetitious, and well defined. The primary questions for plant pattern are: explaining its origin and its propagation. The stability of propagation is an important secondary issue.

We have taken the position that the *de novo* origin of patterns is the more critical process. Surely the primary event, organogenesis, is the same in *de novo* formation and in propagation. Even more surely, any mechanism that can produce pattern *de novo*, without cues, can propagate that pattern once it is present and providing periodic cues. Further, the self-organizing process is likely to be the same as the stabilizing process. These are large scale, or holistic, features that characterize the mechanism abstractly.

In pattern formation in general there is a large gap, with regard to mechanism, between the effect (large scale pattern, i.e. the phenotype) and the ultimate material cause (the genotype). Much effort is currently devoted to the "bottom up" approach. The tactic here is to accumulate details on the many molecular steps that pertain to a given process and then connect them in a network to account for the developmental process (e.g., Loomis & Sternberg 1995). An intricate network to determine the lysogeny-lysis decision in lambda phage illustrates this approach (McAdams & Shapiro 1995) . The strength of this point of departure is molecular biology's process with regard to all questions of essentiality. A process is altered if a pivotal part, a switch, is altered. Specificity of molecular detail about the switches approaches completeness. The binding vs. non-binding of macromolecules and effectors is established as a ubiquitous on-off switch. The question of how the switches "all fit together", however, is still obscure for organ development. For example, lack of the gene product of *deficiens* in snapdragon leads to carpel production in the third whorl of the flower, replacing the normal stamens (Coen & Meyerowitz 1991). The pertinent gene has been sequenced. Thus one has a detailed molecular grip on the control, or switch, for organ identity but no obvious entry to the broader question: how could a given tissue possibly make either? The answer to that (what is being switched?) involves the genome also, but in a very different way. Thus the information currently available accounts well for control, but the connection to the response is not explicit. Sufficiency is lacking. This may be hard to obtain "from below".

The alternative "top down" approach is to start with sufficiency, at a very abstract level, and work down to the details (see Green 1987). In effect, one begins with a holistic view of the phenotype (the responding system) and works backwards toward the control elements (genes). This approach has the tactical advantage that, if development is in reality an integration through space and time, at least the analyst is doing differential, rather than integral, calculus. So, with regard to *de novo* pattern formation, one can ask: of what interesting differential relation could this be the integral? [This contrasts with the question: "What controls it?"] To get started, one needs an analogous process as a model, and theory appropriate to the phenomenon. Fortunately, both are available.

A humble but informative analogue of *de novo* pattern formation in plants is the formation of a potato chip (crisp). Here a flat featureless disk acquires a saddle shape. The final topography resembles that at the tip of an opposite-leaved plant (e.g., a maple or mint). During cooking, the rim of the potato disk hardens first. The center continues to shrink, giving "excess surface" at the periphery. The spontaneous physical solution to deal with the extra rim is to undulate into 3-D. The theory pertinent to this class of phenomena, symmetry breaking, has been expounded by Harrison (1993), for reaction-diffusion theory. The general mechanism requires that the initial uniform object be a wavelength-dependent amplifier. The uniform object has a characteristic wavelength (e.g., half its circumference for

IV.5 Self-organization and the Formation of Patterns in Plants

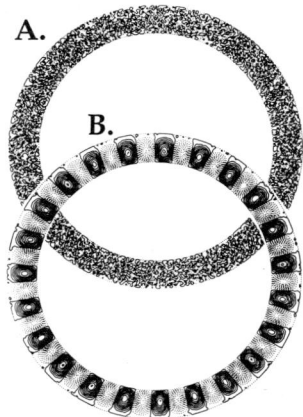

Figure 1: *De novo* whorl formation. Upward contours are drawn in solid lines; depression contours are dashed. A. An initial annulus, flat except for a random pattern of small undulations. The intrinsic wavelength equals the width of the annulus. B. In-plane compression at the margins generates 23 alternating humps and depressions. Compression is physically equivalent to tendency to expand (growth).

the chip). Out of an initial array of random undulations only those close to the intrinsic one will be amplified. These undulations will, to the best approximation, fill the available space with a whole number of districts (crests, troughs). Thus the chip becomes a saddle.

The logic explaining of the "band pass" property is that some feature of the system dampens the growth of short wavelengths, another feature dampens growth of long wavelengths. In the chemical theory, short wavelength pattern is inhibited by diffusion, long wavelength by inhibitor concentration. In the physical model that we use, short wavelengths are suppressed by a solid sheet's reluctance to bend sharply. Long wavelengths are suppressed by an elastic foundation (springs normal to the structure which are reluctant to change length). Two equations are involved. Details are explained in the accompanying box by Rennich & Green (IV.6 this volume).

2 Physical model for buckling undulations

We have chosen the physical version, rather than the chemical, because it involves no turnover of components and because it provides 3-D structures directly. The physical process is one of reaching static elastic equilibrium, so we assume that development is in multiple steps. After each step, the structure is "solidified" in preparation for the next. This is a quasi-static phenomenon. The process which repeatedly puts the system out of equilibrium is a local tendency for expansion of surface (growth) against a constraint. This is equivalent to compression. This input of in-plane stress is countered by the two responses noted above: reluctance of a plate to bend sharply and reluctance of an elastic cushion to change its dimensions. In the plant, the proposed "plate" is a coherent layer of epidermal cells, the tunica. The elastic cushion is thought to be the disorganized cells below the tunica, the corpus.

In the *de novo* formation of a whorl of organs in an annulus, as occurs commonly in flowers, there are no privileged sites. Hence there is no positional information nor, apparently, localized controls of any sort. There are only circular boundary conditions and small random perturbations of initial curvature and other variables. As shown in Fig. 1, compression (growth) gives rise to a ring of 23 undulations of the intrinsic wavelength. The non-linear character of the response insures that these undulations mutually position themselves evenly within the available space. There is some leeway in the natural wavelength, so whole numbers always arise. This sequence fits Beloussov's scenario of no special sites initially and that, after mechanical change, special sites do arise. Here uniform constraint evoked a latent structural periodicity.

When the issue of control elements vs. responding system (so flexible in PI theory) is raised, the well defined contribution of the responding system is abundantly clear in this simulation. The number, and the character, of the undulations can be influenced in three ways. The following examples are from Green *et al.* (1996).

First, the value of a term in the differential equation itself, such as the flexural rigidity of the plate, could be changed. The natural wavelength is proportional to the fourth root of the ratio: (flexural rigidity)/(elastic constant of the foundation). Changing this ratio in the input varies the number of undulations produced in the simulation, as predicted. Thus this type of control can be algebraic.

Second, the dimensions of the formative area may be varied. One way is to enlarge the mean annulus diameter, at constant width of the annulus. Here again, simply more undulations arise, as predicted. This variation is also one where an algebraic interpretation suffices. When, however, annulus width is increased at constant outer diameter, there is an increasing difference in the fit for the undulation at the two margins. Obviously fewer undulations can fit the smaller circumference. The response to change in this dimensional input, or "control", is the production of Y-shaped primordia, or even fused ridges. This provides the annulus with undulations all close to the natural wavelength. This is seen in Fig. 2 A-C.

Third, and especially interesting, is varying the boundary conditions. These are the constants of integration involving initial elevation (here kept constant at zero) and either slope or curvature at the margin. When the slope is fixed, the edge is "clamped". When the curvature is fixed, the edge is "hinged", or in engineering terms "simply supported". In Fig. 2 a left-right comparison of images involves change in boundary conditions only; the dimensions are the same. In the pair A-D the change from clamped to simply supported for the outer margin has drastic

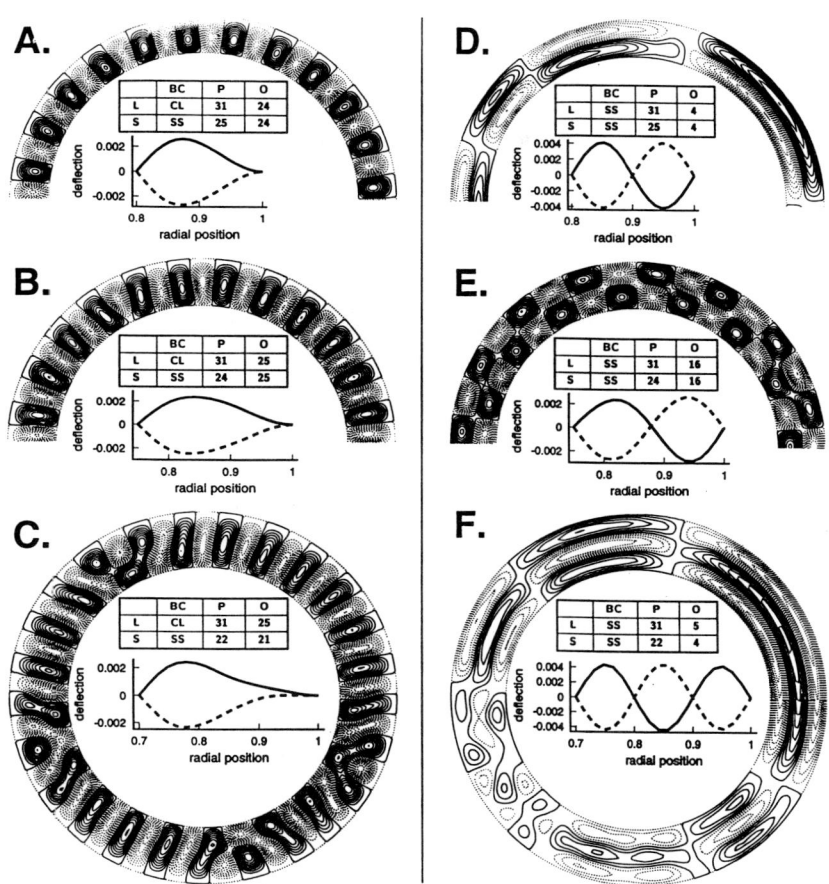

Figure 2: Effects on pattern of varying annulus width (limits of integration) under two different sets of boundary conditions (constants of integration). In each figure the box shows boundary conditions, BC, at the large (L) and small (S) circumference. CL is clamped, SS is simply supported, or hinged. The graph shows the radial profile of humps (solid) and depressions (dashed). Predicted (P) and observed (O) numbers of undulations at the two margins are given. An integral number of half-wavelengths fits radially only in A and D. A-C. As width is increased, some new undulations are "fused" as Y-shaped ridges (depressions), reducing the undulation number at the smaller circumference. D-F. With both margins hinged, the natural undulation is fitted differently to the radial dimension (steep slope at both ends). When the fit of $\frac{1}{2}$ wavelengths is exact (D and F), this promotes formation of circumferential ridges. When the fit is poor, as in E ($2\frac{1}{2}$ wavelengths), the annulus is still subdivided circumferentially, but a checkerboard pattern arises. It includes more circumferential undulation.

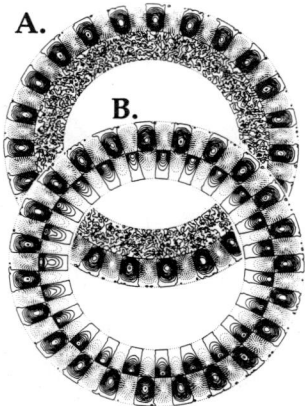

Figure 3: A. Propagation of pattern. An undulating annulus, as in Fig. 1B, is made permanent. It is the outer one. The steep slope at its inner margin becomes the outer clamped boundary condition for a new, naïve annulus (♯2). B. Compression of annulus ♯2 leads to formation of the corresponding undulating pattern, exactly out of phase with ♯1. This is the key cyclic feature of whorled development in plants.

effects. Apparently the flexibility of having both margins hinged now allows the one wavelength fitting across the annulus to provide a node in the center; hence, long ridges, circumferential (trenches) subdivide the annulus. One now adds radial dimensional change to this condition. At the greater width in F, four nodes fit well, and the annulus is largely tri-sectioned circumferentially. However, at an intermediate dimension, in E, the transverse fit in terms of half wavelengths is poor, and much more of the undulation takes place circumferentially. The annulus is still bisected but a "checkerboard" results. The strikingly different responses to simple dimensional changes (compare vertically within each column in Fig. 2) and to boundary conditions (left-right in Fig. 2) show that these parameters can have major, non-intuitive, consequences. The fact that the system is inherently periodic obviously contributes to this diverse behavior as a function of simple progressive change. Thus variation in dimension and boundary state are control elements, or effectors, to be added to the variables already familiar in an algebraic context.

Turning to the issue of propagation of pattern, development at the shoot apex is centripetal. Some feature of an old large annulus must influence the new pattern on a smaller adjacent "naïve" annulus inside it. Most models assume that this influence is a spatially periodic distribution of control molecules acting on a simple responsive tissue. In the present case the responsive region is not considered simple, i.e., it has latent periodicity. We investigated how the physical input of slope periodicity, restricted to the outer boundary of the naïve annulus, could influence pattern development in it. We set (clamped) the slope at the outer boundary of the naïve annulus to be spatially periodic. The inner boundary was clamped at zero slope. The patterned input at the margin sufficed not only to shift the phase of undulations, as is typical in whorled plants, but also to maintain the same number of undulations despite the reduced circumference (Fig. 3). In this case the key feed-forward control was the slope at the periphery, a constant of integration for the master equations.

We thus view the formative annulus of tissue at the plant apex as "an active solid body". It can have a natural wavelength. This latent periodicity can evoke, in *de novo* pattern formation, a whorled pattern without a particular phase. When, of course, the formative area is physically connected to a periodic structure peripheral to itself, as is often the case in plants, that periodicity feeds in at the boundary to set the phase of phenomena in the adjacent naïve area. The phase shift in simple whorled systems is one half wavelength. We obtain it in simulations. We are now applying this approach to spiral patterns (Green 1996a).

3 Summary

The feature of *de novo* pattern by non-patterned mechanical influence on an annulus was illustrated in whorl initiation. Continuous interaction within the system was characteristic of the solution of the non-linear equations. The resulting ring of new undulations, i.e., new mechanical singularities, was able to influence the new-forming pattern in an adjacent annulus. Periodic slope caused the phase shift (alternation) so typical of organs in flowers. In this "top down" modeling, the interplay of control factors and responding system is explicit through the use of equations for buckling. A diversity of controls is available when the mechanism is embodied in differential equations. While the models are abstract, and at present devoid of molecular specificity, they are at least sufficient and explicit, in physical terms, with regard to the origin and propagation of pattern.

Subsequent work has concerned comparing the propagation of whorls and "Fibonacci" spirals, the only two patterns common in nature. In each case the propagation occurs within an annulus. Over time, tissue of its flat inner margin is converted to an undulating profile at the outer margin. The relative work required to do this can be calculated. Thus far, it appears that whorls are of least energy, provided that the intrinsic wavelength of the tissue fits the circumference exactly. Spirals require slightly more energy but are uniquely insensitive to change in annulus circumference. This is an attractive hypothesis to account for the prevalence of the two patterns (Green 1996b).

Acknowledgements The generous contribution of computer time for all the simulations, by the Maui High Performance Computer Center, Kihei, Maui, HI 96572, is gratefully acknowledged. Supported from a grant of the National Science Foundation to P.B.G.

IV.6

The Mathematics of Plate Bending

Steven C. Rennich and Paul B. Green (Stanford)

Recent work (Green IV.5 this Chapter) has investigated the link between patterns seen in plant structures and the physics of solid materials. In this work the plant is modeled as an elastic solid subjected to specific boundary conditions and forces. When stressed, this elastic solid deforms to maintain a state of minimal energy. The equations governing the deflection of such a structure are well defined and can be solved numerically. This work attempts to offer a simple explanation of the plant model, the governing equations and the solution methods. See Green et al. (1996) for further discussion.

1 Model

The plant is modeled by a thin, circular plate attached to a linearly elastic foundation. The thin plate represents the epithelium (tunica) and the elastic foundation represents the inner corpus tissue of the plant. The plant structure surrounding the tunica provides the boundary conditions. The expansion (growth) of the tunica against (relatively) rigid boundaries causes in-plane compression of the tunica. This compression can cause out-of-plane deflection of the tunica and is modeled by applying in-plane compressive forces at the outer boundary of the disk.

2 Linearized equations

If the plate is thin and the deflections are small, the applied stresses and subsequent deflections can be related by a relatively simple equation from small deflection plate theory (Szilard 1974).

$$D\nabla^2\nabla^2 w(x,y) + Q\nabla^2[w(x,y) + w_0(x,y)] + kw(x,y) - p = 0. \qquad (1)$$

Versions of this equation were obtained by Lagrange as early as 1811. This equation captures much of the underlying physics and is easily understood. The constants are D, the flexural rigidity (stiffness) of the plate, Q, the applied in-plane compressive force per unit length, k, the coefficient of elasticity of the linearly elastic foundation, and p, the applied surface pressure. The initial (unstressed) out-of-plane surface deformation, $w_0(x,y)$, is known. The only unknown variable is the out-of-plane surface deflection, $w(x,y)$.

The equation itself represents the sum of the out-of-plane forces acting at any point on the plate. The terms in the equation represent (from left to right) forces due to: plate bending, the applied in-plane compression, the elastic foundation and the surface pressure. The applied forces, Q and p, tend to push a region of the plate away from its unstressed position. The forces due to bending and the elastic foundation tend to force the plate back to its unstressed position. The stressed plate will come to rest when all of the forces sum to zero at every location on the plate. This position is found by solving Eqn. 1 for w. The solution yields the equilibrium deformation of the stressed plate.

Boundary conditions, required for the solution of Eqn. 1 are formed by prescribing the deflection and slope, termed a clamped boundary condition, or the deflection and zero moment, termed a simply supported boundary condition at the plate edges.

The small deflection equation (Eqn. 1) acts as a wavelength dependent amplifier (band pass filter). As the applied in-plane compression, Q, is increased, some undulations present in the initial deflection, w_0, are amplified while others are damped. The most amplified waves are those with wavelength equal to the natural wavelength, λ_n, which depends on the stiffness of the plate and the coefficient of elasticity of the foundation, as follows:

$$\lambda_n = 2\pi \left(\frac{D}{k}\right)^{\frac{1}{4}}. \qquad (2)$$

Undulations with wavelength much smaller than λ_n are damped by the resistance of the plate to bend sharply. Undulations with wavelength much larger than λ_n are damped by the resistance of the elastic foundation to large deflections. The natural wavelength can be thought of as the "easiest" way for the plate to bend. It is that wavelength where the potential energy in the bent plate equals the potential energy in the deflected elastic foundation. Since λ_n is proportional to the fourth root of the ratio D/k, it is very insensitive to changes in either D or k.

3 Nonlinear equations

To allow for large deflections and nonlinear interaction, the treatment in section 2 must be supplemented significantly. The two equations governing moderately large deflections of a plate are known as the Von Kármán equations (Szilard 1974):

$$D\nabla^4 w = \mathcal{L}(w + w_0, \Phi) + p - kw \qquad (3)$$

$$\nabla^4 \Phi = -\frac{t}{2} E \left[\mathcal{L}(w + w_0, w + w_0) - \mathcal{L}(w_0, w_0)\right]. \qquad (4)$$

The differential operator $\mathcal{L}$ is defined as

$$\mathcal{L}(\xi, \eta) = \frac{\partial^2 \xi}{\partial x^2} \frac{\partial^2 \eta}{\partial y^2} + \frac{\partial^2 \xi}{\partial y^2} \frac{\partial^2 \eta}{\partial x^2} - 2 \frac{\partial^2 \xi}{\partial x \partial y} \frac{\partial^2 \eta}{\partial x \partial y} \qquad (5)$$

and the constants and variables are defined as in section 2 with the addition of t, the thickness of the plate and $\Phi(x, y)$, the Airy stress function. For small deflections, neglecting nonlinear terms, Eqns. 3 & 4 decouple and Eqn. 3 reduces to Eqn. 1. For large deflections, nonlinear terms that account for the in-plane stretching or compression of the plate become significant. Since the in-plane stresses are no longer a function of just the applied in-plane compression (the deflection of the plate can also generate in-plane stresses) it becomes mathematically convenient to represent all of the in-plane stresses in terms of the Airy stress function, Φ, (see Szilard 1974). Φ provides a relationship between out-of-plane deflection and in-plane stress.

The boundary conditions used for solving Eqn. 3 are the same as those in section 2. The boundary conditions on the Airy stress function, used when solving Eqn. 4) are formed from the applied force Q (see Szilard 1974).

4 Numerical method

When computing deflections of a compressed circular plate, the equations used are the non-linear Von Kármán equations 3 & 4, written in polar coordinates; see Szilard (1974) for the actual transformation. Briefly, the solution process is an iterative one.

1) First, all constants (D, k, p, t), initial conditions (w_0) and boundary conditions are specified. The deflection, w, and in-plane compression, Q, are set to zero.
2) Solve Eqn. 4 for Φ using the latest approximation of w.
3) Solve Eqn. 3 for w using the latest approximation of Φ and treating all nonlinear functions of w explicitly.
4) If w has not converged repeat from step 2.
5) Increment Q by some small value.
6) If Q has not reached the desired value, repeat from step 2

A deflection for a given in-plane compression is found by slowly ramping the compression up from zero, iterating frequently along the way to ensure that the error in the approximate solution is always below a specified tolerance. In this way the deflection of the plate is always kept very close to the equilibrium value.

IV.7

Mechanical Forces and Signal Transduction in Growth and Bending of Plant Roots

Alexander A. Stein (Moskva)
Mechthild Rutz and Hanna Zieschang (Bonn)

1 Cell growth

The tissues of the root elongation zone consist mainly of cell walls that form a solid deformable framework and of a multitude of intracellular volumes containing fluid under turgor pressure. The fluid exchange takes place in different ways: via the apoplasm that includes extracellular space and cell wall pores, and via plasmodesmata providing direct communication between intracellular volumes. The cell-file structure of the tissue permits relatively distinct partitioning of longitudinally and transversely oriented walls. This geometrical organization allows one to postulate simple relations between macro- and microparameters for the tissue.

The growing cell is the elementary unit of the system. The result of growth is an increase of its linear dimensions and hence volume. Cell wall elongation and absorption of water are the two main constituents of the process. The mechanical forces applied to the cell wall are intracellular liquid pressure and forces that owe their origin to the mechanical contact of the wall as the part of the global framework with other cell walls and finally with soil and obstacles. The resulting tension produces elastic stretching and promotes an inelastic one, the latter including growth deformation accompanied by deposition of new material.

The first attempt at a mechanical description of plant cell growth was given by Lockhart (1965). The relationships reflecting Lockhart's idea in its modern form are given below, taking into account some later modifications (Plant 1982, Ortega 1985) and several considerations of our own. We will formulate the relations describing cell elongation in one direction only. This approach is reasonable, as an example, for a long cylindrical cell if its transverse deformation is negligible. As in the papers mentioned above, external forces are assumed to be absent in this section.

The main quantities to be found are the total and irreversible (growth) longitudinal wall strains that are analogous to those traditionally used in growth mechanics (Regirer & Stein 1985, Stein 1995, 1996a), or the total length of the cell L and its length after unloading K (Rutz 1995). The rates of total and irreversible relative elongations (Erickson 1976) can be introduced by the following formulae:

$$g = \frac{1}{L}\frac{dL}{dt}, \qquad g^i = \frac{1}{K}\frac{dK}{dt}. \qquad (1)$$

The incompressibility of water allows to set the total elongation of the cell wall equal to water influx. The dependence of the latter on the turgor pressure p and the osmotic pressure difference between the cell and its surroundings $\Delta\pi$ is given by the relationship

$$g = H(\rho\Delta\pi - p), \qquad (2)$$

where H is a permeability coefficient. In plant biophysics one uses the term "water potential" for a "driving force" supplying water entrance. Here, this term may be applied to the bracketed expression in Eqn. 2. The following equilibrium condition links the turgor pressure p and the longitudinal tensile stress in the cell wall T

$$p = \alpha T, \qquad (3)$$

the coefficient α depending on geometrical parameters only (but not on K or L for simple geometry).

The rheological behaviour of the wall is controlled by the relationships

$$T = \eta \frac{L-K}{K}, \qquad (4)$$

$$g^i = \lambda(T-S) \text{ (if } T > S\text{)}, \qquad g^i = 0 \text{ (if } T < S\text{)}. \qquad (5)$$

According to Eqns. 4 & 5, the strain rate g is determined by elastic and inelastic elongations, the first being governed by a Hook-like law (Eqn. 4) and the latter showing "viscous" behaviour as one can see from Eqn. 5. The ratio $\epsilon_w^e = (L-K)/K$ is the elastic longitudinal strain supposed to be small. There is no inelastic elongation, if the longitudinal tensile stress T is less than a positive threshold stress S.

The system of equations 1–5 allows to determine the cell deformation if $\Delta\pi$ and the coefficients H, ρ, η, λ, α are given. The interrelation of the constituent processes given by the model can be schematically represented as follows. The water influx due to water potential difference leads to the elastic elongation of the cell. The insertion of new material into the wall is stimulated by the tension and lowers this tension as it develops. Inasmuch as the tension is in equilibrium with turgor pressure, the latter also decreases, thus augmenting the water potential. This leads to new water streaming into the cell.

The system of equations 1–5 is insufficient to describe many real effects. The simplest generalization can be achieved by accepting nonlinear relations instead of linear ones or, similarly, allowing the dependence of coefficients in Eqns. 1–5 on variables. Rutz (1995) used the following relationship connecting L and K directly:

$$\frac{dK}{dt} = \tilde{\lambda}(L - K - \tilde{S}) \qquad (6)$$

that is covered by setting $S = \tilde{S}\eta/K$ and $\lambda = \tilde{\lambda}/\eta$ in Eqn. 5. This modification reflects a possible change of the threshold S with cell growth.

2 One-dimensional tissue growth

The relations between the parameters that characterize cell and tissue deformation are most elementary for the elongation-only zone of the root. If, at a location, we take L to be the averaged cell length, its relative elongation rate g can be identified with the tissue strain rate (x is the spatial coordinate along the root axis, v is the axial velocity of the medium):

$$g = \frac{1}{L}\frac{dL}{dt} = \frac{\partial v}{\partial x}. \tag{7}$$

For the unstressed tissue (the mechanical stress equals zero in the tissue as a whole), the system of equations presented in the previous section can be used if supplemented by Eqn. 7. The time derivatives in Eqns. 1 & 6 should now be identified with corresponding total derivatives. The analysis of stationary solutions for the unstressed root was performed by Plant (1982), Logvenkov (1993) and Rutz (1995). Assuming the dependence of mechanical and liquid exchange controlling parameters on either cell length or its position, it is possible to reproduce measured growth curves. In the latter paper, using the relationship Eqn. 6, it is supposed that the appearing rheological coefficients are constant and that $\Delta\pi$ depends linearly on cell length L. Simulated curves representing L and g as functions of x are reproduced in Fig. 1. Taking into account the diffusion of a regulator being produced at the apical end of the root and consumed all over the growth zone, or the additional dependence of rheological coefficients on L and K, one can further modify the form of the curves and fit them to measured values.

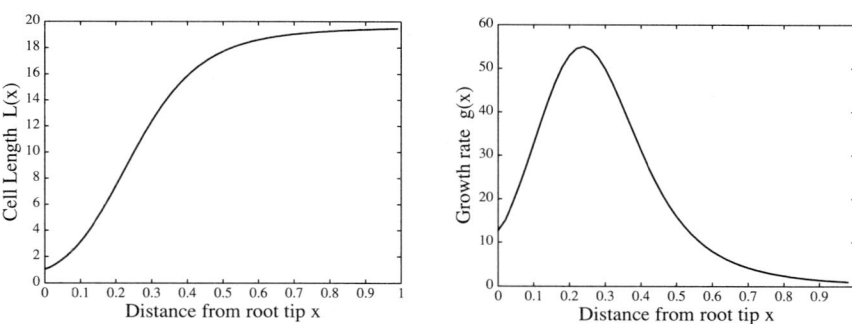

Figure 1: The stationary distributions of cell length $L(x)$ and of relative elongation rate $g(x)$ along the axis of a root (Rutz 1995). The quantities have been rescaled, hence $L(0) = 1$.

A continuum mechanical model of growing file tissue taking into account tissue stress distribution was developed by Stein (1996a). The validity of Lockhart-type rheological relationships for a cell wall is the starting point of the theory. The model allows to describe the participation of an external force in the equilibrium conditions for a tissue element and hence for a cell. This force is generated as a response of the cell wall framework to restrictions and represented by tissue stress in a continuum approach. The model, although allowing for transverse growth, can be reduced to one-dimensional relationships if the transverse extensibility is small as compared to the longitudinal one and the loading is axially symmetrical, or absent. In this case the relationships defining cell wall deformation retain the form of Eqns. 4 & 5. But the equilibrium condition (Eqn. 3) should be replaced by a more complicated one including the turgor pressure p, the longitudinal wall stress T, and the longitudinal tissue stress σ. For small volume concentrations of longitudinal cell walls α, this condition takes the following rough form:

$$\sigma = -p + \alpha T \tag{8}$$

where the geometrical parameter α can be identified with the coefficient α in Eqn. 3.

The total tissue strain is represented as the sum of inelastic (ϵ^i) and elastic (ϵ^e) ones. The tissue elastic strain ϵ^e equals $\epsilon^e_w - p/E$ (E is an elastic coefficient) if one assumes that zero strain corresponds to the unstressed state of the medium as a whole.

The constitutive equations for one-dimensional tissue growth that can be got from the relationships Eqns. 4, 5 & 8 are of the "viscoelastic" type and have the following form:

$$\epsilon^e = \frac{1}{E}\sigma, \tag{9}$$

$$g^i = w + \frac{1}{\theta}\sigma \ (\text{if } \sigma > \sigma_{cr} = -\theta w), \qquad g^i = 0 \ (\text{if } \sigma < \sigma_{cr}) \tag{10}$$

where g^i is the rate of irreversible (growth) strain and can be identified with the variable represented by the same symbol in Eqn. 1. The "intrinsic" growth rate w is a function of turgor pressure p:

$$w = \frac{1}{\theta}(p - Y) \tag{11}$$

Eqns. 9 & 10 include elastic and "viscous" coefficients E and θ (both positive). One can easily see that Eqns. 8 – 11 are equivalent to the relationships Eqns. 3 – 5 if the tissue is unstressed ($\sigma = 0$). It is sufficient to suppose: $E = \alpha\eta$, $\theta = \alpha/\lambda$, $Y = S\alpha$. For the unstressed tissue, $w = g^i$ identically.

The system of equations 7–11 has to be supplemented by the usual equilibrium and kinematic relations but it still remains unclosed. One can use the relationship Eqn. 2 or its modifications but, strictly speaking, this equation is not entirely correct for the tissue. At least three different compartments are involved in the water exchange: intracellular volumes, apoplasm, and the root environment. The parameters characterizing these compartments vary along the root axis. The generalization of Eqn. 2 leads to setting up a broader problem with the relations describing liquid transport included. Examples of models, where some transport phenomena are allowed for, were given by Plant (1982) and Logvenkov (1993).

3 Bending of unrestricted roots

The models discussed in the previous section can be used in the analysis of root deformation both for free growth processes and for root interaction with obstacles and soil. Although this problem is two- or even three-dimensional (if one is going to describe not only bending in a plane but also nutations), estimates can be done showing that relationships between longitudinal components of stress and strain are only essential for small deviations of the root from the straight line (Stein 1996a).

For the free growth of the root, a two-dimensional model was developed (Rutz 1995). It uses the equations given in the previous sections for the unstressed tissue, cell length characteristics L and K being calculated for two sides of the root separately. The curvature, the lateral displacement, and the apical tip deviation are found from obvious geometrical considerations, see below, in particular Eqn. 12.

A special one-dimensional coordinate system that changes in time is associated with the midline. Let s denote the distance of an arbitrary midline point from the tip measured along the midline arc. In order to know the quantities on both sides but expressed as functions of s, a time dependent variable transformation of the lateral coordinates has to be performed. This leads to a Volterra integral equation for the curvature κ, that can be solved analytically to give the transformation.

Although realistic growth curves have been reproduced without any additional control involved, regular oscillations were not obtained using this model. Such oscillations arise if a chemical regulator moving upwards from the apex, and being governed by the diffusion equation, is included. Inasmuch as the regulator has to cover a longer way along the convex side of the root than along the concave one to reach corresponding points, its concentration on the concave side is higher. Due to its growth promoting effect ($\tilde{\lambda}$ and H increase with the concentration), it leads to the balance of side lengths and, due to the delay, to the reversal of lengths at the two sides of the root. Stable oscillations of the tip around the root axis independent of gravitropic effects are the result. They are depicted in Fig. 2.

Additional influences of mechanical forces that arise from the different tensions in the cell walls at the two sides of the root during initial bending were included in the model. Starting from a curved state, the bending leads to an oscillation that is soon dampened out if no chemical regulator is involved. When the regulator is included, the amplitude is the only parameter that changes when additional mechanical forces are applied. This model supports the idea of a chemical regulator involved in the circumnutation mechanism.

An important and intensely investigated field is the bending of roots due to their gravitropic response. The difference in local growth (differential growth) between the upper and lower sides of the root changes with respect to both time and distance from the root tip. In the vertical orientation the differential growth is absent.

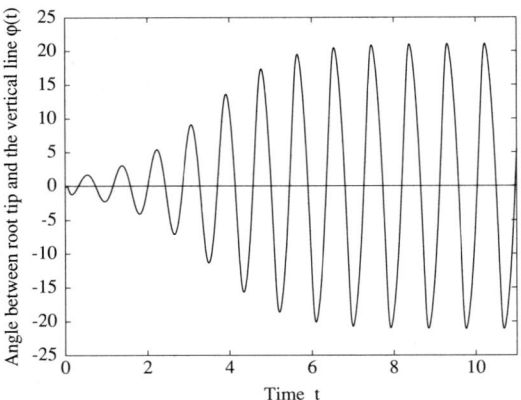

Figure 2: Simulated oscillations of a root elongation zone around the vertical line due to the difference in concentration of a growth promoting regulator on the two sides of the root (Rutz 1995). Plotted is the angle $\varphi(t)$ between the root tip midline and the vertical. Time has been rescaled to the unit of 1.

A kinematic model was developed (Zieschang 1992) that allows to compute the shape of the root midline within a two-dimensional projection. The one-dimensional coordinate system, with a coordinate s as defined above, is associated with this projection. In accordance with the well-known theorem of differential geometry, the curvature distribution generates the unique shape of the curve that can be parameterized by its arc length. In order to obtain the curved shape of the midline, consider g_1 to be the strain rate for the upper side and g_2 to be the analogous quantity for the lower one. Further in this section, the strain rates are identified with their irreversible constituents because elastic constituents are small, being not essential for purely kinematic problems. The local curvature κ is given by the formula

$$\frac{\partial \kappa(s,t)}{\partial t} = \frac{g_1(s,t) - g_2(s,t)}{2\,r} \qquad (12)$$

where r is the radius of the rotationally symmetrical root cross-section. The distance of any individual point from the tip changes with time while the root is growing and can be obtained by solving a set of differential equations. All the parameters characterizing the orientation and the shape of the root, including the angle of tip deviation from the vertical line φ, can be computed if one accepts some simple geometrical assumptions (Zieschang et al. 1996).

Experimental data lead one to expect that the tip deviation angle φ already reached by the root during the bending process is the main parameter responsible for inducing differential growth. A corresponding dependence proposed by Barlow et al. (1991) has the following general form

$$g_\beta(s,t) = R(s)\left(1 + P_\beta\,\varphi(t)^a\,(180 - \varphi(t))^b\right) \qquad (13)$$

where the exponents a and b are constants. The index β equals 1 or 2, the parameter P_β being different for the upper and lower sides. The angle φ is measured in degrees in this section. As a function $R(s)$ defining the growth rate distribution of a vertical root, Zieschang (1992) suggested the von-Bertalanffy function, its behaviour being similar to one displayed in Fig. 1. In contrast to the previous model with expressed oscillations, visualized in Fig. 2, the von-Bertalanffy function leads to a simulation of a slightly oscillating root tip movement occurring during the gravitropic bending process. This movement resembles the circumnutation of a root tip which actually can be observed with gravireacting roots. Some results of the simulation of the kinetics of a tilted primary root are depicted in Fig. 3.

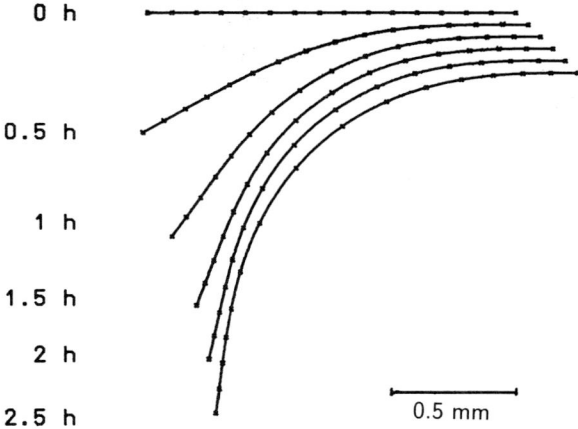

Figure 3: Simulated kinetics of a primary root which was tilted by $\pi/2$. The midlines of the roots are plotted at time intervals of 0.5 h. For each time step the corresponding plot was shifted, so that each single line can clearly be seen. When plotting one midline over the other the simulated root tip at 2.5 h would cross that at 2 h, indicating that the used model function leads to the simulation of a root tip oscillation during the bending process. The parameters were chosen according to experiments with roots of *Phleum pratense* (Zieschang et al. 1997).

A more detailed analysis is needed for the inclusion of a time lag that appears, in particular due to the finite velocity of the signal moving from the tip perceiving the gravity stimulus (the statenchyme cells in the root cap are particularly perceptive) (Wareing & Phillips 1981). Thus g_β should depend on the angle at a preceding instant (Zieschang et al. 1996). This delay mechanism may be another possible reason for growth oscillations and nutations (Israelsson & Johnsson 1967).

4 Growth and Bending of Restricted Roots

For an elementary analysis of restrained growth, the models using the constitutive equations 9 & 10 were accepted. The dependence of their coefficients on concentrations of chemical regulators was taken into account. This approach is reasonable, as an example, if the permeability (in Eqn. 2 characterized by the parameter H) is large enough. Using such models, one can investigate the interplay of mechanical (through growth-rate dependence on stress) and chemical regulation.

The simplest example of restrained growth is a root growing vertically outside the soil (in experiments) and meeting a rigid plane. The restriction leads to the increase of the compressive load P resulting in stability loss and bending (Fig. 4). The problem of this sort for a rod of growing material with the properties described by the Eqns. 9 & 10, if the coefficients are constant or time dependent, was investigated by Stein (1995). Vertical growth is unstable if growth rate decreases under compression and increases under tension.

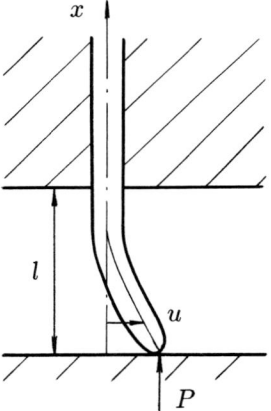

Figure 4: Bending of a root tip growing outside the soil and meeting a rigid plane (l denotes the length of the free part of the root, u the lateral displacement of its tip and P the compressive load).

The regulator coming from the root cap in response to the tip deviation from the vertical line and providing the gravitropic reaction (see the previous section) is a possible stabilizing factor. For an elementary analysis one can neglect the nonuniformity of differential growth along the growth zone produced due to the apex deviation and assume that the spreading of the regulator is rapid compared to the growth process. These assumptions correspond to a relation of the form $w = w(\varphi, y)$ where y is the coordinate measured along a line perpendicular to the midline, or, more schematically, $w_1 - w_2 = \Delta w(\varphi)$ where the indices 1 and 2 indicate "intrinsic" growth rates defined separately for two sides of the root (as the strain rates in the previous section). As in the previous section, φ is the angle of tip deviation from the vertical line (now measured in radians).

The procedure of averaging along the transverse cross-section, based on the constitutive equations 9 & 10, leads, for small deviations from the vertical line, to the following equation which defines the bending moment M of axial forces acting in this cross-section (Stein 1996b):

$$u''_t + \frac{M}{\theta I} + \frac{M_t}{EI} - ku'(0,t) = 0. \qquad (14)$$

Here the dash denotes the derivative with respect to the axial coordinate x and the lower index t means the time derivative, u is the lateral displacement of the root axis. In the linear approach, u'' is the curvature and u' is the angle of deviation from the vertical line, thus $u'(0,t) = \varphi$. The coefficients E and θ have the same meaning as those in Eqns. 9 & 10, I is the moment of inertia of the cross-section. Eqn. 14 represents the rate of curvature change as a sum of elastic and growth constituents taking into account the influence of bending moment and differential growth due to the gravitropic reaction. The last term, including a positive differential growth regulatory parameter k, appears as a result of the assumed regulatory hypothesis. The analysis on the basis of Eqn. 14 uses conventional equilibrium conditions for a bending rod.

It turns out that growth-rate dependence on stress leads to bending before Eulerian buckling takes place. If $k\theta l/E \ll 1$ (l is the length of the free part of the root), vertical growth is always unstable but time for loss of stability is large for small loads and becomes comparable with the growth stopping time for $P \sim P_{cr}/2$ (P_{cr} is Eulerian critical load). The gravitropic reaction can stabilize growth if k is large enough, but it can not stabilize rapid deviations, so the root bends when P_{cr} is achieved.

A root growing in the soil is a fundamentally different mechanical system because the soil restricts lateral movement of the root. The mechanical interaction between the root and the soil is a complicated problem. The root behaviour may vary significantly according to soil rheology and other factors. A model briefly presented below (Stein, in preparation) shows that the joint involvement of mechanical and chemical (gravitropic) controlling mechanisms can stabilize vertical growth.

Let us consider the vertical growth of the root, assuming that this non-perturbed solution exists due to symmetrical conditions. The main part of the root does not grow. The propelling force is generated in the elongation region. The resistance is distributed all over the moving part of the root, but the main load is concentrated at the small apical area where the pushing of the root through the soil takes place. The apex velocity and the load are developed as a result of interplay of different factors. These quantities are assumed to be known during the analysis of perturbations.

The deviation of the apex from the vertical line is characterized by the angle φ (Fig. 5). The lateral movement of the root is assumed to be excluded by soil

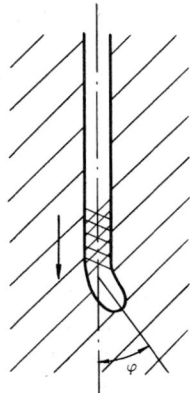

Figure 5: Deviation of a root apex in soil. The growth zone is shown by hatching. The arrow shows the direction of nonperturbed motion.

presence except at the apical end where the root meets and deforms new regions of soil. The derivation of the following two elementary relationships, Eqns. 15 & 16, is based on the possibility of separating the two domains (apex and growth zone) in the model. Supposing that the elastic forces due to bending are small as compared with those produced by growth restriction, one can reduce Eqn. 14 to a rough relationship reflecting the rheology of the growth zone

$$\frac{M}{\theta} + \frac{M_t}{E} - \tilde{k}\varphi = 0 \qquad (15)$$

where $\tilde{k} = Ik$.

The second relationship gives the moment balance for the apex on the assumption that the soil resistance to its advance and rotation depends on the velocities of these movements. Any deviation of the apex breaks the symmetry and, taking into account the axial thrust being produced by the growth zone, leads to the development of a moment that augments the deviation. On the other hand, the moment M being generated due to differential growth restriction and the moment of resistance forces depending on the rotation velocity counteract this process. In linear form that is natural for small deviations, these general assumptions lead to the relationship

$$\nu\varphi - \tau\varphi_t - M = 0. \qquad (16)$$

Both coefficients in Eqn. 16 are positive and depend on the load (or velocity) being developed in nonperturbed movement. The coefficient ν increases with load. In evaluating the results of this analysis, one has to remember that Eqn. 16 is valid for relatively rapid perturbations as compared with the time it takes the root tip to be displaced through a distance equal to the length of the growth zone.

Eqns. 15 & 16, the first being applied to the lower region of the growth zone, form a system of ordinary differential equations that gives both stable and unstable regimes. Stability conditions are the following:

$$\frac{\nu}{\tilde{k}} < \theta < \frac{E\tau}{\nu}. \qquad (17)$$

Increasing the load leads to destabilization due to increasing ν. Oscillatory and exponential regimes of stability loss are possible (and such oscillations are seen in real root shapes). Increasing the gravitropic parameter $\tilde{k}$ only cannot provide vertical growth stability but leads to oscillations if this parameter is large enough. The stable regime is impossible if the growth viscosity θ is too large, so one has a reason to suggest that growth regulation by mechanical stress is an important part of vertical growth stabilization.

5 Conclusion

Modelling of plant tissue growth cannot be properly developed without formulating all the mechanical aspects. There are a lot of experimental data demonstrating the influence of different mechanical factors on this process (Regirer & Stein 1985, Cosgrove 1986). Elementary lumped parameter models like Lockhart's (now 30 years old) do not allow one to analyze spatial growth organization and give way to more complicated models with distributed parameters.

The setting of appropriate problems leads to using methods of continuum mechanics. The first step in this direction is an analysis based on kinematical relations only (Silk & Erickson 1979). This approach may be fruitful in experimental data processing (see also section 3 of this paper) but is not sufficient if one is interested in modelling of controlling mechanisms involved in plant tissue growth. Although some interesting results may be obtained using traditional mechanical models (like elastic body), an advanced analysis is impossible without employment of growing material models and simulating transport phenomena in their interplay with growth. The basic challenge is the interrelation between processes on the cell and tissue levels. The difficulties are not only mathematical but also conditioned by deficiency of experimental data. Nevertheless, there is enough information for following steps and one may anticipate new achievements in this promising direction that will stimulate the performance of new experiments.

Acknowledgements The investigations summarized in this paper have been in part supported by the Russian Fund of Basic Researches, grant 96-01-00956. Other parts (H. Zieschang) have been supported by the "Deutsche Forschungsgemeinschaft". The authors are grateful to W. Alt, P. Barlow, Ph. Brain, P. B. Green, and S. A. Regirer for helpful reading of the manuscript.

IV.8

Growth Field and Cell Displacement within the Root Apex

Jerzy Nakielski (Katowice)

The root apex in angiosperms, structurally divided into the root proper and the root cap, consists of a coherent bundle of cell files radiating from a lens-shaped zone of cells called the quiescent centre (QC) (Clowes 1956). Located at the pole of the root proper in the region where the files of cortical and stelar cells converge, the QC houses mitotically less active cells considered as founders of the "functional" initial cells which occupy the periphery of this zone. From these initials and their derivatives the tissue of the root proper and root cap is developed (Barlow 1994). Though factors regulating the initiation and maintenance of the QC are still being investigated (Kerk & Feldman 1995), it is commonly accepted that the QC functions as a template for cell pattern in the apex and, as a consequence of changes in QC dimensions, the pattern is altered (Feldman 1984).

1 The model

The cell pattern remains approximately constant during steady-state root growth. In the axial longitudinal section through the apex, it can be characterized by two types of mutually orthogonal lines, known as periclines and anticlines (Sachs 1887) which conveniently lend themselves to description by the Root-Natural Coordinate System (R-NCS) (u, v, ϕ) (Hejnowicz & Karczewski 1993). In such a representation the pattern is given by two families of u and v lines for $\phi = const.$ (Fig. 1). If k is adjusted to a diameter of the cylindrical part of the apex, the border between the root proper and root cap is defined by v_0 on one side and $-v_0$ on the opposite side of the root axis. Assuming u_0 as the proximal limit of QC, there are four zones: 1 and 2 describing the root proper, 3 and 4 representing the root cap. The zone 1 which includes the focus of R-NCS, corresponds to the QC. Apical meristems grow symplastically (Priestley 1930, Erickson 1984). During such growth neighbouring cells do not slide, glide or slip, i.e. adjacent cells do not alter position relative to each other. Mathematically, growth can be described by a continuous field of the displacement velocity **V**. The **V** field for roots with QC growing steadily without rotation around the root axis was introduced by Hejnowicz & Karczewski (1993). In R-NCS, for the zonation assumed above, the equations are the following (Nakielski & Barlow 1995): zone 1: $V_u = 0, V_v = 0$; zone 2: $V_u = c(u - u_0), V_v = 0$; zone 3:

$V_u = 0, V_v = -d \cdot \sin(\frac{\pi}{v_0} \cdot v)$; zone 4: $V_u = c(u - u_0), V_v = -d \cdot \sin(\frac{\pi}{v_0} \cdot v)$; where c, d are constants, $V_u = \frac{du}{dt}, V_v = \frac{dv}{dt}, V_\phi = \frac{d\phi}{dt} = 0$. Accordingly, there is no growth in the zone 1 corresponding to the QC. In the other parts of the apex either V_u(zone 2) or V_v(zone 3) or both V_u and V_v (zone 4), are applied.

A knowledge of the **V** field and R-NSC was sufficient to build a 2D simulation model for growth and cell division in the root apex (Nakielski & Barlow 1995). In brief, the model employs the following rules and procedures: (*i*) the apex consists of quadrilateral cells given by the grid of u and v lines; individual cells have vertices located at the grid nodes, whereas cell edges are drawn as chords, (*ii*) to simulate growth, the equations for **V** were solved and, new locations of cell vertices in subsequent Δt, were found, (*iii*) to create a cell divison, cell area A at every timestep of the simulation was calculated and, if some critical value A_c, the same for the whole apex, was exceeded, a mother cell was replaced by two daughter cells having a common wall, (*iv*) the division, always into half in units of u and v, was parallel either to u or v. At a given position it depended upon the value of δ where δ = cell length/cell breadth. For $\delta > 1$, a wall tangent to v line ($u = const.$) was inserted, otherwise the division was along u line ($v = const.$). The model was used to simulate cell patterns in this paper. It was based on the following set of constants: $c = 1.0$, $d = 0.3$, $u_0 = 0.2$, $v_0 = \frac{\pi}{4}$, $\Delta t = 0.002$, and $A_c = 0.9 A_{av}$, where A_{av} was an average cell area at $t = 0$. Progressive cell patterns for growth and cell division generated by the model are shown in Fig. 1. As already reported (Nakielski & Barlow 1995) the apex enlarges maintaining symmetry of its cell pattern. To a first approximation, the pattern can be considered self-perpetuating.

In the model, new locations of cell vertices are calculated from **V**. It should be noticed, however, that individual cells displace during growth under the control of the second rank operator called the growth tensor (GT, Hejnowicz & Romberger 1984). The GT which is equivalent to a covariant derivative of **V** can be calculated either from the absolute derivative of **V** or diadic ($\nabla \cdot \mathbf{V}$). It is considered to generate a growth field within the apex. In such a field a spatial variation of growth occurs and the QC corresponds to the region of minimal growth rates (Nakielski 1991, Hejnowicz & Karczewski 1993). Furthermore, at each point of the field are three mutually orthogonal principal directions of growth (PDG) which, arranged into PDG trajectories, manifest themselves in the cell wall network as periclines and anticlines (Hejnowicz & Romberger 1984, Hejnowicz 1984). In the context of the model, it means that R-NSC built on both these types of cline is a graphical representation of the GT field. This is because in a steady growth, unit vectors of the system coincide with PDGs at every time. Thus, each cell wall lies along one of PDG trajectories in $\phi = const$ and each partition wall is formed along a principal plane defined by two PDGs. The **V** field is subordinated to GT (Hejnowicz & Romberger 1984) in this way the GT operating on it generates the required pattern of PDGs, here described by u and v lines of R-NCS.

Growth and cell division are coupled (Green 1976, Green & Selker 1991) and, it is intuitively clear that any change in the GT field influences cell pattern. Fur-

IV.8 Growth Field and Cell Displacement

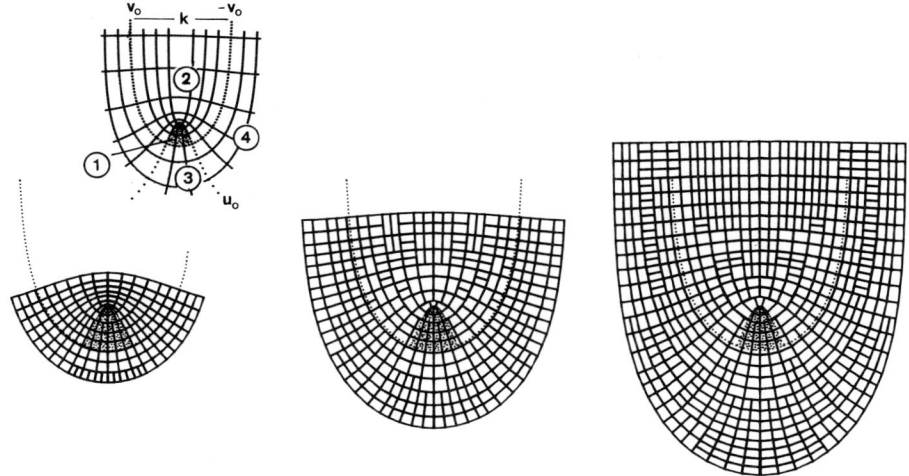

Figure 1: Temporal sequences of growth and cell division in the root apex. The examples $t = 0, t = 0.052, t = 0.066$ are shown. The insert shows the R-NCS (see text). The dotted lines: $v_0, -v_0$ and u_0 divide the apex into zones numbered from 1 to 4; for example, the zone 4 is delineated by v_0 (or $-v_0$) and u_0. The zone 1 (strippled) coincides with the QC.

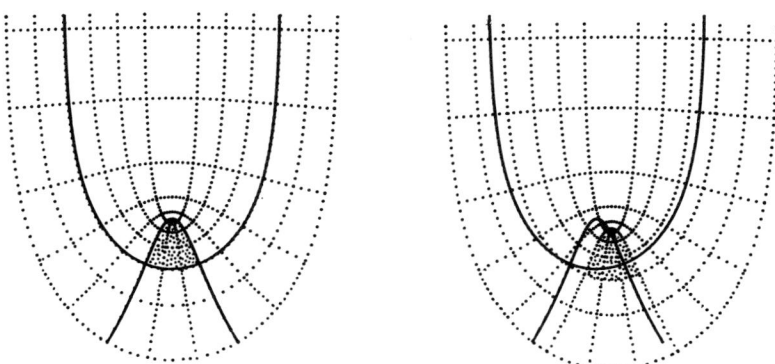

Figure 2: Two applications of GT field: (a) normal, as in Fig. 1, (b) the field displaced forwards and to the right with respect to the existing cell pattern which is indicated by $v_0, -v_0, u_0$ (bold lines). PDG lines are dotted, the actually generated QC is strippled.

thermore, even if the field remains constant, some modifications of the pattern due to the way in which the GT field is applied (in the operational sense) within the apex, are possible (Nakielski & Barlow 1995). To explain the problem, consider Fig. 2. In the normal application (Fig. 2a), PDGs coincide with the existing cell wall network, therefore, the zone 1 (Fig. 1-insert) which is expected to generate the QC, does actually coincide with it. If the QC maintains a stable location, a self-perpetuating cell pattern, as in Fig. 1, is generated. In Fig. 2b, the GT field is displaced and, PDGs do not coincide with the cell pattern. If such a state is maintained, it leads to the modification of cell pattern. Two types of the modification corresponding to the case when the field is displaced along the root axis either proximally, into the interior of the apex, or distally, towards the root tip, are shown in Fig. 3. They show the effects of a collapse and bulge of the grid known from the previous studied (Nakielski & Barlow 1995). Consider now the displacement of the GT field in the lateral direction (Fig. 4). Assuming $q = \frac{1}{20}k$ by which the grids in Fig. 3 were simulated, after $\Delta t = 0.018$ the field is displaced by an amount equivalent to a mean cell length (similarly as in tomato roots, Nakielski & Barlow 1995). Even if the displacement is generally small, a deformation of the grid is significant. The cell pattern becomes asymmetrical; it manifests the extra extension on the left side of the root axis. The final effect is similar to differential growth (Barlow & Rathfelder 1985) but occuring in the meristematic part of the root apex.

Also, a re-distribution of cell files in the dome-like portion of the apex is observed. Now, they radiate from the region actually occupied by the QC. Notice the modification of cell shape, what were previously right angles have become acute or obtuse as the result of a change of the orientation of the cell with respect to PDGs. So specifically deformed grids as shown in Figs. 3 & 4 have been forced to make a combination of the two: lateral and, optionally, distal or proximal displacements. A suitable simulation based on hitherto existing assumptions has revealed the results shown in Fig. 5. Although q for the axial displacements was twice as small as in the simulation used for Fig. 3, the effects of a collapse (Fig. 5a) and bulge (Fig. 5b) embedded into the grid deformed typically for differential growth, are distinguishable. Compared to Fig. 3, these effects now occur near to the peripheral part of the root proper. The position of the focus indicates that in Fig. 5a the dome-like portion of the apex develops faster then the cylindrical one. However, this proportion is inverted in Fig. 5b. As in Fig. 4 the re-distribution of cell files radiating from the QC is observed.

Imagine that the GT field wanders without any directional preference around the mode of its actual application. With reference to the model this corresponds to the situation in which the field is displaced in the direction chosen randomly and the displacement, by $q < \frac{1}{20}k$, occurs as before at every $\Delta t = 0.006$. Two examples of the grid patterns deformed in this way are shown in Fig. 6. In both, a little effect of the differential growth on the opposite flanks in the proximal portion of the apex is observed. It suggests that in one case displacements to the right, whereas in the other case displacements to the left were more frequent. It

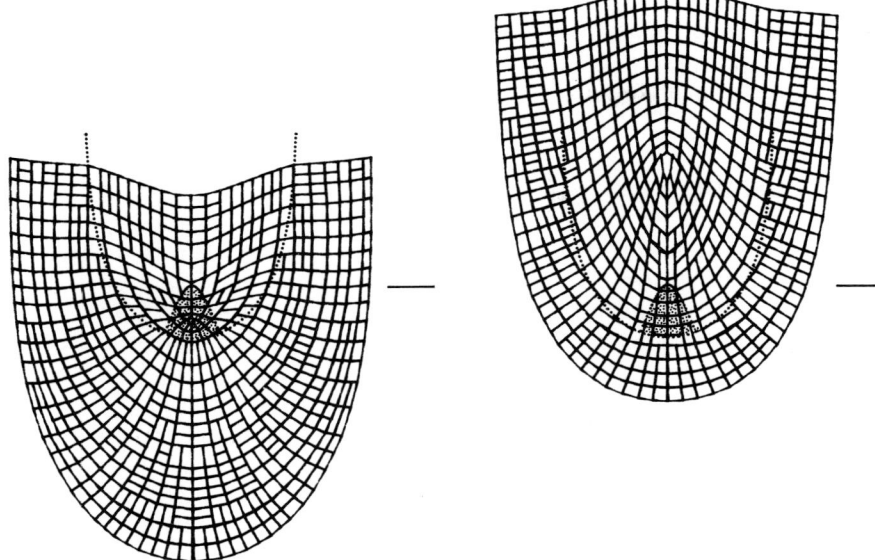

Figure 3: Cell patterns obtained for the GT field displaced along the root axis backwards (a) and forwards (b) by $q = \frac{1}{20}k$, at every $\Delta t = 0.006$ beginning with $t = 0$ (see text); the time step $t = 0.066$ is shown. The focus level is indicated. The QC is strippled. Compare with Fig. 1, $t = 0.066$.

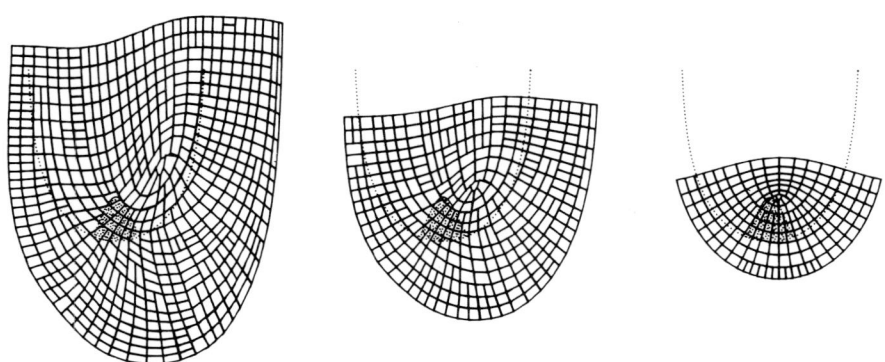

Figure 4: Progressive cell patterns generated for the case when the GT field is displaced laterally to the right by $q = \frac{1}{20}k$, at every $\Delta t = 0.006$. The examples show $t = 0.018$, $t = 0.052$ and $t = 0.066$.

is interesting that the grid as a whole grows more or less as in Fig. 1, except for the fact that re-distribution of cell files inside the apex, mainly in the dome-like region of the root proper, is observed. It is less then in Figs. 4 & 5 but sufficient to induce an asymmetry of cell pattern on the opposite sides of the root axis.

2 Discussion

This work is based on the hypothesis (Nakielski & Barlow 1995) that the GT can operate in various states relating to the spatial application of its field within the growing apex. The modification of cell pattern occurs during the transition from one to the other state as a consequence of an altered position of the QC which is continually changed. The assumption of the existence of two such states has provided a satisfactory explanation of the cellular transformations observed by Barlow (1992) in apices of the wild-type and *gib*-1 mutant of tomato roots. At the pole of wild-type roots, where the cell files of the cortex converge, there are commonly 1-2 tiers of cortical cells sandwiched between the pole of the stele and the cap initials. By contrast, root apices of the *gib*-1 mutant contain 6-8 tiers in this region. If the GT field is displaced forwards, as in Fig. 3a, the cell pattern of wild-type root is modified into one typical for the *gib*-1 mutant. If the field is displaced backward, as in Fig. 3b, a transition in the opposite direction, from the pattern of mutant to the pattern of wild-type root, is described (Nakielski & Barlow 1995). This paper suggests that the number of states in which the GT field can operate during growth is greater and all these states from the basic mode corresponding to Fig. 2a, can be achieved.

The work with tomato roots has shown that the effects of a collapse and bulge of the grid are associated with a tendency to decrease and increase, respectively, of the number of cell tiers in the cortex at the pole of the root. In the light of these results the differential growth due to the lateral displacement of the GT field (Figs. 4 & 5) can relate to some of asymmetries in cell pattern on the opposite sides of the root axis. Such asymmetry can be caused, for example, by instability in the location of the formative division (Barlow 1994) initiating cortical cell files in the apex. The differential growth, if maintained longer, would reflect in the distribution of the rate of elongation on opposite flanks of the apex. The distribution is often asymmetrical even during vertical growth of roots (Barlow & Rathfelder 1985, Ishikawa et al. 1991). The GT field which wanders perpendicularly to the root axis can relate to micronutations (Barlow et al. 1994).

Nakielski & Barlow (1995) have shown a link between the level of endogenous gibberelline and the mode of application of the GT field. The various states in which the GT field can be applied do not disturb the link. There can be, however, other factors, which may be related to interaction between a "living system" of the growing root and the external physical environment (Barlow & Zieschang 1994) and which modify the position of the field established for a given level of gibberelline.

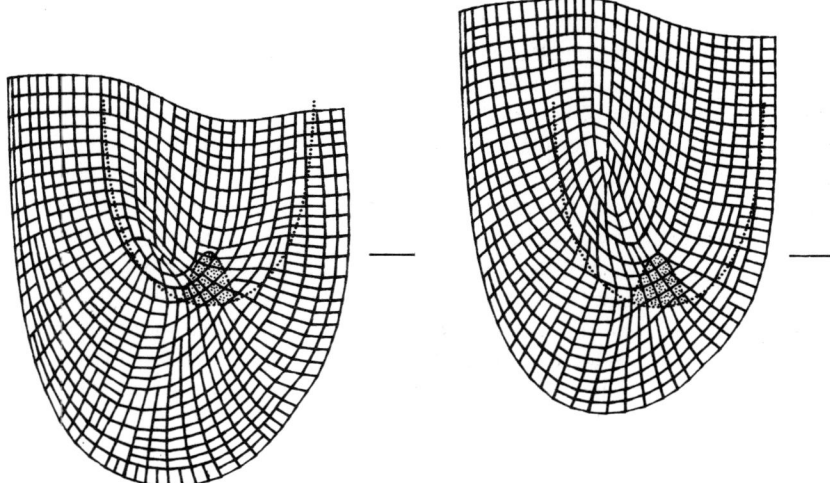

Figure 5: Cell patterns as in Fig. 3 for GT field displaced laterally and backwards (a), and laterally and forwards (b). The displacements were: $q = \frac{1}{20}k$ (lateral) and $q = \frac{1}{40}k$ (axial).

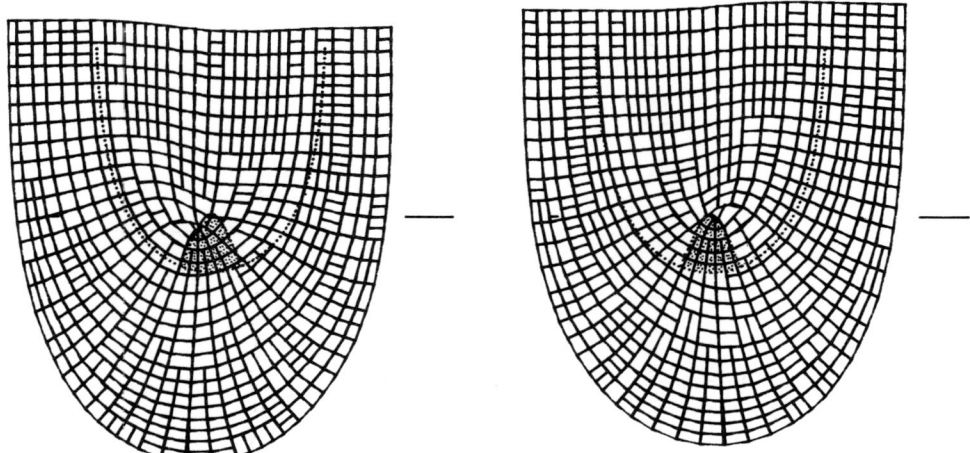

Figure 6: Two cell patterns corresponding to $t = 0.066$ in Fig. 1 but simulated using random wandering of GT field with $q < \frac{1}{20}k$.

The GT field was changed in its application within the apex without any change of the equations for **V**. A steady growth has become quasi-steady in this way. It is obvious that any change in values of constants of the **V** field, for example those defining the QC dimensions, leads to a deformation of the grid pattern, too. Such attempts have not been considered in this paper.

In the modeling based on the GT and PDGs, growth and division of individual cells depend on position of these cells within the GT field. Altered distribution of the field and permanent re-definition of the QC have caused re-specification of positional information in the root apex. Thus, cell fates had been changed during growth. A flexibility of cell fate associated with a change of positional control have been revealed recently by surgical experiments (van der Berg *et al.* 1995).

The model is very straightforward. The "ideal" grid patterns have shown only basic types of modification which can occur in real apices. To establish details of the relationship between the deformation of cell pattern and the mode of application of the GT field, the model with free-shaped cells (Nakielski & Barlow 1995) for real apices is needed. The wandering QC, considered in terms of such a model, will be described in another paper.

IV.9

The Stationary State of Epithelial Tissues

Nicolas Rivier and Benoit Dubertret (Strasbourg)
Gudrun Schliecker (Freiberg)

1 Introduction

We are concerned with the structure and evolution of epithelial tissues. From a geometrical point of view, the tissue fills space at random with cells (Hales 1727, Errera 1886, Lewis 1928, Lewis 1931, Thompson 1942, Matzke 1959, Dormer 1980, Smolyaninov 1980, Weaire & Rivier 1984). "The appreciation that cells are polyhedral figures came with the very first histological report ever published" (Dormer 1980). While the partitioning or filling of space by cells has obvious biological and chemical relevance, the role of disorder is much less clear. In this paper, we will explain why disorder is important, how it is achieved and maintained, and what are its structural characteristics.

Why is disorder important? Because it preserves the local invariance of the tissue in spite of perpetual local topological transformations (cell division and detachment or death). It guarantees the renewal of a tissue which remains globally unchanged. Specifically, cell division or death creates a structural "defect" (Fig. 1). Randomness of the structure ensures that a defect is neither felt nor seen, except in its immediate neighbourhood. By contrast, a dislocation in a crystal is immediately visible and felt afar.

How is disorder achieved and maintained? The structure is in statistical equilibrium under local topological transformations. The state of statistical equilibrium is that which is realized by the largest number of local configurations (maximum entropy). It is the state of a tissue which, even if it does not have many cells, lives sufficiently long to explore, by cell division and death, all possible local configurations. The tissue in statistical equilibrium is invariant over-all, in spite of the renewal of its cells (and also because of this renewal, since it is cell division which explores all the possible local configurations). The analogy with statistical thermodynamics can be pushed further. In one of our papers (Rivier & Dubertret 1995) we have adapted to epithelial tissues a molecular dynamics formalism introduced by Telley *et al.* (1995) to describe the coarsening of froths and polycrystals.

The state of statistical equilibrium has observable characteristics, which are structural relations between the average size of a cell, the number of its neighbours (its topological shape), and the correlations of these characteristics between neighbouring cells. (Some of these characteristics are constrained; see Eqns. 1 & 2.)

For example, in two dimensions, a cell has exactly 6 neighbours on the average. Also, two daughter cells are neighbours and have 4 more neighbours in total than their mother.) These structural characteristics of statistical equilibrium are also diagnostic: If they are not obeyed, this indicates pathologies in cell-renewal such as psoriasis or melanoma.

The tissue is able to respond efficiently and locally to an external constraint disturbing statistical equilibrium, as in the healing of a wound. This is plasticity, achieved through the motion of local topological "defects" called dislocations.

The importance of geometry (form) in biological systems and of geometrical plasticity (growth) for their evolution has been well-known since D'Arcy Thompson (1942). What is new is geometrical randomness, which enables us to obtain new, diagnostic relations. Maximum entropy governs the statistical invariance of the tissue, in spite of local renewal of its cells. It is a principle describing the plasticity of the tissue and its evolution, beyond the specificity and rigidity imposed by genetic programming.

2 Topological representation of biological tissues – cell division and plasticity –

An epithelial tissue can be regarded, in general and at the lowest level of sophistication, as a fluid of polygonal cells, filling space at random. This fluid is made of cells bounded by edges or interfaces and vertices. There are always 3 interfaces and 3 cells incident on each vertex.

Each cell may be characterized by the number n of its neighbours. The random variable n has a distribution p_n. This distribution is constrained by the following relations

$$\sum_n p_n = 1 \qquad \text{normalization}$$

$$\sum_n n p_n = 6 \qquad \text{topology}$$

$$\sum_n A_n p_n = A_{tot}/C \qquad \text{space-filling} \tag{1}$$

$$\sum_n N_{kn} p_n = k \qquad \text{topological correlations}$$

where $A_n \geq 0$ is the average area of n-sided cells, $p_n N_{kn} \geq 0$ is the number of n-sided cells neighbouring any given k-sided cell, and C the total number of cells in the tissue. The second constraint is a topological result (a consequence of Euler's relation). The fourth states that a k-sided cell has k neighbours. Cellular division is the local, elementary topological transformation which shuffles the variable n.

The numbers of sides of the cells directly involved in the division are related,

$$m + 4 = d_1 + d_2 \qquad (2)$$

where m and d_i denote the numbers of sides of the mother and daughter cells, respectively. Daughters should have at least 4 sides, otherwise division would increase the topological size of the cell, and go against the general trend. There are two other cells involved in the mitotic process, at both ends of the dividing membrane. They gain one side each.

Cellular division changes the topological shape (number of sides) of the cells involved. For the tissue to remain in a steady state, its structure must be a random distribution of defects (any long-range order would be visibly offset by the defect) maximizing the entropy which is thereby unchanged by the creation of other defects. The most probable distribution (that which maximizes the entropy) is not only overwhelmingly more probable than any other (it can be achieved by the largest number of macroscopically equivalent microscopic configurations), but it is also robust under cell division. These topological fluctuations establish statistical equilibrium. They are also the agents of growth of the tissue.

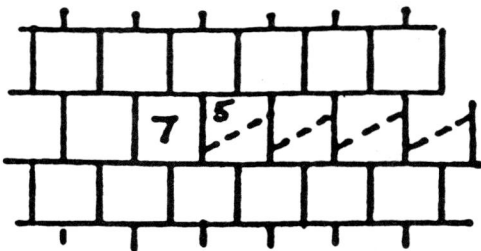

Figure 1: A pair of 5- and 7-sided cells (labelled) is a topological dislocation. All unlabelled cells are hexagonal. The figure shows three layers of hexagonal cells on the left, and four on the right, the signature of a dislocation. The additional layer can be produced by successive division of the cells (dashed lines).

A hexagonal tissue, even random, is flat. A 5- (7-) sided cell is a source ("charge") of positive (negative) curvature. It makes the flat tissue buckle into a cone (saddle). A pair pentagon-heptagon is a (topological) dislocation (Fig. 1). Accordingly, there is an energy (Eqn. 5) associated with local cellular configurations.

Cell division is the source of two dislocations (Fig. 2) which can climb apart through further divisions (Pyshnov 1980) (a local and internal mechanism for adding material (Figs. 2 & 3), and for invariant or pathological (Fig. 3) growth of the tissue).

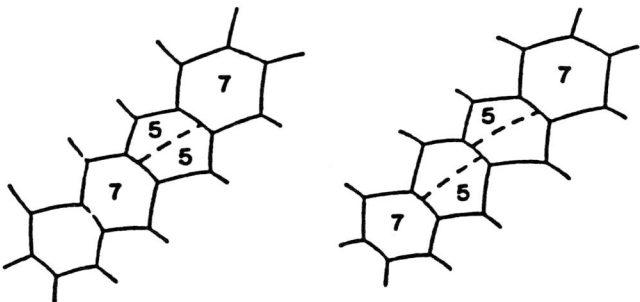

Figure 2: Climb of dislocation by successive cellular divisions. On the left, a pair of dislocations (each consisting of a 5- and a 7- sided cell) results from the division of a hexagonal cell. On the right, a second division makes the two dislocations climb away from each other, leaving in their wake an additional layer of cells. Unlabelled cells are hexagonal.

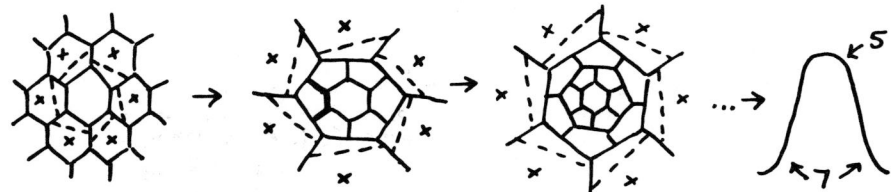

Figure 3: Amplification of local curvature: growth of an appendix by successive cellular divisions

3 Structural characteristics of disorder: Lewis' and Aboav's laws

Statistical equilibrium, characterized by maximum entropy, is the over-all cellular space-filling which can be realized by the largest number of local configurations of cells and their neighbours. This over-all geometry is characterized by a distribution of cells and their neighbours. This over-all geometry is characterized by a distribution of cells p_n and by equations of state, familiar in statistical mechanics (Boltzmann distribution and ideal gas law, respectively).

Entropy is decreased by the imposition of a physical constraint. Conversely, removal of a constraint increases the entropy: the over-all structure is more likely, because it can be realized by many more local microscopic configurations. The same result is achieved by forcing a necessary constraint (Eqn. 1) to duplicate others, thereby making it redundant (Rivier & Lissowski 1982, Peshkin et al. 1991). The duplication condition is a structural relation for tissues in statistical equilibrium (equation of state in thermodynamics). There are two such relations, Lewis' law (Eqn. 3, Lewis 1930) and Aboav's law (Eqn. 4, Aboav 1970). They are diagnostic: A tissue which does *not* obey them is restricted by other forces. Furthermore, *how* they are violated may reveal the nature of these forces. Lewis' law indicates how topological space is most likely to be filled by cells. Aboav's law gives the most probable correlations between neighbouring cells.

If the size-shape relation A_n has the linear form (with some characteristic parameter λ)

$$A_n = (A_{tot}/C)\lambda(n - (6 - 1/\lambda)) \qquad \text{(Lewis' law)} \qquad (3)$$

then the third constraint in Eqn. 1 is a linear combination of the first two and duplicates them (Rivier & Lissowski 1982). The tissue has taken advantage of the arbitrariness in the functional form of the relation $A_n = A(n)$ to increase the entropy further. Eqn. 3, a relation between average sizes and shapes of cells discovered empirically by Lewis (1930), has been observed throughout the biological world (Smolyaninov 1980, Mombach et al. 1990). λ is a parameter characteristic of the tissue imposing the linear relationship.

The same redundancy argument gives the functional dependence for topological correlations (fourth constraint in Eqn. 1). Both N_{kn} and the total number of sides of all neighbours to a n-cell, $n \cdot m(n) = \sum_k k N_{kn} p_k$,

$$\begin{aligned} N_{kn} &= (k-6)\sigma(n-6) + (n+k-6) \\ n \cdot m(n) &= (6 - \sigma\mu_2)n + 6\mu_2(1/6 - \sigma) \qquad \text{(Aboav's law)} \end{aligned} \qquad (4)$$

are linear in n (Peshkin et al. 1991). Here $\mu_2 = \sum_n p_n(n-6)^2$ is the variance of the distribution p_n, and $(1/6-\sigma)$, measuring the correlation strength, is the structural parameter imposing redundancy. Smaller cells surrounding larger cells, and vice-versa, impose $\sigma < 1/6$. In nature, $\sigma < 0$; local invariance under mitosis yields Aboav's law with slope $6 + \sigma\mu_2 = 5$ (Rivier et al. 1995b). $m(6)$ is independent of σ, as A_6 is of λ. The linearity of Aboav's law is impressive, both experimentally (Mombach et al. 1990) and in simulations (Peshkin et al. 1991, Godrèche et al. 1992).

4 Topological energy

Physically, the energy is carried by interfaces (surface tension). Its contribution to the elasticity of the tissue can be given topologically: A n-sided cell is a source of curvature, of charge $(6-n)$ in an elastic medium. A pair of neighbouring, opposite charges is a dislocation, a defect costing less energy. Cells have correlation U_{corr}

and self-energy U_{self} (Rivier et al. 1995b),

$$U_{corr} + U_{self} = \varepsilon \sum_{nk} p_n p_k (n-6) N_{kn}(k-6) + \eta \sum_n p_n (n-6)^2$$
$$= \sum_n p_n [\xi(n-6)^2 + \varepsilon \mu_2(n-6)] = \eta \mu_2. \tag{5}$$

Here, $\xi = \varepsilon \sigma \mu_2 + \eta$ ($\varepsilon, \eta \geq 0$). $U_{corr} = 0$ for hexagonal froths ($\mu_2 = 0$) and is largest for uncorrelated ($\sigma = 1/6$) froths. The topological energy of the tissue or froth is thus measured by the variance μ_2 of the distribution of cell shapes. Maximum entropy yields a relation between μ_2 and p_6 (Fig. 4) as a structural equation of state (Le Caër & Delannay 1993, Rivier 1994) (the equivalent in thermodynamics of the (virial) expansion of pressure over temperature in powers of the density of a real gas):

$$\begin{aligned} \mu_2 p_6^2 &= 1/(2\pi) \simeq 0.159 \quad \text{for} \quad 0.3 \leq p_6 \leq 0.7 \\ \mu_2 &= 1 - p_6 \quad \text{for} \quad p_6 > 0.7. \end{aligned} \tag{6}$$

It was discovered experimentally and found to be universal in froths by Lemaître et al. (1993). The energy (Eqn. 5) enables us to describe how cell division can proceed in random tissues, and to define directions (geodesics) along which successive divisions may occur at no additional energy cost (or gain). Consider the change in the energy of the cells involved in cellular division, without taking correlations into account. The cells involved are the dividing cell (m sides), the daughters (d_1 and $d_2 = 4 + m - d_1$ sides) and the two neighbouring cells affected by the division which change from a, b sides to $a+1$, $b+1$ after division. The difference in energy (second expression in Eqn. 5) is,

$$\Delta U = 2\xi[-(d_1 - 4)(m - d_1) + a + b - 9]. \tag{7}$$

Symmetric division $d_1 \approx d_1$ costs the least energy. Moreover, the smaller cells a, b attract the new dividing edge, and the larger cells repel it. Larger cells can also lose their energy by dividing spontaneously (and offsetting the steady state). It costs energy for the smaller cell to divide. Steady state (constant energy) implies that it is chiefly the 6-, 7-, 8- and 9-sided cells which divide. This is indeed the case in cucumber (Lewis 1928, Rivier et al. 1995a).

Cells with an odd number of neighbours divide as symmetrically as possible (100%, 95%, 100% for $m = 5, 7, 9$, resp. in cucumber) whereas cells with an even number of neighbours sometimes select the second most symmetric alternative instead of the most symmetric (32%, 24% for $m = 6, 8$, resp. in cucumber (Lewis 1928, p. 361)). This is because the difference in energy (Eqn. 7) between the most symmetric division and the next is -2ξ for m odd and only $-\xi$ for m even. Accordingly, the range of daughter cells is narrower for m odd than for m even (Rivier et al. 1995a).

Geodesics ($\Delta U = 0$) trace the path of the successive divisions described in Figs. 2 & 3: Fig. 2 shows a straight climb of the dislocation as a long scar of softer, new tissue in an ordered, hexagonal tissue, with only the initial division costing energy. Fig. 3 shows the growth of an appendix in a region of positive curvature (a ring of 6 pentagons), with the energy cost of the first division recovered at the sixth.

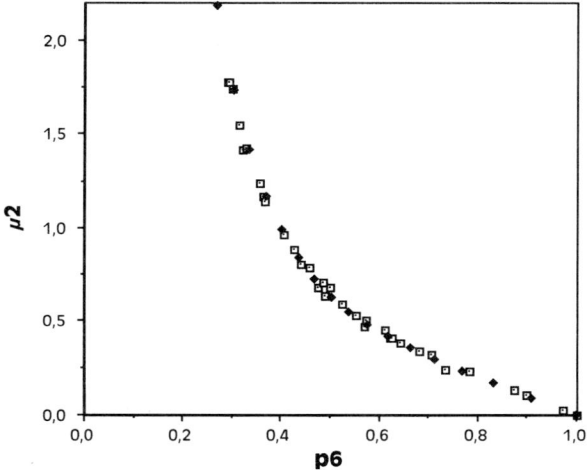

Figure 4: Lemaître's law: μ_2 as a function of p_6. Full diamonds: Maximum entropy theory (Eqn. 6). Squares: experimental (disks on air table, botanical tissues) and numerical data (Lemaître et al. 1993).

To avoid long scars of soft tissue or growth of appendices, geodesics must be shuffled. This is probably why the basal layer in the epidermis of mammals is corrugated (this contention is still to be biologically proven). The hyperbolic points mix the geodesics and ensure that the tissue remains invariant, without scars or appendices, and random.

5 Invariant distribution of cell shapes – coupled rate equations –

The stationary distribution p_s is the solution of coupled, linear, integro-difference rate equations (Eqn. 8) (Rivier et al. 1995a, Dubertret & Rivier 1996). The tissue has N_s s-sided cells, distributed as $p_s = N_s/\sum N_s$, in statistical equilibrium, $0 = dp_s/dt = (1/N)[dN_s/dt - p_s dN/dt]$. The population N_s of s-sided cells is affected by cell division as follows [with the corresponding probabilities in brackets]: One cell is lost if the mother has s sides ($[p_s P(\text{mother}\,|\,s)]$, where $P(\text{mother}\,|\,s)$ is the conditional probability that an existing s-cell divides), or if a s-sided neighbour gains a side ($[\sum_k p_k P(\text{mother}\,|\,s)(2/k)p_s N_{sk}]$, $2/k$ is the proportion of neighbour cells affected by the division). Conversely, one cell is gained if the affected neighbour had $s-1$ cells before the division or if one of the daughters has s sides ($[\sum_k p_k P(\text{mother} \mid s)\Gamma(k \to s)]$). The division kernel $\Gamma(k \to s)$ is the conditional probability that a k-sided dividing cell has a s-sided daughter (the other has $k+4-s$ sides). Thus,

$$0 = \sum_k p_k P(\text{mother} \mid s)[-\delta_{ks} + \Gamma(k \to s) + (2/k)(-p_s N_{sk} + p_{s-1} N_{s-1,k}) - p_s]. \quad (8)$$

Mother and daughter cell distributions are related to p_s by the kernels $P_m(k)$ and $\Gamma(k \to s)$. These equations include also the local topological correlations, or "diffusion" terms N_{sk}. These diffusion terms strongly restrict the range of parameters giving a mathematical solution ($0 \leq p_s \leq 1$) of Eqn. 8 (and of their generalization (Dubertret et al. 1996) to include cell detachment or death).

If the division kernel $\Gamma(k \to s)$ is independent of s or weakly dependent on s, (i.e. $\Gamma(k \to s) = 2/(k-1)$), which describes physical fragmentation (Delannay & Le Caër 1994) better than cellular division) the rate equations become second-order difference equations (coupling p_s, p_{s-1} and p_{s-2}), and their solution can be discussed in general. In some cases, the difference equations are even first-order (coupling p_s and p_{s-1} in detailed balance with p_{s-1} and p_{s-2}). Then, the resulting distribution is very broad; it exists over a wide range of the parameter, unrestricted by local correlations. These cases include fragmentation with $P(\text{mother} \mid k) = 1$ (Delannay & Le Caër 1994) and the rate equations for soap froth, where cells are allowed to disappear without affecting their direct neighbours (global mean-field approximation) (Stavans et al. 1991). By contrast, the distribution resulting from the second-order difference equations is much narrower and only normalized and nonnegative for sharply defined values of the parameters.

6 Conclusion

We have presented a topological mechanism for the renewal of epithelial tissues, which can be represented as random, topological partitions of a surface. Because neighbours are defined topologically, the description allows for gaps and voids between real cells. Randomness is essential for the tissue to remain invariant in spite of cellular division and detachment. There is, however, a great deal of organisation, forced by the topological and the minimal energy constraints and expressed in the relations 3–6 imposed by maximum entropy. Illustration of the mechanism can be seen in the dynamics of biological tissues, for example in budding; see Green IV.5 and Nakielski IV.8 this volume.

Acknowledgements This work is supported by the EC Network, Physics of Foam, CHRXCT 94 0542, and by La Société Dior (BD).

Discussion and Open Problems

Mechanical stresses are important for functioning and regulation at all levels of an organism. They are inseparable from motion, which causes and is produced by mechanical stresses. They are a vital part, at the highest levels, of sensation, e.g. in animal proprioception, touch, balance, and hearing, and plant graviperception and touch. At the level of tissues, with which this Chapter has been concerned, mechanical stresses are vital to our topics of interest, i.e. homeostasis, growth, and morphogenesis. Homeostasis has not been addressed in the papers included in this Chapter, but living creatures exhibit remarkable dynamic adaptation to ambient stress levels (see, e.g., Fung 1990). For this adaptation to occur, there are sensory systems at the cellular and tissue level, and it is these stress- or strain-receptors which ultimately govern stress-responsive tissue remodelling and repair, and the mechanical direction of growth and morphogenesis.

Growth alone can lead to interesting morphogenesis if there is some gradient in growth rates. In the case of uniform growth rates and lack of differentiation, we see only simple forms such as algal thalli and tumors. Where there are growth rate gradients, we see emergent behavior such as the pattern formation discussed in the Green paper (IV.5). But by far the most common mechanisms of morphogenesis are growth coupled with deformation from a reference state. The deformation is proximally caused by mechanical stresses, some of which are generated by the growth and/or deformation themselves, and it is these feedback and control loops which govern much of morphogenesis.

We cannot, of course, separate mechanical regulation of growth and deformation from chemical and biochemical regulation, since it is only through biochemical pathways that mechanical signals are received, transduced, amplified, responded to, and regulated. Whatever occurs at the tissue level is ultimately due to phenomena at the cellular level, as discussed in Chapter I, plus any emergent properties of ensembles of similar cells. It is thus perhaps artificial to present a Chapter on tissue motion and mechanical aspects thereof, divorced from the cellular and biochemical context, or indeed, from the intermediate level of molecular machinery (cf. Chapter I, papers by Lendowsky & Mogilner (I.11); Civelekoglu & Geigant (I.12)). As discussed briefly in Green's contribution in this Chapter (IV.5), science currently has an excellent understanding of gene expression, signal transduction, and morphogenetic mechanisms, but very little of integration of these levels.

All of the models presented in this Chapter are necessarily simplistic; simplification is the essence of modeling, since real biological systems are incomprehensibly complex. For example, the lung morphogenesis model discussed in Lubkin's paper (IV.3) treats lung epithelium as a passive structure responding to its mechanical environment, neglecting any biochemical aspects of its behavior, and neglecting any self-directed deformation and growth. The model in the paper by Rivier, Dubertret & Schlieker (IV.9) abstracts cells to elements in a foam, eliminating all complicating features such as the cytoskeleton. The model in the paper by Green (IV.5) necessarily omits the irreversibility of phyllotactic morphogenesis, in the interests of simplicity and understanding. In fact, the cell model of biological tissues, while so widespread and useful as to be almost universal, nonetheless ignores symplastic behavior which is important in a number of contexts, such as in neural tissue, slime molds, and plant tissues. Indeed, modeling is not possible without simplification, and the simpler a model is, while still adequately representing real phenomena, the more valuable and useful is that model.

However, a real organism is always more complicated than a model can be, and real biological phenomena contain more mechanisms than can be simultaneously tracked with a model. Even one level of a biological system, such as the entirety of ion channels on a single cell, and their interactions, or the entirety of growth factors in developing embryonic lung tissue, and their interactions, cannot possibly be comprehended, and we must simplify. Once we have an understanding, through simple models, of phenomena at one level of an organism (such as in feedback loops of a small number of ion channels or growth factors), we can begin to integrate our understanding across different levels and kinds of organization (such as the interaction between ion flux, actin polymerization, gene expression, cellular motion, tissue deformation, and morphogenesis). It is essential to integrate across these levels of understanding, as much as we are capable. This is the major open problem.

References

Abercrombie M and Heaysman JEM (1953) Observations on the social behaviour of cells in tissue culture. I. Speed of movement of chick heart fibroblasts in relation to their mutual contacts. *Exp. Cell Res.* **5** 111–131

Abercrombie M, Heaysman JEM and Pegrum SM (1970a) The locomotion of fibroblasts in culture. I. Movements of the leading edge. *Exp. Cell Res.* **59** 393–398

Abercrombie M, Heaysman JEM and Pegrum SM (1970b) The locomotion of fibroblasts in culture. II. Ruffling. *Exp. Cell Res.* **60** 437–444

Aboav DA (1970) The arrangement of grains in a polycrystal. *Metallogr.* **3** 383–390

Achenbach F, Achenbach U and Wohlfarth-Bottermann KE (1979) Plasmalemma invaginations, contraction and locomotion in normal and caffeine-treated protoplasmic drops of *Physarum*. *Eur. J. Cell Biol.* **20** 12–23

Achenbach U and Wohlfarth-Bottermann KE (1981a) Synchronization and signal transmission in protoplasmic strands of *Physarum*. Effects of externally applied substances and mechanical influences. *Planta* **151** 574–583

Achenbach U and Wohlfarth-Bottermann KE (1981b) Synchronization and signal transmission in protoplasmic strands of *Physarum*. The endoplasmic streaming as a pacemaker and the importance of phase deviations for the control of streaming reversal. *Planta* **151** 584–594

Achenbach F and Wohlfarth-Bottermann KE (1981c) Morphogenesis and disassembly of the circular plasmalemma invagination system in *Physarum polycephalum*. *Differentiation* **19** 179–188

Adzick NS and Longaker MT (1992) *Fetal Wound Healing*. Elsevier, New York

Agarwal P (1995) The cell programming language. *Artificial Life* **2** 37–77

Agutter PS (1994) Models for solid-state transport; messenger RNA movement from nucleus to cytoplasm. *Cell Biol. Intern.* **18** 849–858

Alcantara F and Monk M (1974) Signal propagation during aggregation in the slime mold *Dictyostelium discoideum*. *J. Gen. Microbiol.* **85** 321–334

Alt W (1980) Biased random walk models for chemotaxis and related diffusion approximations. *J. Math. Biol.* **9** 147–177

Alt W (1987) Mathematical models in actin-myosin interaction. *Fortschr. Zool.* **34** 219–230

Alt W (1994) Cell motion and orientation: theories of elementary behavior between environmental stimulation and autopoietic regulation. In: *Frontiers in Mathematical Biology*, ed. SA Levin *Lecture Notes in Biomath.*, Vol. 100, pp. 79–101. Springer, Berlin

Alt W (1996) Biomechanics of actomyosin-mediated motility in keratinocytes. *Biofizika (Biophysics)* **41** 169–177

Alt W, Brosteanu O, Hinz B and Kaiser HW (1995) Patterns of spontaneous motility in videomicrographs of human epidermal keratinocytes (HEK). *Biochem. Cell Biol.* **73** 441–459

Alt W and Hoffmann G (1990) (eds.) *Biological Motion*. Lect. Notes in Biomath., Vol. 89, Springer, Berlin

Alt W and Tranquillo RT (1995) Basic morphogenetic system modeling shape changes of migrating cells: how to explain fluctuating lamellipodial dynamics. *J. Biol. Systems* **3** 905–916

André E, Brink M, Gerisch G, Isenberg G, Noegel A, Schleicher M, Segall JE and Wallraff E (1989) A *Dictyostelium* mutant deficient in severin, an F-actin fragmenting protein, shows normal motility and chemotaxis. *J. Cell Biol.* **108** 985–995

Baranowski Z (1976) Connection of the wave phenomena with myxomycete plasmodium migration. In: *Nonmuscle Forms of Motility*, ed. GM Frank, Biol. Res. Centre USSR, pp. 47–50. Acad. Sci. Press, Pushchino. (In Russian)

Baranowski Z (1978) The contraction-relaxation waves in *Physarum polycephalum* plasmodia. *Acta Protozool.* **17** 387–397

Baranowski Z and Teplov VA (1992) Endoplasmic streaming mediates integrity of autooscillations in *Physarum polycephalum*. *Cell Biol. Int. Reports* **16** 1091–1096

Baranowski Z and Wohlfarth-Bottermann KE (1982) Endoplasmic veins from *Physarum polycephalum* plasmodia : A new strand model with defined age, structure, and behaviour. *Eur. J. Cell Biol.* **27** 1–9

Barbosa MS and Schlegel R (1989) The E6 and E7 genes of HPV-18 are sufficient for inducing two-stage in vitro transformation of human keratinocytes. *Oncogene* **4** 1529–1532

Barlow PW (1992) The meristem and quiescent centre in cultured root apices of the *gib*-1 mutant of tomato (*Lycopersicon esculentum* Mill.). *Ann. Bot.* **69** 533–543

Barlow PW (1994) The cellular and molecular biology of the quiescent centre in relation to root development. In: *Plant Molecular Biology. Molecular Genetic Analysis of Plant Development and Metabolism*, eds. G Coruzzi and P Puigdomenech, NATO ASI Series, Vol. H81, pp. 17–30. Kluwer Academic Publishers, Dordrecht

Barlow PW, Brain P, Butler R and Parker JS (1991) A model for root gravitropism. In: *Root ecology and its practical application*, eds. L Kutschera, E Hübl, H Persson, E Lichtenegger and M Sobotik, pp. 335–338. Verein für Wurzelforschung, Klagenfurt

Barlow PW, Parker JS and Brain P (1994) Oscillations in axial plant organs. *Adv. Space Res.* **14** 149–158

Barlow PW and Rathfelder EL (1985) Distribution and redistribution of extension growth along vertical and horizontal gravireacting maize roots. *Planta* **165** 134–141

Barlow PW and Zieschang HE (1994) Root movements: towards an understanding through attempts to model the processes involved. *Plant Soil* **165** 293–300

Barocas VH and Tranquillo RT (1994) Biphasic theory and in vitro assays of cell-fibril mechanical interactions in tissue-equivalent collagen gels. In: *Cell Mechanics and Cellular Engineering*, eds. VC Mow, F Guilak, R Tran-Son-Tay and RM Hochmuth, pp. 185–209. Springer, New York

Barocas VH and Tranquillo RT (1996) An anisotropic biphasic theory of tissue-equivalent mechanics: the interplay among cell traction, fibril network deformation, and contact guidance. *J. Biomech. Eng.* To appear

Belchetz PE, Plant TM, Nakai Y, Keogh EJ and Knobil E (1978) Hypophysial responses to continuous and intermittent delivery of hypothalmic gonadotrophin-releasing hormone. *Science* **202** 631–633

Belintsev BN, Beloussov LV and Zaraisky AG (1987) Model of pattern formation in epithelial morphogenesis. *J. Theor. Biol.* **129** 369–394

Bell GI, Dembo M and Bongrand P (1984) Cell adhesion: Competition between nonspecific repulsion and specific bonding. *Biophys. J.* **45** 1051–1064

Bell LGE and Jeon KW (1963) Locomotion of *Amoeba proteus*. *Nature* **198** 675–676

Beloussov LV (1994) The interplay of active forces and passive mechanical stresses in animal morphogenesis. In: *Biomechanics of Active Movement and Division of Cells*, ed. N Akkas, NATO ASI Series, pp. 131–180. Springer, Berlin

Beloussov LV, Dorfman JG and Cherdantzev VG (1975) Mechanical stresses and morphological patterns in amphibian embryos. *J. Embr. Exp. Morphol.* **34** 559–574

Beloussov LV, Labas JA and Kazakova NI (1993) Growth pulsations in hydroid polyps: kinematics, biological role and cytophysiology. In: *Oscillations and Morphogenesis*, ed. L Rensing, pp. 183–193. Marcel Dekker, New York

Beloussov LV, Labas JA, Kazakova NI and Zaraisky AG (1989) Cytophysiology of growth pulsations in hydroid polyps. *J. Exp. Zool.* **249** 258–270

Beloussov LV and Lakirev AV (1988) Self-organization of biological morphogenesis: general approaches and topo-geometrical models. In: *Thermodynamics and Pattern Formation in Biology*, eds. I Lamprecht and AI Zotin, pp. 321–336. W. de Gruyter, New York

Beloussov LV, Lakirev AV and Naumidi II (1988) The role of external tensions in differentiation of *Xenopus laevis* embryonic tissues. *Cell Diff. & Devel.* **25** 165–176

Beloussov LV, Lakirev AV, Naumidi II and Novoselov VV (1990) Effects of relaxation of mechanical tensions upon the early morphogenesis of *Xenopus laevis* embryos. *Int. J. Dev. Biol.* **34** 409–419

Beloussov LV and Luchinskaia NN (1995a) Mechanodependent heterotopies of the axial organs in *Xenopus laevis* embryos. *Ontogenez (Russ. J. Devel. Biol.)* **26** 213–222

Beloussov LV and Luchinskaia NN (1995b) Biomechanical feedback in morphogenesis, as exemplified by stretch responses of amphibian embryonic tissues. *Biochem. Cell Biol.* **73** 555–563

Beloussov LV and Mittenthal J (1992) Hyperrestoration of mechanical stresses as a possible driving mechanisms of morphogenesis. *Zhurn. Obsch. Biol.* **53** 797–807. (In Russian)

Beloussov LV, Saveliev SV, Naumidi II and Novoselov VV (1994) Mechanical stresses in embryonic tissues: patterns, morphogenetic role and involvement in regulatory feedback. *Int. Rev. Cytol.* **150** 1–34

Ben-Jacob E, Shochet O, Tenenbaum A, Cohen I, Czirók A and Vicsek T (1994) Generic modelling of cooperative growth patterns in bacterial colonies. *Nature* **368** 46–49

Ben-Jacob E, Shumeli H, Shochet O and Tenenbaum A (1992) Adaptive self-organisation during growth of bacterial colonies. *Physica A* **187** 378–424

Bereiter-Hahn J (1987) Mechanical principles of architecture of Eukaryota cells. In: *Cytomechanics*, eds. J Bereiter-Hahn, OR Anderson and WE Reif, pp. 3–30. Springer, Berlin

Bereiter-Hahn J and Lüers H (1994) The role of elasticity in the motile behaviour of cells. In: *Biomechanics of Active Movement and Division of Cells.*, ed. N Akkas, NATO ASI Series Vol H84, pp. 181–230. Springer, Berlin

Bereiter-Hahn J, Strohmeier R, Kunzenbacher I, Beck K and Vöth M (1981) Locomotion of *Xenopus* epidermis cells in primary culture. *J. Cell Sci.* **52** 289–311

Bereiter-Hahn J , Karl I, Lüers H, Vöth M (1995) Mechanical basis of cell shape: investigations with the scanning acoustic microscope. *Biochem. Cell Biol.* **73** 337–348

Bernardo JM and Smith AFM (1994) *Bayesian Theory*. John Wiley and Sons, Chichester UK

Bernfield MR and Banerjee SD (1982) The turnover of basal lamina glycosaminoglycan correlates with epithelial morphogenesis. *Dev. Biol.* **90** 291–305

Bernfield MR, Banerjee SD, Koda JE and Rapraeger AC (1984) Remodeling of the basement membrane as a mechanism of morphogenetic tissue interaction. In: *The Role of Extracellular Matrix in Development*, ed. RL Trelstad, pp. 542–572. Alan R. Liss, New York

Bershadsky A, Chausovsky A, Becker E, Lyubimova A and Geiger B (1996) Involvement of microtubules in the control of adhesion-dependent signal transduction. *Current Biol.* **6** 1279–1289

von Bertalanffy L (1932) *Theoretische Biologie. Band I: Allgemeine Theorie, Physikochemie, Aufbau und Entwicklung des Organismus*. Gebr. Borntraeger, Berlin

Bestehorn M, Friedrich R, Fuchs A, Haken H, Kuhn A and Wunderlin A (1988) Synergetics applied to pattern formation and pattern recognition. In: *Optimal Structures in Heterogeneous Reaction Systems*, ed. PJ Plath *Springer Series in Synergetics*, Vol. 44, pp. 164–194. Springer, Berlin

Beylina SI, Matveeva NB, Priezzhev AV, Romanovsky YM, Sukhorukov AP and Teplov VA (1984) Plasmodium of the myxomycete *Physarum polycephalum* as an autowave self-organizing system. In: *Self-Organization. Autowaves and Structures far from Equilibrium*, ed. VI Krinsky, pp. 218–221. Springer, Berlin

Biler P (1995) Growth and accretion of mass in an astrophysical model. *Appl. Math.* **23** 179–189

Bishop JE, Mitchell JJ, Absher PM, Baldor L, Geller HA, Woodcock-Mitchel J, Hamblin MJ, Vacek P and Low RB (1993) Cyclic mechanical deformation stimulates human lung fibroblast proliferation and autocrine growth factor activity. *Am. J. Respir. Cell Mol. Biol.* **9** 126–133

Blechschmidt E and Gasser RF (1978) *Biokinetics and Biodynamics of Human Differentiation*. Thomas Publisher Springfield, Illinois

Blume-Jensen P, Claesson-Welsh L, Siegbahn A, Zsebo KM, Westermark B and Heldin CH (1991) Activation of the human c-kit product by ligand-induced dimerization mediates circular actin reorganization and chemotaxis. *EMBO J.* **10** 4121–4128

Bonner JT (1952) *Morphogenesis*. Princeton University Press, Clinton, Mass.

Bottino D (1996) *An immersed boundary model of amoeboid deformation and locomotion*. Ph.D. thesis, Dept. Mathematics, Tulane University

Braam J and Davis RW (1990) Rain-, wind-, and touch-induced expression of calmodulin and calmodulin-related genes in *Arabidopsis*. *Cell* **60** 357–364

Bretscher A (1993) Microfilaments and membranes. *Curr. Opin. Cell Biol.* **5** 653–660

Brix K, Reinecke A and Stockem W (1990) Dynamics of the cytoskeleton in *Amoeba proteus*. III. Influence of microinjected antibodies on the organization and function of the microfilament system. *Eur. J. Cell Biol.* **51** 279–284

Brosteanu O (1994) *Methoden zur Analyse der Lamellipodienaktivität von Leukozyten*. Ph.D. thesis, University of Bonn

Brown AF and Dunn GA (1989) Microinterferometry of the movement of dry matter in fibroblasts. *J. Cell Sci.* **92** 379–389

Burström HG, Uhrström I and Wunscher R (1967) Growth, turgor, water potential, and Youngs modulus in pea internodes. *Physiol. Plant.* **20** 213–231

Bussemaker H, Deutsch A and Geigant E (1997) Mean-field analysis of a dynamical phase transition in a microscopic model for swarming behaviour. *Phys. Rev. Lett.* To appear

Cao L, Fishkind DJ and Wang YL (1993) Localization and dynamics in nonfilamentous actin in cultured cells. *J. Cell Biol.* **123** 173–181

Cao L and Wang Y (1990) Mechanism of the formation of contractile ring in dividing cultured animal cells. II. Cortical movement of microinjected actin filaments. *J. Cell Biol.* **111** 1905

Carrington WA and Fogarty KE (1987) 3-D molecular distribution in living cells by deconvolution of optical sections using light microscopy. In: *Proc of the 13th Annual Northeast Bioeng. Conf.*, ed. K Foster, pp. 108–111

Casciari JJ, Sotirchos SV and Sutherland RM (1992) Mathematical modeling of microenvironment and growth in EMT6/Ro multicellular tumor spheroids. *Cell Prolif.* **25** 1–22

Chaplain MAJ (1993) The development of a spatial pattern in a model for cancer growth. In: *Experimental and Theoretical Advances in Biological Pattern Formation*, eds. HG Othmer, PK Maini and JD Murray, pp. 45–59. Plenum Press, New York

Chen WT (1979) Induction of spreading during fibroblast movement. *J. Cell Biol.* **81** 684–691

Chen WT (1981) Mechanism of retraction of the trailing edge during fibroblast movement. *J. Cell Biol.* **90** 187–200

Cheng TPO (1992) Minipodia, novel structures for extension of the lamella: A high-spatial-resolution video microscopic study. *Exp. Cell Res.* **203** 25–31

Cherdantzeva EV and Cherdantzev VG (1985) Determination of dorsoventral polarity in the embryos of *Brachidanio rerio* (Teleostei). *Ontogenez (Sov. J. Devel. Biol.)* **16** 270–280

Civelekoglu G (1994) *Modelling the dynamics of actin in cells*. Ph.D. thesis, University of Vancouver

Civelekoglu G and Edelstein-Keshet L (1994) Modelling the dynamics of F-actin in the cell. *Bull. Math. Biol.* **56** 587–616

Clark M and Spudich JA (1974) Biochemical and structural studies of actomyosin-like proteins from nonmuscle cells: isolation and characterization of myosin from amoeba of *Dictyostelium discoideum*. *J. Mol. Biol.* **86** 209–222

Clark RAF and Henson PM (1988) *The Molecular and Cellular Biology of Wound Repair*. Plenum Press, New York

Cline CA, Schatten H, Balczon R and Schatten G (1983) Actin-mediated surface motility during sea urchin fertilization. *Cell Motil.* **3** 513–524

Clowes FAL (1956) Localization of nucleic acid synthesis in root meristems. *J. Exp. Bot.* **7** 397–312

Coates TD, Watts RG, Hartman RS and Coates TC (1992) Relationship of F-actin distribution to development of polar shape in human polmorphonuclear neutrophils. *J. Cell Biol.* **117** 765–774

Coen ES and Meyerowitz EM (1991) The war of the whorls; genetic interactions controlling flower development. *Nature* **353** 31–37

Cohen K, Diegelmann RF and Lindblad WJ (1992) *Wound Healing : Biochemical and Clinical Aspects*. W. B. Saunders Co., Philadelphia

Cohen MH and Robertson A (1971a) Wave propagation in the early stages of aggregation of cellular slime molds. *J. Theor. Biol.* **31** 101–118

Cohen MH and Robertson A (1971b) Chemotaxis and the early stages of aggregation of cellular slime molds. *J. Theor. Biol.* **31** 119–130

Combettes PL (1995) *The Convex Feasibility Problem in Image Recovery. Advances in Imaging and Electron Physics*, Vol. 97, Academic Press, New York

Condeelis J (1993) Life at the leading edge: the formation of cell protrusions. *Ann. Rev. Cell Biol.* **9** 411–444

Condeelis J, Hall A, Bresnick A, Warren V, Hock R, Bennett H and Ogihara S (1988) Actin polymerization and pseudopod extension during amoeboid chemotaxis. *Cell Motil. Cytoskel.* **10** 77–90

Cook J (1995) Waves of alignment in populations of interacting, oriented individuals. *Forma* **10** 171–203

Cosgrove DJ (1986) Biophysical control of plant cell growth. *Ann. Rev. Plant Physiol.* **37** 377–405

Coulombre AJ (1956) The role of intraocular pressure in the development of the chick eye. *J. Exp. Zool.* **133** 211–226

Cramer L, Mitchison TJ and Theriot JA (1993) Actin dependent motile forces and cell motility. *Curr. Opin. Cell Biol.* **6** 82–85

Crystal RG, West JB, Barnes PJ, Cherniack NS and Weibel ER (1991) *The Lung: Scientific Foundations*. Raven Press, New York

Curtis ASG (1964) Mechanism of adhesion of cells to glass. *J. Cell Biol.* **20** 199–215

Curtis ASG, Forrester JV and Clark P (1986) Substratum hydroxylation and cell adhesion. *J. Cell Sci.* **86** 9–24

Curtis ASG and Seehar GM (1978) The control of cell division by tension or diffusion. *Nature* **274** 52–53

Dallon JC and Othmer HG (1996) A discrete cell model with adaptive signalling for aggregation of *Dictyostelium discoideum*. Submitted to Phil. Trans. Roy. Soc. Lon.

Darcy PK and Fisher PR (1990) Pharmacological evidence for a role for cyclic AMP signalling in *Dictyostelium discoideum* slug behaviour. *J. Cell Sci.* **96** 661–667

Darwin C (1875) *Insectivorous Plants*. John Murray, London

Davis B (1990) Reinforced random walks. *Prob. Thy. Rel. Fields* **2** pp. 203–229

De Bruyn PPH (1946) The amoeboid movement of the mammalian leukocyte in tissue cultures. *Anat. Rec.* **95** 177–192

De Chastellier C and Ryter A (1977) Changes of the cell surface and the digestive apparatus of *Dictyostelium discoideum* during the starvation period triggering aggregation. *J. Cell Biol.* **75** 218–236

De Hostos EL, Bradtke B, Lottspeich F, Guggenheim R and Gerisch G (1991) Coronin, an actin binding protein of *Dictyostelium* localized to cell surface projections, has sequence similarities to G protein β subunits. *EMBO J.* **10** 4097–4104

Delannay R and Le Caër G (1994) Topological characteristics of 2D cellular structures generated by fragmentation. *Phys. Rev. Lett.* **73** 1553–1556

Dembo M (1989a) Field theories of the cytoplasma. *Comments on Theoretical Biology* **1** 159–177

Dembo M (1989b) Mechanics and control of the cytoskeleton in *Amoeba proteus*. *Biophys. J.* **55** 1053–1080

Dembo M (1994) On free boundary problems and amoeboid motion. In: *Biomechanics of Active Movement and Division of Cells*, ed. N Akkas, NATO ASI Series, pp. 231–283. Springer, Berlin

Dembo M and Harlow F (1986) Cell motion, contractile networks, and the physics of interpenetrating reactive flow. *Biophys. J.* **50** 109–121

Dembo M, Harlow F and Alt W (1984) The biophysics of cell surface motility. In: *Cell Surface Dynamics: Concepts and Models*, eds. A Perelson, C DeLisi and F Wiegel, pp. 495–541. Marcel Dekker, New York

Dembo M, Maltrud M and Harlow F (1986) Numerical studies of unreactive contractile networks. *Biophys. J.* **50** 123–137

Dembo M, Oliver T, Ishihara A and Jacobson K (1996) Imaging the traction stresses exerted by locomoting cells with the elastic substratum method. *Biophys. J.* **70** 2008–2022

Deneubourg JL, Gregoire JC and Fort EL (1990) Kinetics of larval gregarious behavior in the bark beetle *Dendroctonus micans* (Coleoptera: Scolytidae). *J. Insect Behavior* **3** 169–182

Desmond ME and Jacobson AG (1977) Embryonic brain enlargement requires cerebrospinal fluid pressure. *Dev. Biol.* **57** 188–198

Deuling HJ and Helfrich W (1976) The curvature elasticity of fluid membranes: a catalogue of vesicle shapes. *J. Phys. (Paris)* **37** 1335–1345

Deutsch A (1995) Towards analyzing complex swarming patterns in biological systems with the help of lattice-gas cellular automata. *J. Biol. Syst.* **3** 947–955

Deutsch A (1996a) Orientation-induced pattern formation: swarm dynamics in a lattice-gas automaton model. *Int. J. Bifurc. Chaos* **6** 1735–1752

Deutsch A (1996b) Lattice-gas approach to collective transport phenomena in biological pattern formation. In: *Nonlinear Physics of Complex Systems – Current Status and Future Trends*, eds. J Parisi, SC Müller and W Zimmermann. Springer, Berlin

Devreotes PN (1989) *Dictyostelium discoideum*: a model system for cell-cell interactions in development. *Science* **245** 1054–1059

Diaspro A, Adami M, Sartpre M and Nicolini C (1990) IMAGO: a complete system for acquisition, processing, two/three-dimensional and temporal display of microscopic bio-images. *Comp. Meth. Progr. Biomed.* **31** 225–236

Dickinson RB, Guido S and Tranquillo RT (1993) Biased cell migration of fibroblasts exhibiting contact guidance in oriented collagen gels. *Ann. Biomed. Eng.* **22** 343–356

Dickinson RB and Tranquillo RT (1993) A stochastic model for cell random motility and haptotaxis based on adhesion receptor fluctuations. *J. Math. Biol.* **31** 563–600

Dickinson RB and Tranquillo RT (1995) Transport equations and cell movement indices based on single cell properties. *SIAM J. Appl. Math.* **55** 1419–1454

Dietl H, Prast H and Philippu A (1993) Pulsatile release of catecholamines in the hypothalamus of conscious rats. *Naunyn - Schmiedebergs Arch. Pharm.* **347** 28–33

Diggle PJ (1990) *Time Series: a Biostatistical Introduction.* Clarendon Press, Oxford

DiMilla PA, Barbee K and Lauffenburger DA (1991) Mathematical model for the effects of adhesion and mechanics on cell migration speed. *Biophys. J.* **60** 15–37

Dormer KJ (1980) *Fundamental Tissue Geometry for Biologists.* Cambridge Univ. Press, Cambridge, UK

Doroszewski J, Nowak-Wyrzykowska M and Stolowska L (1993) The method of moments as applied to the study of granulocytes' shape and movement. *Medica Materia Polona* **25** 87–92

Driesch H (1891) *Die mathematisch-mechanische Betrachtung morphologischer Probleme der Biologie.* Gustav Fischer, Jena

Driesch H (1921) *Philosophie des Organischen.* Engelmann, Leipzig

Dubertret B and Rivier N (1996) The renewal of the epidermis: a topological mechanism. *Science* Submitted

Dufort PA and Lumsden CJ (1993a) Cellular automaton model of the actin cytoskeleton. *Cell Motil. Cytoskel.* **25** 87–104

Dufort PA and Lumsden CJ (1993b) High microfilament concentration results in barbed-end caps. *Biophys. J.* **65** 1757–1766

Dunn GA (1980) Mechanisms of fibroblast locomotion. In: *Cell Adhesion and Motility*, eds. ASG Curtis and JD Pitts, pp. 409–423. Cambridge University Press, Cambridge, UK

Dunn GA (1983) Characterizing a kinesis response: time averaged measures of cell speed and directional persistence. *Agents and Actions Suppl.* **12** 14–33

Dunn GA and Brown AF (1986) Alignment of fibroblasts on grooved surfaces described by a simple geometric transformation. *J. Cell Sci.* **83** 313–340

Dunn GA and Brown AF (1987) A unified approach to analysing cell motility. *J. Cell Sci. Suppl.* **8** 81–102

Dunn GA and Brown AF (1990) Quantifying cellular shape using moment invariants. In: *Biological Motion*, eds. W Alt and G Hoffmann *Lecture Notes in Biomath.*, Vol. 89, pp. 10–34. Springer, Berlin

Dunn GA and Ebendal T (1978) Contact guidance on oriented collagen gels. *Exp. Cell Res.* **111** 475–479

Dunn GA and Zicha D (1995) Dynamics of fibroblast spreading. *J. Cell Sci.* **108** 1239–1249

Durston AJ (1973) *Dictyostelium discoideum* aggregation fields as excitable media. *J. Theor. Biol.* **42** 483–504

Durston AJ (1974) Pacemaker activity during aggregation in *Dictyostelium discoideum*. *Develop. Biol.* **37** 225–235

Dvorak HE (1986) Tumors: wounds that do not heal. *New Eng. J. Med.* **315** 1650–1659

Dworkin M and Kaiser D (eds.) (1993) *Myxobacteria II*, Washington. American Society for Microbiology

Edelstein-Keshet L and Ermentrout B (1990) Models for contact mediated pattern formation. *J. Math. Biol.* **29** 33–58

Efron B and Tibshirani R (1986) Bootstrap methods for standard errors, confidence intervals, and other measures of statistical accuracy. *Statistical Science* **1** 54–77

Ehrengruber MU, Deranleau DA and Coates TD (1996) Shape oscillations of human neutrophil leukocytes: characterization and relationship to cell motility (Review). *J. Exp. Biol.* **199** 741–747

Ehrlich HP and Rajaratnam JB (1990) Cell locomotion forces versus cell contraction forces for collagen lattice contraction: an in vitro model of wound contraction. *Tissue Cell* **22** 407–417

Elsdale TR and Bard JBL (1972) Collagen substrata for studies on cell behavior. *J. Cell Biol.* **54** 626–637

Erickson RO (1976) Modeling of plant growth. *Annu. Rev. Plant Physiol.* **27** 407–434

Erickson RO (1986) Symplastic growth and symplasmic transport. *Plant Physiol.* **82** 1153

Ermakov VG and Priezzhev AV (1984) Wave regimes of the contractile activity of the strands of myxomycete plasmodium *Physarum* and their connection with the transport of the protoplasm. *Biophysics* **29** 106–112. (In Russian)

Ermentrout B and Edelstein-Keshet L (1993) Cellular automata approaches to biological modeling. *J. Theor. Biol.* **160** 97–133

Errera L (1886) Sur une condition fondamentale d'équilibre des cellules vivantes. *Comptes Rendus Acad. Sci.* **103** 822–824

Europe-Finner GN, Gammon B, Wood CA and Newell PC (1989) Inositol tris and polyphosphate formation during chemotaxis of *Dictyostelium*. *J. Cell Sci.* **93** 585–592

Euteneuer U and Schliwa M (1984) Persistent, directional motility of cells and cytoplasmic fragments in the absence of microtubules. *Nature* **310** 58–61

Evans E (1993) New physical concepts for cell amoeboid motion. *Biophys. J.* **64** 1306–1322

Felder S and Elson E (1990) Mechanisms of fibroblast locomotion: analysis of forces and motions at the leading lamellas of fibroblasts. *J. Cell Biol.* **111** 2513–2526

Feldman LJ (1984) The development and dynamics of the root apical meristem. *Amer. J. Botany* **71** 1308–1314

Fisher PR, Merkl R and Gerisch G (1989) Quantitative analysis of cell motility and chemotaxis in *Dictyostelium discoideum* by using an image processing system and a novel chemotaxis chamber providing stationary chemical gradients. *J. Cell Biol.* **108** 973–984

Fishkind DJ and Wang Y (1995) New horizons for cytokinesis. *Curr. Opin. Cell Biol.* **7** 23–31

Folkman J and Moscona A (1978) Role of cell shape in growth control. *Nature* **273** 345–349

Forgacs G (1995) On the possible role of cytoskeletal filamentous networks in intracellular signalling; an approach based on percolation. *J. Cell Sci.* **108** 2131–2143

Forscher P, Chi HL and Thompson C (1992) Novel form of growth cone motility involving site- directed actin filament assembly. *Nature* **357** 515–518

Fraenkel GS and Gunn DL (1961) *The Orientation of Animals: Kinesis, Taxis and Compass Reactions*, p. 334. Dover, New York

Freitas I and Baronzio GF (1994) Neglected factors in cancer treatment: cellular interactions and dynamic microenvironment in solid tumours. *Anticancer Res.* **14** 1097–1102

Friedrich R and Uhl C (1992) Symmetric analysis of human electroencephalograms: petit-mal epilepsy. In: *Evolution of Dynamical Structures in Complex Systems*, eds. R Friedrich and A Wunderlin, pp. 249–265. Springer, Berlin

Frisch U, Hasslacher B and Pomeau Y (1986) Lattice-gas automata for the Navier-Stokes equation. *Phys. Rev. Lett.* **56** 1505–1508

Fuchs A, Friedrich R, Haken H and Lehmann D (1988) Spatio-temporal analysis of multichannel alpha EEG map series. In: *Computational Systems - Natural and Artificial*, ed. H Haken *Springer Series in Synergetics*, Vol. 38, pp. 74–83. Springer, Berlin

Fukui Y (1990) Actomyosin organization in mitotic *Dictyostelium* amoebae. In: *Cytokinesis: Mechanisms of Furrow Formation during Cell Division*, eds. G Conrad and T Schroeder, pp. 156–165. New York

Fukui Y and Inoué S (1991) Cell division in *Dictyostelium* with special emphasis on actomyosin organization in cytokinesis. *Cell Motil. Cytoskel.* **18** 41–54

Fung YC (1990) *Biomechanics: Motion, Flow, Stress, and Growth*. Springer, Berlin

Futrelle RP (1982) *Dictyostelium* chemotactic response to spatial and temporal gradients: theories to the limits of chemotactic sensitivity and of pseudochemotaxis. *J. Cell. Biochem.* **18** 197–212

Futrelle RP, Traut J and McKee WG (1982) Cell behavior in *Dictyostelium discoideum*: preaggregation response to localized cyclic AMP pulses. *J. Cell. Biol.* **92** 807–821

Gabbiani G (1994) Modulation of fibroblastic cytoskeletal features during wound healing and fibrosis. *Path. Res. Pract.* **190** 851–853

Gajewski H and Zacharias K (1996) Global behaviour of a reaction-diffusion system modelling chemotaxis. Preprint No. 232, Weierstraß-Institut für Angewandte Analysis und Stochastik, Berlin

Gardiner CW (1983) *Handbook of Stochastic Methods*. Springer, Berlin

Geigant E (1997) *Nichtlineare Integro–Differential–Gleichungen zur Modellierung interaktiver Musterbildungsprozesse auf S^2*. Doctoral thesis, University of Bonn

Geigant E, Ladizhanski K and Mogilner A (1997) Analysis of a differential integral equation modelling the orientational distribution of F-actin in cells. *SIAM J. Appl. Math.* To appear

Gerisch G and Hess B (1974) Cyclic AMP controlled oscillations in suspended *Dictyostelium* cells: their relation to morphogenetic cell interactions. *Proc. Nat. Acad. Sci. (USA)* **71** 2118–2122

Gingell D and Vince S (1982) Substratum wettability and charge influence the spreading of *Dictyostelium* amoeba and the formation of ultrathin cytoplasmic lamellae. *J. Cell Sci.* **54** 255–285

Glazier JA and Graner F (1993) Simulation of the differential driven rearrangement of biological cells. *Phys. Rev. E* **47** 2128–2154

Godrèche C, Kostov I and Yekutieli I (1992) Topological correlations in cellular structures and planar graph theory. *Phys. Rev. Lett.* **69** 2674–2677

Grebecka L (1977) Behaviour of anucleate anterior and posterior fragments of *Amoeba proteus*. *Acta Protozool.* **16** 87–105

Grebecka L and Grebecki A (1975) Morphometric study of moving *Amoeba proteus*. *Acta Protozool.* **14** 337–361

Grebecki A (1976) Co-axial motion of the semi-rigid cell frame in *Amoeba proteus*. *Acta Protozool.* **15** 221–248

Grebecki A (1977) Non-axial cell frame movements and the locomotion of *Amoeba proteus*. *Acta Protozool.* **16** 53–85

Grebecki A (1984) Relative motion in *Amoeba proteus* in respect to the adhesion sites. I. Behavior of monotactic forms and the mechanism of fountainphenomenon. *Protoplasma* **123** 116–134

Grebecki A (1990) Dynamics of the contractile system in pseudopodial tips of normally locomoting amoebae, demonstrated in vivo by video-enhancement. *Protoplasma* **54** 98–111

Grebecki A and Cieslawska M (1978a) Plasmodium of *Physarum polycephalum* as a synchronous contractile system. *Cytobiol.* **17** 335–342

Grebecki A and Cieslawska M (1978b) Dynamics of the ectoplasmic walls during pulsation of plasmodial veins of *Physarum polycephalum*. *Protoplasma* **97** 365–371

Green PB (1976) Growth and cell pattern formation on an axis: Critique of concepts, terminology and modes of study. *Bot. Gaz.* **137** 187–202

Green PB (1987) Inheritance of pattern: analysis from phenotype to gene. *Amer. Zool.* **27** 657–73

Green PB (1996a) How plants produce pattern: a review and a proposal that undulating field behavior is the mechanism. In: *Symmetry in Plants*, eds. RU Jean and D Barabé. World Scientific Press, Singapore

Green PB (1996b) Transduction to generate plant form and pattern: an essay on cause and effect. *Ann. Bot.* **78** 269–281

Green PB and Selker JML (1991) Mutual alignments of cell walls, cellulose, and cytoskeletons: their role in meristem. In: *The Cytoskeletal Basis of Plant Growth and Form*, ed. CW Lloyd, pp. 303–322. Academic Press, London

Green PB, Steele CS and Rennich SC (1996) Phyllotactic patterns: a biophysical mechanism for their origin. *Ann. Bot.* **77** 515–527

Gross J, Farinelli W, Sadow P, Anderson R and Bruns R (1995) On the mechanism of skin wound "contraction": a granulation tissue "knockout" with a normal phenotype. *Proc. Natl. Acad. Sci.* **92** 5982–5986

Gross JD, Peacey MJ and Trevan DJ (1976) Signal emission and signal propagation during early aggregation in *Dictyostelium discoideum*. *J. Cell Sci.* **22** 645–656

Gruler H and De Boisfleury-Chevance A (1994) Directed cell movement and cluster formation: physical principles. *J. Phys. I (France)* **4** 1085–1105

Grünbaum D (1996) Population-level movements with internal state dynamics. Submitted to Proc. Nat. Acad. Sci. US

Guido S and Tranquillo RT (1993) A methodology for the systematic and quantitative study of cell contact guidance in oriented collagen gels: Correlation of fibroblast orientation and gel birefringence. *J. Cell Sci.* **105** 317–331

Gurdon JB, Harger P, Mitchell A and Lemaire P (1994) Activin signalling and response to a morphogen gradient. *Nature* **371** 487–492

Gurwitsch AG (1922) Über den Begriff des embryonalen Feldes. *Arch. Entwicklungsmech.* **52** 167–181

Gurwitsch AG (1923) *Versuch einer synthetischen Biologie*. Gebr. Borntraeger, Berlin

Gurwitsch AG (1930) *Die histologischen Grundlagen der Biologie*. Fischer, Jena

Haberey M (1971) Bewegungsverhalten und Untergrundkontakt von *Amoeba proteus*. *Mikroskopie* **27** 226–234

Hales S (1727) *Vegetable Staticks*. Innis and Woodwark

Hall AA, Warren V, Dharmawardhane S and Condeelis J (1989) Purification of actin nucleating activity and polymerization inhibitor in amoeboid cells: their regulation by chemotactic stimulation. *J. Cell Biol.* **109** 2207–2213

Happel J and Brenner H (1991) *Low Reynolds Number Hydrodynamics*. Kluwer Academic Publishers, Boston, 5th edn.

Harris A (1973) Behavior of cultured cells on substrata of variable adhesiveness. *Exp. Cell Res.* **77** 285–297

Harris AK (1982) Traction and its relations to contraction in tissue cell locomotion. In: *Cell Behaviour*, eds. R Bellairs, A Curtis and G Dunn, pp. 77–108. Cambridge University Press, New York

Harris AK (1984) Tissue culture cells on deformable substrata: Biomechanical implications. *J. Biomech. Engin.* **106** 19–24

Harris AK (1986) Cell traction in relationship to morphogenesis and malignancy. In: *Developmental Biology*, ed. MS Steinberg, pp. 339–357. Plenum Publ. Corp., New York

Harris AK (1994a) Locomotion of tissue culture cells considered in relation to amoeboid locomotion. *Int. Rev. Cytol.* **150** 35–68

Harris AK (1994b) Cytokinesis: the mechanism of formation of the contractile ring in animal cell division. In: *Biomechanics of Active Movement and Division of Cells*, ed. N Akkas, NATO ASI Series, pp. 17–66. Springer, Berlin

Harris AK, Stopak D and Warner P (1984) Generation of spatially periodic patterns by a mechanical instability: a mechanical alternative to the Turing model. *J. Embryol. Exp. Morphol.* **80** 1–20

Harris AK, Wild P and Stopak D (1980) A new wrinkle in the study of locomotion. *Science* **208** 177–179

Harrison LG and Lacalli TC (1993) Controversy over concepts: long-range communication versus "no crosstalk". In: *Oscillations and Morphogenesis*, ed. L Rensing, pp. 43–56. Marcel Dekker, New York

Hartman RS, Lau K, Chou W and Coates TC (1993a) Development of a shape vector that identifies critical forms assumed by human polmorphonuclear neutrophils during chemotaxis. *Cytometry* **14** 832–839

Hartman RS, Lau K, Chou W and Coates TC (1994) The fundamental motor of the human neutrophil is not random: evidence for local non-markov movement in neutrophils. *Biophys. J.* **67** 2535–2545

Hartman RS, Yi D and Coates TC (1993b) Analysis of multi-paramter video measurements of human neutrophil movement and its relation to cell shape and cytosolic calcium. *Comp. Meth. Progr. Biomed.* **39** 195–201

Haupt W (1977) *Bewegungsphysiologie der Pflanzen.* Georg Thieme Verlag, Stuttgart

Hearmon RFS (1948) The elasticity of wood and plywood. Forest Products Research, Special Report no. 7 (pp. 87), Department of Scientific and Industrial Research, His Majesty's Stationary Office, London

Hedberg K, Bengtsson T, Safiejko-Mroczka B, Bell PB and Lindroth M (1993) PDGF and neomycin induce similar changes in the actin cytoskeleton in human fibroblasts. *Cell Motil. Cytoskel.* **24** 139–149

Heinrich V, Svetina S and Žekš B (1993) Nonaxisymmetric shapes in a generalized bilayer-couple model and the transition between oblate and prolate axisymmetric shapes. *Phys. Rev. E* **48** 3112–3123

Hejnowicz Z (1984) Trajectories of principal growth directions. Natural coordinate system in plant growth. *Acta Soc. Bot. Pol.* **53** 29–42

Hejnowicz Z (1989) Differential growth resulting in the specification of different types of cellular architecture in root meristems. *Environ. Exp. Bot.* **29** 85–93

Hejnowicz Z and Hejnowicz K (1991) Modeling the formation of root apices. *Planta* **184** 1–7

Hejnowicz Z and Karczewski J (1993) Modeling of meristematic growth of root apices in a natural coordinate system. *Amer. J. Botany* **80** 309–315

Hejnowicz Z and Romberger JA (1984) Growth tensor of plant organs. *J. Theor. Biol.* **110** 93–114

Hejnowicz Z and Sievers A (1995a) Tissue stresses in organs of herbaceous plants. I. Poisson ratios of tissues and their role in determination of the stresses. *J. Exp. Bot.* **46** 1035–1043

Hejnowicz Z and Sievers A (1995b) Tissue stresses in organs of herbaceous plants. II. Determination in three dimensions in the hypocotyl of sunflower. *J. Exp. Bot.* **46** 1045–1053

Hejnowicz Z and Sievers A (1996a) Tissue stresses in organs of herbaceous plants. III. Elastic properties of the tissues of sunflower hypocotyl and origin of tissue stresses. *J. Exp. Bot.* **47** 519–528

Hejnowicz Z and Sievers A (1996b) Gravitropic response of the stem of *Reynoutria*. Involvement of tissue stresses. *J. Plant Physiol.* In press

Hejnowicz Z and Wohlfarth-Bottermann KE (1980) Propagated waves induced by gradients of physiological factors within plasmodia of *Physarum polycephalum*. *Planta* **150** 144–152

Helfrich W (1973) Elastic properties of lipid bilayers: theory and possible experiments. *Z. Naturforsch.* **28c** 693–703

Hernandez L and Green PB (1993) Transductions for the expression of structural pattern. *Plant Cell* **5** 1723–1738

Herrero MA and Velasquez JJL (1995) A mechanism of blowup for the Keller-Segel model of chemotaxis. Preprint. Dept. de Matem. Aplicada, Univ. Complutense, Madrid

Herrero MA and Velasquez JJL (1996a). Chemotactic collapse for the Keller-Segel model. *J. Math. Biol.* To appear

Herrero MA and Velasquez JJL (1996b). Singularity patterns in a chemotaxis model. *Math. Ann.* **306** 583–623

Hiramoto Y (1956) Cell division without mitotic apparatus in sea urchin eggs. *Exp. Cell Res.* **11** 630–636

Hiramoto Y (1963) Mechanical properties of sea urchin eggs. ii. changes in mechanical properties from fertilization to cleavage. *Exp. Cell Res.* **32** 76–89

Hiramoto Y (1965) Further studies on cell division without mitotic apparatus in sea urchin eggs. *J. Cell Biol.* **25** 161–168

Hiramoto Y (1968) The mechanics and mechanisms of cleavage in the sea urchin egg. *22nd Symposium of the Society for Experimental Biology* pp. 311–327

Hiramoto Y (1969a) Mechanical properties of the protoplasm of the sea urchin egg. I. Unfertilized egg. *Exp. Cell Res.* **56** 201–208

Hiramoto Y (1969b) Mechanical properties of the protoplasm of the sea urchin egg. II. Fertilized egg. *Exp. Cell Res.* **56** 209–218

Hiramoto Y (1978) Mechanical properties of the dividing sea urchin egg. In: *Cell Motility: Molecules and Organization*, eds. S Hatano, H Ishikawa and H Sato, pp. 653–663. Univ. Park Press, Baltimore

His W (1874) *Unsere Körperform und das physiologische Problem ihrer Entstehung*. Leipzig

Hitt AL, Hartwig JH and Luna EJ (1994) Ponticulin is the major high affinity link between the plasma membrane and the cortical actin network in *Dictyostelium*. *J. Cell Biol.* **126** 1433–1444

Hodick D and Sievers A (1989) On the mechanism of trap closure of Venus flytrap *Dionaea muscipula* Ellis. *Planta* **179** 32–42

Höfer T, Maini PK, Sherratt JA, Chaplain MAJ, Chauvet P, Metevier D, Montes PC and Murray JD (1994) A resolution of the chemotactic wave paradox. *Appl. Math. Lett.* **7** 1–5

Höfer T, Sherratt JA and Maini PK (1995a) *Dictyostelium discoideum*: cellular self-organization in an excitable biological medium. *Proc. Roy. Soc. Lond. B* **259** 249–257

Höfer T, Sherratt JA and Maini PK (1995b) Cellular pattern formation during *Dictyostelium* aggregation. *Physica D* **85** 425–444

Houliston E and Elinson RP (1991) Evidence for the involvement of microtubules, ER and kinesis in the cortical rotation of fertilized frog eggs. *J. Cell Biol.* **114** 1017–1028

Howard TH and Meyer WH (1984) Chemotactic peptide modulation of actin assembly and locomotion in neutrophils. *J. Cell Biol.* **98** 1265–1271

Ingber DE, Dike E, Hansen L, Karp S, Liley H, Maniotis A, McNamee H, Mooney D, Plopper G, Sims J and Wang N (1994) Cellular tensegrity: exploring how mechanical changes in the cytoskeleton regulate cell growth, migration and tissue pattern during morphogenesis. *Int. Rev. Cytol.* **150** 173–224

Isaeva VV and Presnov EV (1990) *Topological structure of morphogenetic fields.* Nauka, Moskva. In Russian

Ishijima A, Doi T, Sakurada K and Yanagida T (1991) Subpiconewton force fluctuations of actomyosin in vitro. *Nature* **352** 301–306

Ishikawa H, Hasenstein KH and Evans ML (1991) Computer-based video digitizer analysis of surface extension in maize roots. *Planta* **183** 381–390

Israelsson D and Johnsson A (1967) A theory of circumnutations in *Helianthus annuus*. *Physiol. Plant.* **22** 1251–1262

Jäger W and Luckhaus S (1992) On explosions of solutions to a system of partial differential equations modelling chemotaxis. *Trans. AMS* **329** 819–824

Janmey PA (1994) Phosphoinositides and calcium as regulators of cellular actin assembly and disassembly. *Ann. Rev. Physiol.* **56** 169–191

Johnsson A and Israelsson D (1969) Phase-shift in geotropical oscillations: a theoretical and experimental study. *Physiologia Plantarum* **25** 35–42

Kadanoff L (1985) Simulating hydrodynamics: a pedestrian model. *J. Stat. Phys.* **39** 267–283

Kamiya N (1968) The mechanism of cytoplasmic movement in a myxomycete plasmodium. In: *Aspects of Cell Motility. Symp. Soc. Exp. Biol.* **22**, pp. 199–214. Cambridge Univ. Press, Cambridge

Kamiya N (1981) Physical and chemical basis of cytoplasmic streaming. *Ann. Rev. Plant Physiol.* **32** 205–236

Käs J, Sackmann E, Podgornik R, Svetina S and Žekš B (1993a) Thermally induced budding of phospholipid vesicles - a discontinuous process. *J. Phys. II France* **3** 631–645

Käs J, Strey H, Baermann M and Sackmann E (1993b) Direct Measurements of the wave-vector-dependent bending stiffness of freely flickering actin filaments. *Europhys. Lett.* **21** 865–870

Keller EF and Segel LE (1970) Initiation of slime mold aggregation viewed as an instability. *J. Theor. Biol.* **26** 399–415

Keller H and Bebie H (1996) Protrusive activity quantitatively determines the rate and direction of cell locomotion. *Cell Motil. Cytoskel.* **33** 241–251

Keller R (1987) Cell rearrangement in morphogenesis. *Zool. Sci.* **4** 763–779

Kerk NM and Feldman LJ (1995) A biochemical model for the initiation and maintenance of the quiescent center: implications for organization of root meristems. *Development* **121** 2825–2833

Kerr JP and Bartlett EB (1995) Medical image processing utilizing neural networks trained on a massively parallel computer. *Comput. Biol. Med.* **25** 393–403

Kessler DA and Levine H (1993) Pattern formation in *Dictyostelium* via the dynamics of cooperative biological entities. *Phys. Rev. E* **48** 4801–4804

Killich T, Plath PJ, Haß EC, Xiang W, Bultmann H, Rensing L and Vicker MG (1994) Cell movement and shape are non-random and determined by intracellular, oscillatory, rotating waves in *Dictyostelium* amoebae. *BioSystems* **33** 75–87

Killich T, Plath PJ, Xiang W, Bultmann H, Rensing L and Vicker MG (1993) The locomotion, shape and pseudopodial dynamics of unstimulated *Dictyostelium* cells are not random. *J. Cell Sci.* **106** 1005–1013

King CA, Davies AH and Preston TM (1981) Lack of substrate specificity on the speed of amoeboid locomotion in *Naegleria gruberi*. *Experientia* **37** 706

King CA, Preston TM and Miller RH (1983) Cell-substrate interactions in amoeboid locomotion - a matched reflection interference and transmission electron microscopy study. *Cell Biol. Int. Rep.* **7** 641–649

Kitanishi-Yumura T and Fukui Y (1989) Actomyosin organization during cytokinesis: reversible translocation and differential redistribution in *Dictyostelium*. *Cell Motil. Cytoskel.* **12** 78–89

Klopocka W, Kolodziejczyk J, Lopatowska A and Grebecki A (1996) Relationship between pinocytosis and adhesion in *Amoeba proteus*. *Cell Biol. Int.* **20**. In press

Kocks C, Hellio R, Gounon P, Ohayon H and Cossart P (1993) Polarized distribution of *Listeria monocytogenes* surface protein ActA at the site of directional actin assembly. *J. Cell Sci.* **105** 699–710

Kolega J (1986) Effects of mechanical tension on protrusive activity and microfilament and intermediate filament organization in an epidermal epithelium moving in culture. *J. Cell Biol.* **102** 1400–1411

Kolodziejczyk J, Klopocka W, Lopatowska A, Grebecka L and Grebecki A (1995) Resumption of locomotion by *Amoeba proteus* readhering to different substrata. *Protoplasma* **189** 180–186

Korn ED, Carlier MF and Pantolini D (1987) Actin polymerisation and ATP hydrolysis. *Science* **238** 638–644

Kowalczynska HM and Kaminski J (1991) Adhesion of L1210 cells to modified styrene copolymer surfaces in the presence of serum. *J. Cell Sci.* **99** 587–593

Kralj-Iglič V, Svetina S and Žekš B (1996) Shapes of bilayer vesicles with membrane embedded molecules. *Eur. Biophys. J.* **24** 311–321

Kreitmeier M, Gerisch G, Heizer C and Müller-Taubenberger A (1995) A talin homologue of *Dictyostelium* rapidly assembles at the leading edge of cells in response to chemoattractant. *J. Cell Biol.* **129** 179–188

Krischer K, Rico-Martinez R, Kevrekidis IG, Rotermund G, Ertl G and Hudson JL (1993) Model identification of a spatio-temporally varying catalytic reaction. *AICHE-J* **39** 89–98

Krüger J and Wohlfarth-Bottermann KE (1978) Oscillating contractions in protoplasmic strands of *Physarum*: stretch-induced phase shift and their synchronization. *J. Interdiscip. Cycle Res.* **9** 61–71

Kubo R, Toda M and Hashitsume N (1991) *Statistical Physics II: Nonequilibrium Statistical Mechanics*. Chapter 5. Springer, Berlin

Kucera P and Monnet-Tschudi F (1987) Early functional differentiation in the chick embryonic disc: interactions between mechanical activity and extracellular matrix. *J. Cell Sci.* Suppl. **8** 415–432

Kulikowski JL (1985) Basic problems of image analysis based on methods of moments. Reports of the institute of biocybernetics and biomedical engineering, Polish Academy of Sciences 1-15. (In Polish)

Kuroda R, Hatano S, Hiramoto Y and Kuroda H (1988) Change of cytosolic Ca-ion concentration in the contraction and relaxation cycle of *Physarum* microplasmodia. *Protoplasma, Suppl.* **1** 72–80

Lackie (1986) *Cell Movement and Cell Behavior*. Allen & Unwin, London

Ladizhanski K (1994) *Distribution of generalized aspect with applications to actin fibers and social interactions*. Master's thesis, Weizmann Institute of Science, Rehovot

Landau LD and Lifshitz EM (1986) *Theory of Elasticity*. Pergamon Press, Oxford UK. J.B. Sykes and W.H. Reid Translators

Latushkin OA, Netrebko NB, Romanovsky YM and Teplov VA (1988) Mathematical two-dimensional model of autowave processes in living cell. In: *Collective Dynamics of Excitations and Pattern Formation in Biological Tissue*, ed. VG Yakhno, pp. 109–119. Inst. Appl. Phys. USSR Acad. Sci. Press, Gorky. (In Russian)

Lauffenburger D and Linderman JJ (eds.) (1993) *Receptors: Models for Binding, Trafficking, and Signaling*. Oxford University Press, New York

Lauffenburger DA (1991) Models for receptor mediated cell phenomena. *Ann. Rev. Biophys. Chem.* **20** 387–414

Le Caër G and Delannay R (1993) Correlations in topological models of 2D random cellular structures. *J. Phys. A* **26** 3931–3954

Lechleiter JD and Clapham DE (1992) Spiral waves and intercellular calcium signalling. *J. Physiol.* **86** 123–128

Lee J, Ishihara A and Jacobson K (1993) How do cells move over surfaces? *Trends in Cell Biology* **3** 366–370

Lee J, Leonard M, Oliver TN, Ishihara A and Jacobson K (1994) Traction forces generated by locomoting cells. *J. Cell Biol.* **127** 1957–1964

Lee KJ, Cox EC and Goldstein RE (1996) Competing patterns of signalling activity in *Dictyostelium discoideum*. Preprint

Lemaître J, Gervois A, Troadec JP, Rivier N, Ammi M, Oger L and Bideau D (1993) Arrangements of cells in Voronoi tesselations of monosize packings of discs. *Phil. Mag. B* **67** 347–363

Lendowski V (1996) *Modellierung und Simulation der Bewegung eines Körpers in reaktiven Zweiphasenflüssigkeiten*. Doctoral thesis, Universität Bonn. To appear in *Bonner Mathematische Schriften* **296** (1997)

Levin SA and Segel LA (1985) Pattern generation in space and aspect. *SIAM Review* **27** 45–67

Levine H and Sleeman BD (1997) A system of reaction diffusion equations arising in the theory of reinforced random walks. *SIAM J. Appl. Math.* To appear

Lewis FT (1928) The correlation between cell division and the shapes and sizes of prismatic cells in the epidermis of the *Cucumis. Anat. Rec.* **38** 341–376

Lewis FT (1930) A volumetric study of the growth and cell division in two types of radially prismatic epidermal cells of Cucumis. *Anat. Rec.* **47** 59

Lewis FT (1931) A comparison between the mosaic of polygons in a film of artificial emulsion and the pattern of simple epithelium in surface view (cucumber epidermis and human amnion). *Anat. Rec.* **50** 235–265

Li GH, Qin CD and Li MH (1994) On the mechanisms of growth cone locomotion: modeling and computer simulations. *J. Theor. Biol.* **169** 335–362

Li SL, Miyata Y, Yahara I and Fujita-Yamaguchi Y (1993) Insulin induced circular membrane ruffling on rat 1 cells expressing a high number of human insulin receptors: circular ruffles caused by rapid actin reorganization exhibit high density of insulin receptors and phosphotyrosines. *Exp. Cell Res.* **205** 353–360

Li YX and Goldbeter A (1992) Pulsatile signaling in intercellular communications - periodic stimuli are more efficient than random or chaotic signals in a model based on receptor desensitization. *Biophys. J.* **61** 161–171

Liang S (1986) Random-walk simulations of flow in Hele Shaw cells. *Phys. Rev. A* **33** 2663–2674

Linde H and Engel H (1991) Autowave propagation in heterologous media. *Physica D* **49** 13–20

Lipp P and Niggli E (1993) Microscopic spiral waves reveal positive feedback in subcellular calcium signalling. *Biophys. J.* **65** 2272–2276

Liu M, Skinner SJM, Xu J, Han RNN, Tanswell AK and Post M (1992) Stimulation of fetal rat lung cell proliferation in vitro by mechanical stretch. *Am. J. Physiol.* **263** L363–L383

Lockhart JA (1965) An analysis of irreversible plant cell elongation. *J. Theor. Biol.* **8** 264–275

Logvenkov SA (1993) Mechanism of differentiation of plant root cells. *Biofizika (Biophysics)* **38** 860–864

Loomis WF and Sternberg PW (1995) Genetic networks. *Science* **269** 649

Lorch IJ (1969) The rate of the attachment to the substratum: A study of nuclear-cytoplasmic relationship. *J. Cell. Physiol.* **73** 171–177

Lubkin SR and Murray JD (1995) A mechanism for early branching in lung morphogenesis. *J. Math. Biol.* in press

Mabuchi I (1986) Biochemical aspects of cytokinesis. *Intern. Rev. Cytology* **101** 175–213

MacKay SA (1978) Computer simulation of aggregation in *Dictyostelium discoideum*. *J. Cell Sci.* **33** 1–16

MacKintosh FC, Käs J and Janmey PA (1995) Elasticity of semiflexible biopolymer networks. *Phys. Rev. Lett.* **75** 4425–4429

Mandelkow E, Mandelkow EA, Hotani H, Hess B and Müller SC (1989) Spatial patterns from oscillating microtubules. *Science* **246** 1291–1293

Marchand JB, Moreau P, Paoletti A, Cossart P, Carlier MF and Pantaloni D (1995) Actin-based movement of *Listeria monocytogenes*: actin assembly results from the local maintenance of uncapped filament barbed ends at the bacterium surface. *J. Cell Biol.* **130** 331–343

Margulis L and Schwartz KV (1988) *Five Kingdoms*. W. H. Freeman, New York

Martiel L and Goldbeter A (1987) A model based on receptor desensitization for cyclic AMP signaling in *Dictyostelium* cells. *Biophys. J.* **52** 807–828

Matthes T and Gruler H (1988) Analysis of cell locomotion. Contact guidance of human polymorphonuclear leukocytes. *Europ. Biophys. J.* **15** 343–357

Matthiessen K and Müller SC (1995) Global flow waves in chemically induced convection. *Phys. Rev. E* **52** 492–495

Matzke EB (1950) In the twinkling of an eye. *Bull. Torrey Bot. Club* **27** 222–227

McAdams HH and Shapiro L (1995) Circuit simulation of genetic networks. *Science* **269** 650–656

McRobbie SJ and Newell PC (1984) Chemotaxis-mediated changes in cytoskeletal actin of cellular slime moulds. *J. Cell Sci.* **68** 139–151

Mescheryakov VN (1991) How genes detect right and left? In: *Analytical Aspects of Differentiation*, eds. EV Presnov, VM Maresin and AI Ivanov, pp. 137–166. Nauka, Moskva. (In Russian)

Miao L, Seifert U, Wortis M and Döbereiner HG (1994) Budding transitions of fluid-bilayer vesicles: the effect of area-difference elasticity. *Phys. Rev. E* **49** 5389–5407

Michelson S and T. LJ (1994) Dormancy, regression and recurrence: Towards a unifying theory of tumor growth control. *J. Theor. Biol.* **169** 327–338

Miike H, Müller SC and Hess B (1989) Hydrodynamic flows traveling with chemical waves. *Phys. Lett. A* **141** 25–30

Mittenthal JE and Mazo RM (1983) A model for shape generation by strain and cell-cell adhesion in the epithelium of an arthropod leg segment. *J. Theor. Biol.* **100** 443–483

Moghe P and Tranquillo RT (1994) Stochastic model of chemoattractant receptor dynamics in leukocyte chemosensory movement. *Bull. Math. Biol.* **56** 1041–1093

Moghe P and Tranquillo RT (1995) Stochasticity in membrane-localized ligand-receptor-G protein binding - consequences for leukocyte movement behavior. *Ann. Biomed. Eng.* **23** 257–267

Mogilner A (1996) Models for angular self-organization of cytoskeleton. In preparation

Mogilner A and Edelstein-Keshet L (1995) Selecting a common direction: How orientational order can arise from simple contact responses between interacting cells. *J. Math. Biol.* **33** 619–660

Mogilner A and Edelstein-Keshet L (1996) Selecting a common direction: II. Peak-like solutions representing total alignment of cell clusters. *J. Math. Biol.* **34** 811–842

Mogilner A and Edelstein-Keshet L (1996) Spatio-angular order in populations of self-aligning objects: formation of oriented patches. *Physica D* **89** 346–367

Mogilner A and Oster GF (1996) Cell motions driven by actin polymerization. *Biophys. J.* In press

Mombach JCM, Vasconcellos MAZ and de Almeida RMC (1990) Arrangement of cells in vegetable tissues. *J. Phys. D* **23** 600–606

Monk PB and Othmer HG (1989) Cyclic AMP oscillations in suspensions of *Dictyostelium discoideum*. *Phil. Trans. Roy. Soc. Lond. B.* **323** 185–224

Monk PB and Othmer HG (1990) Wave propagation in aggregation fields of the cellular slime mould *Dictyostelium discoideum*. *Proc. R. Soc. Lond. B.* **240** 555–589

Mori Y, Yoshida K, Morita T and Nakanishi Y (1994) Branching morphogenesis of mouse embryonic submandibular epithelia cultured under three different conditions. *Dev. Growth and Diff.* **36** 529–539

Müller SC (1988) Spatial patterns in (bio)chemical reactions. In: *From Chemical to Biological Organization*, eds. M Markus, SC Müller and G Nicolis, pp. 85–98. Springer, Berlin

Murray J, Vawter-Hugart H, Voss E and Soll DR (1992) Three-dimensional motility cycle in leukocytes. *Cell Motil. Cytoskeleton* **22** 211–223

Murray JD and Oster GF (1984) Generation of biological pattern and form. *IMA J. Math. Appl. Med. Biol.* **1** 51–75

Nagai T (1996) Blow-up of radially symmetric solutions to a chemotaxis system. *Math. Meth. Appl. Sci.* To appear

Nakanishi Y, Morita T and Nogawa H (1987) Cell proliferation is not required for the initiation of cleft formation in mouse embryonic submandibular epithelium *in vitro*. *Development* **99** 429–438

Nakielski J (1991) Distribution of linear growth rates in different directions in root apical meristems. *Acta Soc. Bot. Pol.* **60** 77–86

Nakielski J (1992) Regeneration of the root apex: modelling study by means of the growth tensor. In: *Mechanics of Swelling*, ed. TK Karalis, NATO ASI Series H 64, pp. 179–191. Springer, Berlin

Nakielski J and Barlow PW (1995) Principal directions of growth and the generation of cell patterns in wild-type and *gib*-1 mutant roots of tomato (*Lycopersicon esculentum* Mill.) grown in vitro. *Planta* **196** 30–39

Nanjundiah V (1973) Chemotaxis, signal relaying and aggregation morphology. *J. Theor. Biol.* **42** 63–105

Newell PC, Europe-Finner GN, Liu G, Gammon B and Wood CA (1990) Chemotaxis of *Dictyostelium discoideum*: The signal transduction pathway to actin and myosin. In: *Biology of the Chemotactic Response*, eds. JP Armitage and JM Lackie, pp. 241–272. Cambridge University Press, Cambridge UK

Noble P (1990) Images of cells changing shape: pseudopods, skeletons and motile behaviour. In: *Biological Motion*, eds. W Alt and G Hoffmann *Lecture Notes in Biomath.*, Vol. 89, pp. 35–41. Springer Verlag, Berlin

Nogawa H and Ito T (1995) Branching morphogenesis of embryonic mouse lung epithelium in mesenchyme-free culture. *Development* **121** 1015–1022

Nogawa H and Takahashi Y (1991) Substitution for mesenchyme by basement-membrane-like substratum and epidermal growth factor in inducing branching morphogenesis of mouse salivary epithelium. *Development* **112** 855–861

Nowakowska G and Grebecki A (1978) Attachment of *Amoeba proteus* to the substrate during upside-down crawling. *Acta Protozool.* **17** 353–358

Nüsslein-Volhard C, Frohnhofer HG and Lehmann R (1987) Determination of anteroposterior polarity in *Drosophila*. *Science* **238** 1675–1681

Odell GM (1984) A mathematically modelled cytogel cortex exhibits periodic Ca^{2+}-modulated contraction cycles seen in *Physarum* shuttle streaming. *J. Embryol. Exp. Morphol., Suppl.* **83** 261–287

Odell GM, Oster GF, Alberch P and Burnside B (1981) The mechanical basis of morphogenesis. I. Epithelial folding and invagination. *Dev. Biol.* **85** 446–462

Oliver T, Dembo M and Jacobson K (1995) Traction forces in locomoting cells. *Cell Motil. Cytoskel.* **31** 225–240

Olsen L, Sherratt JA and Maini PK (1995) A mechanochemical model for adult dermal wound contraction and the permanence of the contracted tissue displacement profile. *J. Theor. Biol.* **177** 113–128

Olsen L, Sherratt JA and Maini PK (1996a) Simple modelling of extracellular matrix alignment in dermal wound healing. I. Cell flux induced alignment. Preprint

Olsen L, Sherratt JA and Maini PK (1996b) Simple modelling of extracellular matrix alignment in dermal wound healing. II. Stress induced alignment. In preparation

Olsen L, Sherratt JA and Maini PK (1996c) A mathematical model for fibro-proliferative wound healing disorders. *Bull. Math. Biol.* **58** 787–808

Omann GM, Porasik MM and Sklar LA (1989) Oscillating actin polymerization/depolymerization in human polymorphonuclear leukocytes. *J. Biol. Chem.* **264** 16355–16358

Opas M (1978) Interference reflection microscopy of adhesion of *Amoeba proteus*. *J. Microscopy* **112** 215–221

Opas M (1981) Effects of induction of endocytosis on adhesiveness of *Amoeba proteus*. *Protoplasma* **107** 161–169

Opas M (1994) Substratum mechanics and cell differentiation. *Int. Rev. Cytol.* **150** 119–138

Ortega JKE (1985) Augmented growth equation for cell wall expansion. *Plant Physiol.* **78** 318–320

Oster GF and Moore HPH (1989) The budding of membranes. In: *Cell to Cell Signalling: From Experiments to Theoretical Models*, ed. A Goldbeter, pp. 171–187. Academic Press, London

Oster GF, Murray JD and Harris AK (1983) Mechanical aspects of mesenchymal morphogenesis. *J. Embryol. Exp. Morphol.* **78** 83–125

Oster GF, Murray JD and Odell GM (1985) The formation of microvilli. In: *Molecular Determinants of Animal Form*, ed. GM Edelman, pp. 365–384. Alan R. Liss, New York

Oster GF and Odell GM (1984) Mechanics of cytogels I: Oscillations in *Physarum*. *Cell Motil.* **4** 469–503

Othmer HG, Dunbar SR and Alt W (1988) Models of dispersal in biological systems. *J. Math. Biol.* **26** 263–298

Othmer HG, Monk PB and Rapp PE (1985) A model for signal-relay adaptation in *Dictyostelium discoideum*. 2. Analytical and numerical results. *Mathem. Biosc.* **77** 79–139

Othmer HG and Stevens A (1997) Aggregation, blowup and collapse: the ABCs of taxis in reinforced random walks. *SIAM J. Appl. Math.* **57** 1044–1082

Panfilov AV and Winfree AT (1985) Dynamical simulations of rotational scroll rings in 3D excitable media. *Physica D* **17** 323–330

Parnas H and Segel LA (1977) Computer evidence concerning the chemotactic signal in *Dictyostelium discoideum*. *J. Cell Sci.* **25** 191–204

Parnas H and Segel LA (1978) A computer simulation of pulsatile aggregation in *Dictyostelium discoideum*. *J. Theor. Biol.* **71** 185–207

Patlak CS (1953) Random walk with persistence and external bias. *Bull. Math. Biophys.* **33** 311–338

Pavlov DA and Potapov MM (1994) Continuum model of intracellular motility and its projection-difference approximation. In: *Calcul. Math. Cybern. 1*, Ser. 15, pp. 10–16. Proc. Mosc. Univ. (In Russian)

Peshkin M, Strandburg K and Rivier N (1991) Entropic prediction for cellular networks. *Phys. Rev. Lett.* **67** 1803–1806

Peskin CS, Odell GM and Oster GF (1993) Cellular motions and thermal fluctuations: the Brownian rachet. *Biophys. J.* **65** 316–324

Peterson MD and Titus MA (1994) F-actin distribution of *Dictyostelium* myosin I double mutants. *J. Eukaryotic Microbiology* **41** 652–657

Plant RE (1982) A continuum model for root growth. *J. Theor. Biol.* **98** 45–59

Podolski JL and Steck TL (1990) Length distribution of F-actin in *Dictyostelium discoideum*. *J. Cell Biol.* **265** 1312–1318

Pollard TD (1981) Cytoplasmic contractile proteins. *J. Cell Biol.* **91** 156s–165s

Pollard TD and Cooper JA (1986) Actin and actin-binding proteins. A critical evaluation of mechanisms and functions. *Ann. Rev. Biochem.* **55** 987–1035

Preston TM and King CA (1978) An experimental study of the interaction between the soil amoeba *Naegleria gruberi* and a glass substrate during amoeboid locomotion. *J. Cell Sci.* **34** 1245–158

Priestley JH (1930) Studies in the physiology of cambial activity. II. The concept of sliding growth. *New Phytol.* **29** 96–140

Probine MC (1963) Cell growth and the structure and the mechanical properties of the wall in the internodal cells of *Nitella opaca*. III. Spiral growth and wall structure. *J. Exp. Bot.* **14** 101–113

Pyshnov MB (1980) Topological solution for cell proliferation in intestinal crypt. I - Elastic growth without cell loss. *J. Theor. Biol.* **87** 189–200

Rapp PE, Monk PB and Othmer HG (1985) A model for signal-relay adaptation in *Dictyostelium discoideum*. 1. Biological processes and the model network. *Mathem. Biosc.* **77** 35–78

Rappaport R (1961) Experiments concerning the cleavage stimulus in sand dollar eggs. *J. Exp. Zool.* **148** 81–89

Rappaport R (1967) Cell division: direct measurement of maximum tension exerted by furrow of echinoderm eggs. *Science* **156** 1241–1243

Rappaport R (1969) Division of isolated furrows and furrow fragments in invertebrate eggs. *Exp. Cell Res.* **56** 87–91

Rappaport R (1986) Establishment of the mechanism of cytokinesis in animal cells. *Intern. Rev. Cytology* **105** 245–281

Rappaport R and Ebstein RP (1965) Duration of stimulus and latent period preceeding furrow formation in sand dollar eggs. *J. Exp. Zool.* **158** 373–382

Rascle M and Ziti C (1995) Finite time blow-up in some models of chemotaxis. *J. Math. Biol.* **33** 388–414

Rashevsky N (1938) *Mathematical Biophysics*. Univ. of Chicago Press, rev. ed.

Redmond T and Zigmond SH (1993) Distribution of F-actin elongation sites in lysed polymorphonuclear leukocytes parallels the distribution of endogenous F-actin. *Cell Motil. Cytoskel.* **26** 7–18

Regirer SA and Stein AA (1985) Mechanical aspects of growth, development, and remodelling processes in biological tissues. *Itogi Nauki i Tekhniki VINITI (Advances in Science and Technology of VINITI)* **1** 3–142. (In Russian)

Reinke J (1901, 2nd ed. 1911) *Einleitung in die theoretische Biologie*. Gebr. Paetel, Berlin

Reinke J (1922) *Grundlagen einer Biodynamik*. Gebr. Borntraeger, Berlin

Rivier N (1994) Maximum entropy for random cellular structures. In: *From Statistical Physics to Statistical Inference and Back*, eds. P Grassberger and JP Nadal, pp. 77–93. Kluwer, Dordrecht

Rivier N, Arcenegui Siemens X and Schliecker G (1995a) Cell division and evolution of biological tissues. In: *Fragmentation physics*, eds. D Beysens, X Campi and E Pefferkorn, pp. 266–282. World Scientific Press, Singapore

Rivier N and Dubertret B (1995) Why does skin stay smooth? *Phil. Mag. B* **72** 311–322

Rivier N and Lissowski A (1982) On the correlation between sizes and shapes of cells in epithelial mosaics. *J. Phys. A* **15**

Rivier N, Schliecker G and Dubertret B (1995b) The stationary state of epithelia. *Acta Biotheor.* **43** 403–423

Romanovsky YM, Chernyaeva EB, Kolin'ko VG and Khors NP (1981) Mathematical models of the protoplasmic movement. In: *Autowave Processes in Systems with Diffusion*, ed. MT Grekhova, pp. 202–209. Inst. Appl. Phys. USSR Acad. Sci. Press, Gorky (In Russian)

Romanovsky YM and Teplov VA (1988) Autowave mechanochemical model for *Physarum* shuttle streaming. In: *Thermodynamics and Pattern Formation in Biology*, eds. I Lamprecht and AI Zotin, pp. 395–414. W. de Gruyter, New York

Romanovsky YM and Teplov VA (1995) The physical bases of cell movement. The mechanisms of self-organization of amoeboid motility. *Physics Uspekhi* **38** 521–542

Rutz M (1995) *Theorien und Simulationen des differentiellen Flankenwachstums von Wurzelspitzen*. Diploma thesis, Botanical Institute, University Bonn

Ryabova LV (1995) Two-component cytoskeletal system as a basis of cortical contractility in *Xenopus laevis* eggs. *Ontogenez (Russ. J. Devel. Biol.)* **26** 236–247

Ryter A and Brachet P (1978) Cell surface changes during early developmental stages of *Dictyostelium discoideum*: a scanning electron microscopic study. *Biol. Cellulaire* **31** 265–270

Sachs J (1887) Lecture XXVII. Relations between growth and cell division in the embryonic tissues. In: *Lectures in Plant Physiology*, pp. 431–459. Clarendon Press, Oxford. (Translated by H. Marshal Ward)

Samans KE, Hinz I, Hejnowicz Z and Wohlfarth-Bottermann KE (1984) Phase relations of oscillatory contraction cycles in *Physarum* plasmodia: I. A serial infrared registration device and its application to different plasmodial stages. *J. Interdiscipl. Cycle Res.* **15** 241–250

Satterwhite LL and Pollard TD (1992) Cytokinesis. *Curr. Opin. Cell Biol.* **4** 43–52

Savill NJ and Hogeweg P (1997) Modeling morphogenesis: from single cells to crawling slugs. *J. Theor. Biol.* Accepted for publication

Schaaf R (1985) Stationary solutions of chemotaxis systems. *Trans. AMS* **292** 531–556

Schienbein M and Gruler H (1995) Chemical amplifier, self-ignition mechanism, and amoeboid cell migration. *Phys. Rev. E* **52** 4183–4197

Schimansky-Geier L, Schweitzer F and Mieth M (1996) Interactive structure formation with Brownian particles. In: *Self-Organization of Complex Structures: From Individual to Collective Dynamics*, ed. F Schweitzer, pp. 101–118, Gordon and Breach, London

Schlage W, Kuhn C and Bereiter-Hahn J (1981) Established *Xenopus* tadpole heart endothelium (XTH) cells exhibiting selected properties of primary cells. *Europ. J. Cell Biol.* **24** 342

Schleicher M and Noegel AA (1992) Dynamics of the *Dictyostelium* cytoskeleton during chemotaxis. *New Biologist* **4** 461–472

Schmidt C, Horwitz A, Lauffenburger D and Sheetz M (1993) Integrin-cytoskeletal interactions in migrating fibroblast are dynamcis, asymmetric, and regulated. *J. Cell Biol.* **123** 977–991

Schroeder E (1968) Cytokinesis: filaments in the cleavage furrow. *Exp. Cell Res.* **53** 272–276

Schroeder E (1972) The contractile ring: II. determining its brief existence, volumetric changes, and vital role in cleaving arbacia eggs. *J. Cell Biol.* **53** 419–434

Schroeder TE (1978) Microvilli on sea urchin eggs: A second burst of elongation. *Develop. Biol.* **64** 342–346

Schweitzer F, Lao K and Family F (1997) Active random walkers simulate trunk trail formation by ants. *BioSyst.* In press

Schweitzer F and Schimansky-Geier L (1994) Clustering of "active" walkers in a two-component system. *Physica A* **206** 359–379

Segel LA, Goldbeter A, Devreotes PN and Knox BE (1986) A mechanism for exact sensory adaptation based on receptor modification. *J. Theor. Biol.* **120** 151–179

Seifert U, Berndl K and Lipowsky R (1991) Shape transformations of vesicles: phase diagram for spontaneous-curvature and bilayer-coupling models. *Phys. Rev. A* **44** 1182–1202

Sellen DB (1983) The response of mechanically anisotropic cylindrical cells to multiaxial stress. *J. Exp. Bot.* **34** 681–687

Shaffer BM (1975) Secretion of cyclic AMP induced by cyclic AMP in the cellular slime mold *Dictyostelium discoideum*. *Nature* **255** 549–552

Sheetz MP (1994) Cell migration by graded attachment to substrates and contraction. *Seminars in Cell Biology* **5** 149–155

Sherratt JA and Lewis J (1993) Stress induced alignment of actin filaments and the mechanism of cytogel. *Bull. Math. Biol.* **55** 637–654

Siegert F and Weijer CJ (1989) Digital-image processing of optical-density wave propagation in *Dictyostelium discoideum* and analysis of the effects of caffeine and ammonia. *J. Cell Sci.* **93** 325–335

Siegert F and Weijer CJ (1995) Spiral and concentric waves organize multicellular *Dictyostelium* mounds. *Curr. Biol.* **5** 937–943

Siegert F, Weijer C J Siegert F and Weijer CJ (1992) Three-dimensional scroll waves organize *Dictyostelium* slugs. *Proc. Natl. Acad. Sci. USA.* **89** 6433–6437

Silk WK and Erickson RO (1979) Kinematics of plant growth. *J. Theor. Biol.* **76** 481–501

Simons P (1992) *The Action Plant: Movement and Nervous Behaviour in Plants*. Blackwell Publishers, Oxford, UK

Skierczynski BA, Usami S and Skalak R (1994) A model of the leukocyte migration through solid tissue. In: *Biomechanics of Active Movement and Division of Cells*, ed. N Akkas, Nato ASI Ser. H 84, pp. 285–328. Springer, Berlin

Sklar LA and Omann GM (1990) Kinetics and amplification in neutrophil activation and adaptation. *Semin. Cell Biol.* **1** 115–123

Slack JMW (1987) Morphogenetic gradients - past and present. *Trend. Biol. Sci. (TIBS)* **12** 200–205

Small JV, Rinnerthaler G and Hinssen H (1982) Organization of actin meshworks in cultured cells: the leading edge. *Spring Harb. Symp. Quant. Biol.* **46** 599–611

Small V (1994a) Introduction: actin and cell crawling. *Seminars in Cell Biology* **5** 137–138

Small V (1994b) Lamellipodia architecture: actin filament turnover and the lateral flow of actin filaments during motility. *Seminars in Cell Biology* **5** 157–163

Smith DA and Saldana R (1992) Model of the Ca^{2+} oscillator for shuttle streaming in *Physarum polycephalum*. *Biophys. J.* **61** 368–380

Smolyaninov VV (1980) *Mathematical Models of Tissues*. Nauka, Moscow. (In Russian)

Smolyaninov VV and Bliokh ZL (1976) Mechanics of fibroblast movement in vitro. In: *Nonmuscle Forms of Motility*, ed. GM Frank, pp. 5–31. Biol. Res. Centre USSR Acad. Sci. Press, Pushchino. (In Russian)

Soll DR, Voss E, Varnum-Finney B and Wessels D (1988) The "dynamic morphology system": A method for quantitating changes in shape, pseudopod formation and motion in normal and mutant amoebae of *Dictyostelium discoideum*. *J. Cell Biochem.* **37** 177–192

Soll DR, Wessels D and Sylwester A (1993) The motile behavior of amoebae in the aggregation wave in *Dictyostelium discoideum*. In: *Experimental and Theoretical Advances in Biological Pattern Formation*, eds. HG Othmer, PK Maini and JD Murray. Plenum, London

Solyanik GI, Bulkiewicz RI and Kulik GI (1995) One of the mechanisms of the metastatic cells dominance in heterogeneous tumor. *Exp. Oncol.* **17** 158

Spooner BS (1973) Microfilaments, cell shape changes, and morphogenesis of salivary epithelium. *Amer. Zool.* **13** 1007–1022

Spooner BS and Wessells NK (1972) An analysis of salivary gland morphogenesis: role of cytoplasmic microfilaments and microtubules. *Dev. Biol.* **27** 38–54

Stavans J, Domany E and Mukamel D (1991) Universality and pattern selection in two dimensional cellular structures. *Europhys. Letters* **15** 479–484

Steeg PS, Alley MC and Grever MR (1994) An added dimension: will three-dimensional cultures improve our understanding of drug resistance? *J. Natl. Cancer Inst.* **86** 953–955

Stein AA (1995) Deformation of a rod of growing biological material under longitudinal compression. *Prikl. Mat. Mekh. (J. Appl. Maths Mechs)* **59** 149–157

Stein AA (1996a) A mathematical model of file plant tissue at the stage of primary growth. *Biofizika (Biophysics)* In print

Stein AA (1996b) The interaction of a root growing vertically with a rigid barrier. *Biofizika (Biophysics)* **41** N5 1097–1101

Steinbock O and Müller SC (1995) Spatial attractors in aggregation patterns of *Dictyostelium discoideum*. *Z. Naturforschung* **50** 275–281

Stern CD (1984) A simple model for early morphogenesis. *J. Theor. Biol.* **107** 229–242

Stevens A (1992) *Mathematical Modeling and Simulations of the Aggregation of Myxobacteria. Chemotaxis-Equations as Limit Dynamics of Moderately Interacting Stochastic Processes*. Ph.D. thesis, Universität Heidelberg

Stockem W and Brix K (1994) Analysis of microfilament organization and contractile activities in *Physarum. Int. Rev. Cytol.* **149** 145–215

Stocum L (1995) *Wound Repair, Regeneration and Artificial Tissues*. R. G. Landes Co., Austin

Stossel TP (1994) The machinery of blood cell movements. *Blood* **84** 367–379

Stossel TP, Chaponnier C, Ezzel RM, Hartwig JH and Janmey PA (1985) Nonmuscle actin–binding proteins. *Ann. Rev. Cell Biol.* **1** 353–402

Strohmeier R and Bereiter-Hahn J (1987) Hydrostatic pressure in epidermal cells is dependent on Ca-mediated contractions. *J. Cell Sci.* **88** 631–640

Sussman MM (1987) Cultivation and synchronous morphogenesis of *Dictyostelium* under controlled experimental conditions. *Meth. Cell Biol.* **28** 9–29

Svetina S and Žekš B (1989) Membrane bending energy and shape determination of phospholipid vesicles and red blood cells. *Eur. Biophys. J.* **17** 101–111

Svetina S and Žekš B (1991) Mechanical behavior of closed lamellar membranes as a possible common mechanism for the establishment of developmental shapes. *Int. J. Develop. Biol.* **35** 359–365

Svetina S and Žekš B (1992) The elastic deformability of closed multilayered membranes is the same as that of a bilayer membrane. *Eur. Biophys. J.* **21** 251–255

Swanson JA and Taylor DL (1982) Locally and spatially coordinated movements in *Dictyostelium discoideum* amoebae during chemotaxis. *Cell* **28** 225–232

Symons MH and Mitchison TJ (1991) Control of actin polymerisation in live and permeabilized fibroblasts. *J. Cell Biol.* **114** 503–513

Szilard R (1974) *Theory and Analysis of Plates. Classical and Numerical Methods*. Prentice-Hall, Engelwood Cliffs NJ

Takahashi Y and Nogawa H (1991) Branching morphogenesis of mouse salivary epithelium in basement membrane-like substratum separated from mesenchyme by the membrane filter. *Development* **111** 327–335

Takeuchi S (1979) Wound healing in the cornea of the chick embryo. IV. Promotion of the migratory activity of isolated corneal epithelium in culture by the application of tension. *Dev. Biol.* **70** 232–240

Tang YH and Othmer HG (1994) A G-protein based model of adaptation in *Dictyostelium discoideum*. *Math. Biosci.* **120** 25–76

Tang YH and Othmer HG (1995) Excitation, oscillations and wave propagation in a G-protein based model of signal transduction in *Dictyostelium discoideum*. *Phil. Trans. R. Soc. Lond.* **349** 179–195

Telley H, Liebling TM and Mocellin A (1995) The Laguerre model of grain growth in two dimensions. I: Cellular structures viewed as dynamical Laguerre tesselations; II: Examples of coarsening simulations. *Phil. Mag. B* **73**

Teplov VA (1988) Autooscillations in *Physarum* plasmodium. Correlation between force generation and viscoelasticity during rhythmical contractions of protoplasmic strand. *Protoplasma, Suppl.* **1** 81–88

Teplov VA (1989) Biomechanics of autooscillatory contraction in the cell. In: *Biomechanics*, ed. A Morecki, Lecture notes of the Biocybernetics seminar, pp. 192–209. ICB Press, Madralin (Warsaw)

Teplov VA and Romanovsky YM (1987) Mechanochemical distributed autooscillations in cell motility. Autowave phenomena in the contractile activity of *Physarum polycephalum* plasmodium. Tech. Rep., Biol. Res. Centre USSR, Acad. Sci. Press, Pushchino. (In Russian)

Teplov VA, Romanovsky YM and Latushkin OA (1991) A continuum model of contraction waves and protoplasmic streaming in strands of *Physarum* plasmodium. *BioSystems* **24** 269–289

Theriot JA and Mitchison TJ (1991) Actin microfilament dynamics in locomoting cells. *Nature* **352** 126–131

Theriot JA, Rosenblatt J, Portnoy DA, Goldschmidt-Tilney LG, DeRosier DJ and Tilney MS (1992) How *Listeria* exploits host cell actin to form its own cytoskeleton. I. Formation of a tail and how that tail might be involved in movement. *J. Cell Biol.* **118** 71–81

Thimann KV and Schneider CL (1938) Differential growth in plant tissue. *Amer. J. Bot.* **25** 627–641

Thompson D'Arcy W (1942) *On Growth and Form*. Cambridge University Press, Cambridge

Tilney LG and Portnoy DA (1989) Actin filaments and the growth, movement, and spread of the intracellular bacterial parasite *Listeria monocytogenes*. *J. Cell. Biol.* **109** 1597–1608

Timoshenko S (1934) *Theory of Elasticity*. McGraw-Hill, New York

Tomchik KJ and Devreotes PN (1981) Adenosine 3',5'-monophosphate waves in *Dictyostelium discoideum*: a demonstration by isotope dilution-fluorography. *Science* **212** 443–446

Tranquillo RT (1990) Theories and models of gradient perception. In: *Biology of the Chemotactic Response*, eds. JP Armitage and JM Lackie, pp. 35–75. Cambridge University Press, Cambridge, UK

Tranquillo RT and Alt W (1990) Glossary of terms concerning oriented movement. In: *Biological Motion*, eds. W Alt and G Hoffman, Lecture Notes in Biomath., p. 510. Springer, Berlin

Tranquillo RT and Alt W (1994) Dynamic morphology of leukocytes: statistical analysis and a stochastic model for receptor-mediated cell motion and orientation. In: *Biomechanics of Active Motion and Division of Cells*, ed. N Akkas, pp. 437–443. Springer, New York

Tranquillo RT and Alt W (1996) Stochastic model of receptor-mediated cytomechanics and dynamic mophology of leukocytes. *J. Math. Biol.* **34** 361–412

Tranquillo RT and Alt W (1997) Simulation of chemotactic receptor-mediated cytomechanics of leukocytes: random and chemotactic movement. In preparation

Tranquillo RT, Durrani MA and Moon AG (1992) Tissue engineering science: consequences of cell traction force. *Cytotechnol.* **10** 225–250

Tranquillo RT and Lauffenburger DA (1987) Stochastic model of leukocyte chemosensory movement. *J. Math. Biol.* **25** 229–262

Tranquillo RT and Murray JD (1992) Continuum model of fibroblast-driven wound contraction: inflammation-mediation. *J. Theor. Biol.* **158** 135–172

Tranquillo T and Lauffenburger DA (1986) Consequences of chemosensory phenomena for leukocyte chemotactic orientation. *Cell Biophys.* **8** 1–46

Tranquillo T and Lauffenburger DA (1987) Stochastic model of leukocyte chemosensory movement. *J. Math. Biol.* **25** 229–262

Tsubaki Y (1981) Some beneficial effects of aggregation in young larvae of *Pryeria sinica* Moore (Lepidoptera: Zygaenidae). *Res. Popul. Ecol.* **23** 156–167

Tsubaki Y and Shiotsu Y (1982) Group feeding as a strategy for exploiting food sources in the Burnet moth *Pryeria sinica. Oecologia (Berlin)* **55** 12–20

Tyson JJ, Alexander KA, Manoranjan VS and Murray JD (1989) Spiral waves of cyclic AMP in a model of slime mold aggregation. *Physica D* **34** 193–207

Tyson JJ and Murray JD (1989) Cyclid AMP waves during aggregation of *Dictyostelium* amoebae. *Development* **106** 421–426

Usami S, Wung SL, Skierczynski BA, Skalak R and Chien S (1992) Locomotion forces generated by a polymorphonuclear leukocyte. *Biophys. J.* **63** 1663–1666

Vaishnav RN and Vossoughi J (1987) Residual stress and strain in aortic segments. *J. Biomechan.* **20** 235–239

Van der Berg C, Willemsen V, Hage W, Weisbeek P and Scheres B (1995) Cell fate in *Arabidopsis* root meristem determined by directional signalling. *Nature* **378** 62–65

Van Haastert PJ, Wang M, Bominaar AA, Devreotes PN and Schaap P (1992) CAMP-induced desensitization of surface cAMP receptors in *Dictyostelium*: different second messengers mediate receptor phosphorylation, loss of ligand binding, degradation of receptor, and reduction of receptor mRNA levels. *Mol. Biol. Cell.* **3** 603–612

Van Oss C, Panfilov AV, Hogeweg P, Siegert F and Weijer CJ (1996) Spatial pattern formation during aggregation of the slime mould *Dictyostelium discoideum. J. Theor. Biol.* **181** 203–213

Varnum-Finney BJ, Voss E and Soll DR (1987) Frequency and orientation of pseudopod formation of *Dictyostelium discoideum* amoebae chemotaxing in a spatial gradient: Further evidence for a temporal mechanism. *J. Cell Motil. Cytoskeleton* **8** 18–26

Vasiev BN, Hogeweg P and Panfilov AV (1994) Simulation of *Dictyostelium discoideum* aggregation via reaction-diffusion model. *Phys. Rev. Lett.* **73** 3173–3176

Vereycken V, Gruler H, Bucherer C, Lacombe C and Lelivre JC (1995) The linear motor in the human neutrophil migration. *J. Phys. III France* **5** 1469–1480

Vicker MG (1981) Ideal and non-ideal concentration gradient propagation in chemotaxis studies. *Exp. Cell Res.* **136** 91–100

Vicker MG (1989) Gradient and temporal signal perception in chemotaxis. *J. Cell Sci.* **92** 1–4

Vicker MG (1994) The regulation of chemotaxis and chemokinesis in *Dictyostelium* amoebae by spatial gradient and temporal signal fields of cyclic-AMP. *J. Cell Sci.* **107** 659–667

Vicker MG, Schill W and Drescher K (1984) Chemoattraction and chemotaxis in *Dictyostelium discoideum*: myxamoebae cannot read spatial gradients of cyclic adenosine monophosphate. *J. Cell Biol.* **98** 2204–2214

Vicker MG, Xiang W, Plath P and Wosniok W (1996) Autowaves of F-actin assembly determine pseudopodium extension and cell locomotion in *Dictyostelium discoideum*. Physica D. In press

Vicsek T, Czirók A, Ben-Jacob E, Cohen I and Shochet O (1995) Novel type of phase transition in a system of self-driven particles. *Phys. Rev. Lett.* **75** 1226

Vincent JFV and Jeronimidis G (1991) The mechanical design of fossil plants. In: *Biomechanics and Evolution*, eds. JMV Rayner and RJ Wooton, pp. 179–194. Cambridge University Press

Wachsstock DH, Schwarz WH and Pollard TD (1994) Cross-linker dynamics determine the mechanical properties of actin gels. *Biophys. J.* **66** 801–809

Waddington C (1940) *Organisers and Genes*. Cambridge University Press, Cambridge

Wanek N, Marcum BA, Lee HT and Campbell RD (1980) Effect of hydrostatic pressure on morphogenesis in nerve-free hydra. *J. Exp. Zool.* **211** 275–280

Wang YL (1985) Exchange of subunits at the leading edge of living fibroblasts: possible role of treadmilling. *J. Cell Biol.* **101** 597–602

Wang YL, Lanni F, Mcneil PL, Ware BR and Taylor DL (1982) Mobility of cytoplasmic and membrane associated actin in living cells. *Proc. Nat. Acad. Sci. USA* **79** 4660–4664

Wang YL and Taylor DL (1979) Distribution of fluorescently labeled actin in living sea urchin eggs during early development. *J. Cell Biol.* **82** 672–679

Wareing P and Phillips I (1981) *Growth and differentiation in plants*. Pergamon Press, Oxford, 3rd edn.

Warrick H and Spudich JA (1987) Myosin structure and function in cell motility. *Ann. Rev. Cell Biol.* **3** 379–421

Weaire D and Rivier N (1984) Soap cells and statistics - random patterns in two dimensions. *Contemp. Phys.* **25** 59–99

Weber I, Wallraff E, Albrecht R and Gerisch G (1995) Motility and substratum adhesion of *Dictyostelium* wild-type and cytoskeletal mutant cells: a study by RICM/brightfield double-view image analysis. *J. Cell Sci.* **108** 1519–1530

Weibel ER (1963) *Morphometry of the Human Lung*. Springer, Berlin

Weibel ER (1991) Fractal geometry: a design principle for living organisms. *Am. J. Physiol.* **261** 361–369

Weiss P (1939) *Principles of Development*. Holt, New York

Weliky M, Minusuk S, Keller R and Oster GF (1991) Notochord morphogenesis in *Xenopus laevis*: simulation of cell behavior underlying tissue convergence and extension. *Development* **113** 1231–1244

Weliky M and Oster GF (1990) The mechanical basis of cell rearrangement. I. Epithelial morphogenesis during *Fundulus* epiboly. *Development* **109** 373–386

Welsh B, Gomatam J and Burgess A (1983) Three-dimensional chemical waves in the Belousov-Zhabotinsky reaction. *Nature* **304** 611–614

Wessels D, Soll DR, Knecht D, Loomis WF, De Lozanne A and Spudich J (1988) Cell motility and chemotaxis in *Dictyostelium* amoebae lacking native myosin heavy chains. *Develop. Biol.* **128** 164–177

White JG (1990) Laterally mobile, cortical tension elements can self-assemble into a contractile ring. In: *Cytokinesis: Mechanics of Furrow Formation during Cleavage Division*, Vol. 582, pp. 50–59. Ann. N. Y. Acad. Sci. (1990)

White JG and Borisy GG (1983) On the mechanism of cytokinesis in animal cells. *J. Theor. Biol.* **101** 289–316

Wilkinson PC (1988) Chemotaxis and chemokinesis: confusion about definitions. *J. Immunol. Methods* **110** 143–149

Winfree AT (1973) Scroll-shaped waves of chemical activity in three dimensions. *Science* **181** 937–939

Winfree AT (1974) Introduction in mathematical problems in biology. In: *Mathematical Problems in Biology*, ed. P van den Driessche *Lecture Notes in Biomath.*, Vol. 2, pp. 243–260. Springer, Berlin

Winfree AT (1990) *The Geometry of Biological Time*. Springer, Berlin

Winfree AT and Strogatz SH (1984) Organizing centers for three dimensional chemical waves. *Nature* **311** 611–615

Winklbauer R and Selchow A (1992) Motile behavior and protrusive activity of migratory mesoderm cells from the *Xenopus* gastrula. *Dev. Biol.* **150** 335–351

Winklbauer R, Selchow A, Nagel M and Angres B (1992) Cell interaction and its role in mesoderm cell migration during Xenopus gastrulation. *Dev. Dyn.* **195** 290–302

Witke W, Schleicher M and Noegel AA (1992) Redundancy in the microfilament system: abnormal development of *Dictyostelium* cells lacking two F-actin cross-linking proteins. *Cell* **68** 53–62

Wohlfarth-Bottermann KE (1979) Oscillatory contractile activity in *Physarum*. *J. Exp. Biol.* **81** 15–32

Wolpert L (1969) Positional information and the spatial pattern of cellular differentiation. *J. Theor. Biol.* **25** 1–47

Wosniok W (1987) Estimation of individual growth curves from aggregation data. In: *Advances in System Analysis: Erwin-Riesch Workshop - System Analysis of Biological Processes*, ed. DPF Möller, pp. 133–139. Vieweg, Braunschweig

Yoshimoto Y and Kamiya N (1978) Studies on contraction rhythm of the plasmodial strand. *Protoplasma* **95** 89–133

Yoshimoto Y and Kamiya N (1982) Ca^{2+} oscillation in the homogenate of *Physarum* plasmodium. *Protoplasma* **110** 63–65

Yoshimoto Y and Kamiya N (1984) ATP- and calcium controlled contraction in a saponin model of *Physarum polycephalum*. *Cell Struct. Funct.* **9** 135–141

Zhu QL and Clark M (1992) Association of calmodulin and an unconventional myosin with the contractile vacuole complex of *Dictyostelium discoideum*. *J. Cell Biol.* **118** 347–358

Zhukarev V, Ashton F, Sanger JM, Sanger JW and Shuman H (1995) Organization and structure of actin filament bundles in *Listeria*-infected cells. *Cell Motil. Cytoskel.* **30** 229–246

Zieschang HE (1992) *Wachstumsrelevante Parameter während der Graviresponse von Primärwurzeln*. Doctoral thesis, Botanical Institute, Bonn

Zieschang HE, Brain P and Barlow PW (1997) Modelling of root growth in two dimensions. *J. Theor. Biol.* **184** 237–246

Zigmond SH (1980) Polymorphonuclear leukocyte chemotaxis: detection of the gradient and development of cell polarity. *Excerpta Medica* **41** 229

Zigmond SH (1993) Recent quantitative studies of actin filament turnover during cell locomotion. *Cell Motil. Cytoskel.* **25** 309–316

Zigmond SH and Sullivan SJ (1979) Sensory adaptation of leukocytes to chemotactic peptides. *J. Cell Biol.* **82** 517–527

Participants in the Workshop "Cell and Tissue Motion", Bonn-Röttgen, March 19–24, 1995

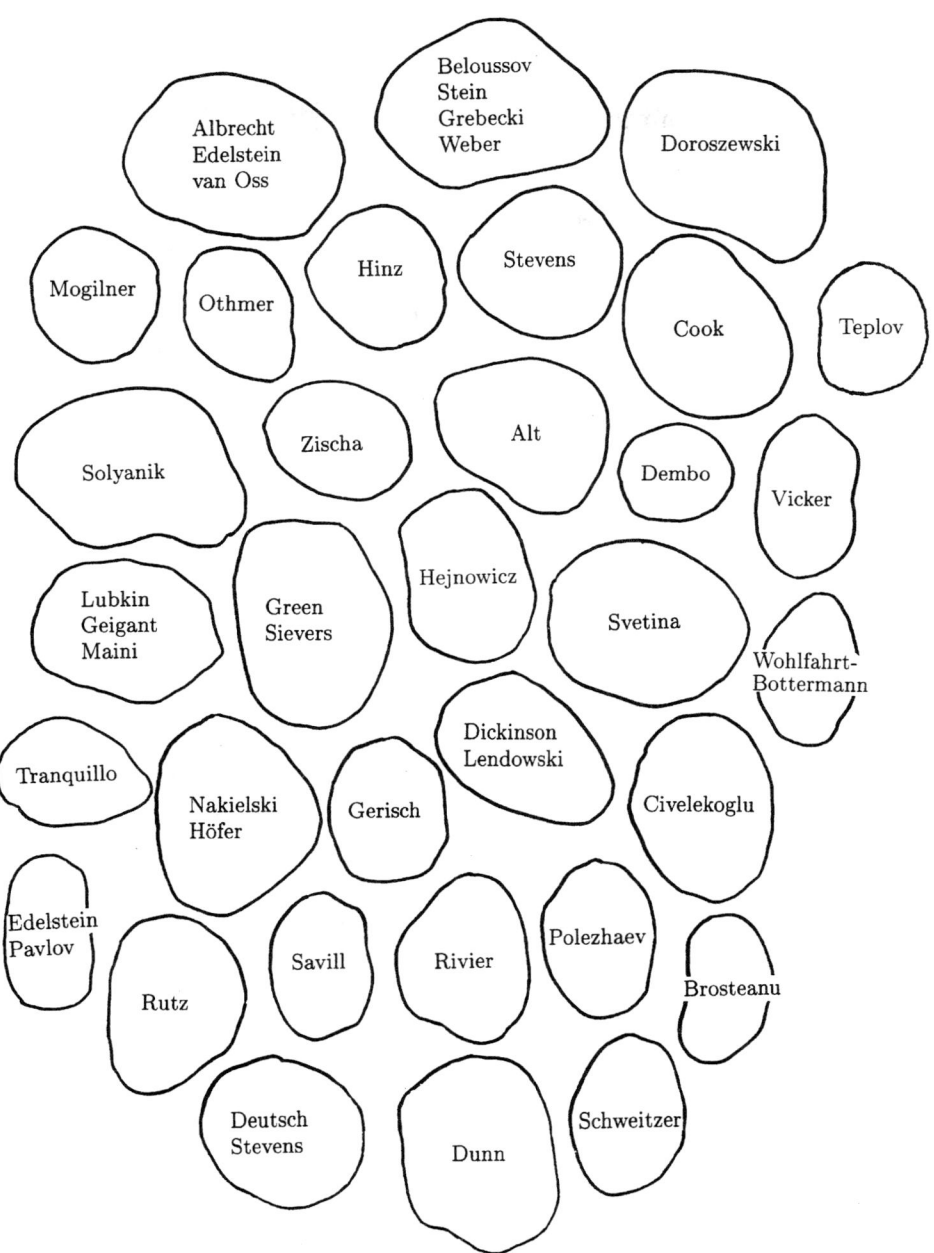

Addresses

Book authors and participants in the SFB-Workshop "Cell and Tissue Motion", Bonn-Röttgen, March 19–24, 1995

Richard Albrecht
albrecht@vms.biochem.mpg.de
Abt. Zellbiologie, MPI für Biochemie
Am Klopferspitz 18a
D-82152 Martinsried, Germany

Wolfgang Alt
wolf.alt@uni-bonn.de
Abt. Theor. Biologie, Univ. Bonn
Kirschallee 1
D-53115 Bonn, Germany

Lev V. Beloussov
lbelous@ecol.msu.ru
Dept. of Embryology, Faculty of Biology
Lomonosov State University
RUS-119899 Moskwa, Russia

Jürgen Bereiter-Hahn
bereiter-hahn
@zoology.uni-frankfurt.de
AK Kinematische Zellforschung
FB Biologie, Univ. Frankfurt
Marie-Curie-Str. 9
D-60439 Frankfurt, Germany

Philip Brain
phil.brain@bbsrc.ac.uk
IARC Long Ashton Research Station
Dept. of Agric. Sci., Univ. Bristol
GB-BS18 9AF Bristol, Great Britain

Oana Brosteanu
oana@imise.uni-leipzig.de
Inst. Med. Informatik, Statistik u.
Epidemiologie (IMISE),
UNIV. lEIPZIG
Liebigstr. 27
D-04103 Leipzig, Germany

Gul Civelekoglu
gulc@prizma.net.tr eldpa.epfl.ch
Lab. de Génie Médical,
EPFL – Ecublens
CH-1015 Lausanne, Helvetia

Julian Cook
Biomathematics Department
UCLA School of Medicine
10833 Le Conte Avenue
Los Angeles, CA 90095-1766, U.S.A.

Micah Dembo
mxd@bu.edu
Biomed. Eng., Boston University
44 Cummington St.
Boston, MA 02215, U.S.A.

Andreas Deutsch
deutsch@io.bota.uni-bonn.de
Abt. Theoretische Biologie, Univ. Bonn
Kirschallee 1
D-53115 Bonn, Germany

Richard B. Dickinson
dickinso@che.ufl.edu
Dept. of Chemical Engineering
University of Florida
Gainesville, FL 32611-6005, U.S.A.

Jan Doroszewski
cmkp1@warman.com.pl
Biophysics and Biomath.,
Medical Center of Postgraduate Education
ul. Marymoncka, 99
PL-01-813 Warszawa, Poland

Graham A. Dunn
gad@helios.rai.kcl.ac.uk
M.R.C. Muscle& Cell Motility Unit
The Randall Institute
King's College London
26-29 Drury Lane
GB-WC2B 5RL London, U.K.

Edith Geigant
geigant@sn-bota-1.bota.uni-bonn.de
Abt. Theoretische Biologie, Univ. Bonn
Kirschallee 1
D-53115 Bonn, Germany

Günther Gerisch
gerisch@vms.biochem.mpg.de
Abt. Zellbiologie, MPI für Biochemie
D-82152 Martinsried, Germany

Andrej Grebecki
grebecki@nencki.gov.pl
Cell Biology
Nencki Institute for Experimental Biology
ul. Pasteura 3
PL-02093 Warszawa, Poland

Paul Green
pbgreen@popserver3.stanford.edu
Dept. of Biological Sciences
Stanford University
Stanford, CA 94305, U.S.A.

Zygmunt Hejnowicz
khejn@usctoux1.cto.us.edu.pl
Dept. of Biophysics and Cell Biology
Silesian University
ul. Jagiellonska 28
PL-40-032 Katowice, Poland

Boris Hinz
hinz@sn-bota-1.bota.uni-bonn.de
Abt. Theoretische Biologie, Univ. Bonn
Kirschallee 1
D-53115 Bonn, Germany

Thomas Höfer
idefix@mpipks-dresden.mpg.de
MPI für Physik komplexer Systeme
Bayreuther Str. 40
D-01187 Dresden, Germany

Hans Wilhelm Kaiser
hwkaiser@mailer.meb.uni-bonn.de
Dermatologie, Universitäts-Hautklinik
Sigmund-Freud-Str. 25
D-53127 Bonn, Germany

Leah Keshet
keshet@math.ubc.ca
Dept. of Mathematics
University of British Columbia
Mathematics Road
Vancouver, B.C. V6T 1Z2, Canada

Volker Lendowski
lendowsk@io.bota.uni-bonn.de
Abt. Theoretische Biologie, Univ. Bonn
Kirschallee 1
D-53115 Bonn, Germany

Sharon Lubkin
lubkin@amath.washington.edu
Dept. Applied Mathematics
University of Washington
Seattle, WA 98195 , U.S.A.

Philip Maini
maini@maths.ox.ac.uk
Centre for Mathematical Biology
Mathematical Institute
24-29 St. Giles
GB-OX1 3LB Oxford, U.K.

Alex Mogilner
mogilner@ucdmath.ucdavis.edu
Dept. of Mathematics
University of California
Davis, CA 95616, U.S.A.

Jerzy Nakielski
nakiel@usctoux1.cto.us.edu.pl
Dept. of Biophysics and Cell Biology
Silesian University
ul. Jagiellonska 28
PL-40-032 Katowice, Poland

Hans Othmer
othmer@math.utah.edu
Dept. of Mathematics, University of Utah
102 Widstoe Building
Salt Lake City, UT 84112, U.S.A.

Dmitri A. Pavlov
dapavlov@cs.msu.su
Faculty of Computational
Mathematics and Cybernetics
Lomonosov State University
RUS-119899 Moskwa, Russia

Beate Pfistner
b.pfistner@uni-bonn.de
Abt. Theoretische Biologie, Univ. Bonn
Kirschallee 1
D-53115 Bonn, Germany

Andrey A. Polezhaev
apol@lpi.ac.ru
Lebedev Physical Institute
Russian Academy of Sciences
Leninskiy prosp. 53
RUS-117924 Moskwa, Russia

Nicolas Rivier
nick@fresnel.u-strasbg.fr
Laborat. Physique Théorique
Université Louis Pasteur
3, rue de l'Université
F-67084 Strasbourg, France

Mechthild Rutz
thhesse@x4u2.desy.de
Abt. Theoretische Biologie, Univ. Bonn
Kirschallee 1
D-53115 Bonn, Germany

Nicholas Savill
njs@behold.biol.ruu.nl
Dept. Theor. Biol. Bioinform
Univ. Utrecht
Padualaan 8
NL-3584 CH Utrecht, The Netherlands

Frank Schweitzer
frank@summa.physik.hu-berlin.de
Institute of Physics
Humboldt University
D-10099 Berlin, Germany

Andreas Sievers
Abt. Zellbiologie, Botanisches Institut
Universität Bonn
Venusbergweg 22
D-53115 Bonn, Germany

Galina I. Solyanik
farmm@iepor.kiev.ua
Dept. Pharmacodynamics
Ukrainian Academy of Sciences
Vasilkovskaya ul. 45
UKR-252022 Kiev, Ukraina

Aleksandr A. Stein
stein@inmech.msu.su
Institute of Mechanics
Lomonosov State University
Michurinsky prosp. 1
RUS-117192 Moskwa, Russia

Angela Stevens
stevens@iwr.uni-heidelberg.de
Institut für Angewandte Mathematik
Universität Heidelberg
Im Neuenheimer Feld 294
D-69120 Heidelberg, Germany

Sasha Svetina
sasa@biofiz.mf.uni-lj.si
Institute of Biophysics
Medical Faculty
University of Ljubljana
Lipiceva 2
SI-61105 Ljubljana, Slovenia

Vladimir A. Teplov
teplov@venus.iteb.serpukhov.su
Inst. of Theoretical and
Experimental Biophysics
Russian Academy of Sciences
RUS-142292 Pushchino, Russia

Robert T. Tranquillo
tranquillo@cems.umn.edu
Dept. of Chem. Engineering
and Materials Science
University of Minnesota
421 Washington Avenue SE
Minneapolis, MN 55455, U.S.A.

Catelijne Van Oss
cvo@binf.biol.ruu.nl
Dept. of Theor. Biol. Bioinform.
Univ. Utrecht
Padualaan 8
NL-3584 Utrecht, The Netherlands

Pavel Vesely
pvy@img.cas.cz
Inst. of Molecular Genetics
Acad. of Science
Flemingovo nam 2
CS-16637 Prague 6 – Dejvice
Czech Republic

Michael G. Vicker
vicker@zfn.uni-bremen.de
Fachbereich Biologie, Universität Bremen
Postfach 33 04 40
D-28334 Bremen, Germany

Igor Weber
iweber@biochem.mpg.de
Abt. Zellbiologie, MPI für Biochemie
Am Klopferspitz 18a
D-82152 Martinsried, Germany

Rudolf Winkelbauer
agwinkel@biolan.uni-koeln.de
Zoologisches Institut, Universität Köln
Weyertal 119
D-50931 Köln, Germany

Karl-Ernst Wohlfarth-Bottermann
Lotharstraße 113
D-53115 Bonn, Germany

Daniel Zicha
dan@helios.rai.kcl.ac.uk
M.R.C. Cell Biophysics Unit
26-29 Drury Lane
GB-WC2B 2RL London, U.K.

Hanna Zieschang
reitz.zieschang@t-online.de
Botanisches Institut, Univ. Bonn
Kirschallee 1
D-53115 Bonn, Germany

Index

Aboav's law 279
actin 10, 20, 117
   -assembly 20
   -based motility 141
   -binding protein (ABP) 20, 22, 46, 71, 94, 99, 102
   cortex 10, 20, 147
   F- (filamentous) 11, 20, 25, 73, 89, 91, 101, 139
   flow 20
   G- (globular) 28, 76, 91, 94, 101
   motor 112
   -myosin 10, 22, 55, 68, 71, 84, 102
      complexes 86
      network 112, 115
   network 20, 94, 96
   polymerization 76, 284
   waves 28
active pressure component 89
activin 218
adaptation 115, 142, 143, 172
adenylate cyclase 194, 202
adhesion 7, 34, 47, 73, 117, 122, 167, 203
   -dependent motor functions 118
   differential 207
   dynamics 147
   focal 36
   friction 74
   molecules 122, 149
   substrate 8
adhesivity 209
aggregation 171, 183, 187, 193, 200, 209
Airy stress function 252
alignment 149
α-actinin 22, 94, 102
Amoeba proteus 80, 115, 117
angiogenesis 159
anisotropic

biphasic theory 160
   cell traction 162
   diffusion model 116
approximation
   closure 177
   continuity 188
   diffusion 168, 188
   finite-difference 200
   Hele-Shaw 232
   mean-field 180
   pseudo-steady state 144
assembly
   F-actin 111
   network 59
   self- 6
asters 62
attachment zones 120
attractants 165
auto-oscillatory processes 83
autocorrelation 44
   analysis 15, 16, 32
   function 31, 158
autowave 21, 28, 80, 83–85, 139
   excitation 86
   pattern 92
auxin 218
AX-2 cells 22

bacterium 93
Beloussov-Zhabotinsky (BZ) reaction 4, 27, 133
bending
   fluctuation 95
   plant roots 255
bifurcation 176
   analysis 107, 189
   diagram 107
   Hopf 80, 89
biological time 139

blastocoel 221, 227
blastoderm 223
blastomere 215
blastula 227
blebs 73
blow-up 172, 189
Boltzmann
    distribution 278
    equation 180
boundary conditions
    free moving 92
    Neumann 97
    no-slip 97
brightness correction 48
Brownian
    motion 95
    particle 189
buckling 214, 243, 277
budding 282
    of membranes 112

caffeine 134, 202
calcium 22, 28, 80
    control 86
    influx 194
    oscillation 87, 91
calmodulin 85
cAMP 22, 85, 133–135, 172, 193, 207
    internal 198
    oscillator 28
    wave 198
cancer
    cells 165
    growth 210
    -ogenic 20
cardiocytes 28
cell
    aggregation 133, 171
    alignment 174
    area opaca 223
    AX-2 22
    cancer 165
    -cell communication 209
    -cell contact 171, 225

-cell interaction 165, 171, 203, 209
cleavage 70
clustering 193
cortex 55
diffusion 200
dispersal 204
division 55, 204, 268, 275
embryonic mesoderm 7
equator 55
eucaryotic 101
geometry 213
growth 204
locomotion 3, 21, 33, 123
L1210 47, 53, 54, 119
mesoderm 8
migration 149, 213
morphology 167
motility 38, 135
movement 157, 159, 194
orientation 151
path 141, 149
pattern 267, 268
polarization 195
prespore 208
prestalk 208
programming language 203
shape 7, 29, 67, 144, 219, 270
sorting 172, 203, 204
streaming 172, 194
-substratum interaction 117
surface complexity 49
translocation 30
velocity 150
Walker carcinosarcoma 42
wall 213, 242, 255, 268
    elongation 214
XTH-2 7
cellular automaton 171, 174, 203
    lattice-gas 172, 179
centroid 29, 125
cGMP pathway 196
Chaos carolinensis 20
Chapman-Kolmogorov equation 157
chemical

regulation 262
signals 173
chemoattractant 22, 149, 172, 182, 206
waves 193
chemokinesis 133
chemotactic 182, 198
coefficient 200
factor (CF) 141
gradient 134
response 200
signal 136
chemotaxis 116, 133, 141, 154, 163, 168, 183, 193, 198, 205
-driven instability 202
equation 188
system 183
chick heart fibroblast 37
circadian rhythm 28
circular
maps 24
travelling waves 80
cleavage 55, 61, 63
furrow 61
CME (condition of a mechanical equilibrium) 222
cohesive forces 59
colcemid 10, 38
colchicine 43
collagen 160, 167
gels 149
collagenase 173
collapse 172
collenchyma 236
competition process 191
complexity scale 129
compression 98, 245, 251, 262
compressive stress 235
confocal laser scan microscopy 12
constriction force 70
contact
guidance 149, 159, 168, 209
model 115
inhibition 231
of motion 165

continuity approximation 188
contractile
cortex 73, 115
cytogel 231
cytoskeletal filaments 55
forces 56
ring 56, 71
stress 59
contractility 20, 57, 80
contraction 34, 59, 73, 121
auto-oscillatory 6
force 163
peristaltic 83
-relaxation model 90
coronin 22
correlation
coefficient 44
angular-temporal 111
matrix 31
topological 282
cortex 10
actin 147
contraction 73, 115
layer 55, 73
tension 80
crawling 7, 33
cross-correlation
function 43
structure 34
cross-linking proteins 20, 59, 102
cucumber 280
curvature 277
-dependent motion 232
local 260
cytochalasin B 230
cytokinesis 5, 55, 56, 60, 61, 63, 67, 70
cytomechanical model 143, 147
cytoplasmic flux 7
cytoskeletal viscosity 59
cytoskeleton 34, 55, 80, 101, 117, 284
cytosol flow 76, 80

Dictyostelium 4, 21, 33, 39, 80, 115, 119, 133, 137, 172, 193, 209

discoideum 205
   lacteum 194
   minutum 194
Darcy's law 56, 58, 85, 232
deadhesion 34
death of cells 206
deconvolution method 48
deformation 284
   cyclic mechanical 231
   field 124
degradation 160, 200
delocalization rule 215
Dendroctomus micans 183
depolymerization 20, 56, 94
   rate 95
descriptor analysis 47
desensitization 194
determination 215
development 213, 233, 243
dichotomous branching 231
DIC (differential interference contrast)
   microscopy 33
differential
   adhesion 207
      energy 231
   growth 263, 270
   interference contrast microscopy 33
differentiation 219, 283
diffusion 171, 184, 187, 195, 257, 282
   approximation 168, 188
   coefficient 149
   equation 150
   gradient 218
   limited growth 112, 205
   rotational 104, 175
   tensor 150, 157
dimensional analysis 129
Dionaea 242
directional persistence 195
   tensor 153
   time 153
disassembly 34, 59, 73
disease
   fibrocontractive 165

fibromatoses 162
   fibroplasia 159
disorder 275
division 203
DNA 168
   synthesis 21
drag 71, 74
drift
   correction 127
   vector 157
   velocity, 150
DRIMAPS (digitally recorded interference microscopy with automatic phase-shifting) 33
dynamic reciprocity 168
dynamics
   angular-temporal 32
   F-actin 111, 145
   temporal 173

echinoderm egg 55
ECM (extracellular matrix) 20, 159, 168, 213
ecological applications 209
ectoplasm 83
efficiency
   of cell locomotion 42
   tracks 45
egg maturation 223
elastic
   cushion 246
   energy 67
   equilibrium 245
   extension 236
   foundation 214, 251
   medium 279
   solid 251
   stretching 255
   substrata 123
elasticity 214
   function 89
   moduli 236
electrostatic forces 121
embryology 171, 215

embryonic
  mesoderm cells 7
  regulation 217
endocytosis 35, 142
endoplasm 83
  flow 85
energy
  bonds 203
  bending 67, 69, 231
  differential-adhesion 231
  elastic 67
  potential 70
  topological 279
engulfment 204
enterocytes 93
enthalpy 69
epiboly 230
epidermis 236
epigenetical landscape 216
epithelium 214, 229, 243, 251
  models 230
  tissues 275
equation(s)
  Boltzmann 180
  Chapman-Kolmogorov 157
  chemotaxis 188
  diffusion 150
  Fitzhugh Nagumo 207
  Fokker-Planck 95
  integro
    -differential 94, 104
    -partial differential 171, 173, 209
  Langevin 189
  Laplace's 232
  Martiel-Goldbeter (MG) 197
  mass transport 75
  Navier-Stokes 232
  partial integro-differential 104
  reaction-diffusion-advection 172, 176
  rheological 164
  selection (of Eigen-Fisher type) 191
  transport-diffusion 109
  Volterra integral 259
  von Kármán 252

equilibrium
  elastic 245
  mechanical 222
Escherichia coli 134
eucaryotic cell 101
excitability 172, 202, 230
excitable media 194, 207
excitatory waves 81
expansion 251
extension 9, 77, 143, 147, 226
external stress field 112
extracellular
  matrix (ECM) 20, 159, 168, 213
  proteins 147
  messenger 193

F-actin 11, 20, 22, 25, 73, 89, 91, 101, 139
  assembly 111
  dynamics 111, 145
  turnover 22
fascin 102
fertilization 223
fetal breathing 229
fiber orientation tensor 160
fibrin 159
fibroblast 26, 33, 39, 77, 93, 118, 132, 139, 149, 159, 173, 179
fibrocontractive disease 165
fibromatoses 162
fibroplasia 159
field(s)
  interacting 65
  of mechanical stress (FMS) 218
  stimulus 154
  traction density 125
filament(s)
  anisotropy 57
  intermediate 81
  reversible actin 22
filamin 20, 76, 102
filopodia (-pods) 15, 73, 93
fimbrin 102
fish keratinocytes 42

fitness 191
Fitzhugh Nagumo equations 207
flexural rigidity 246, 251
flow
   cytosol 76, 80
   endoplasmic 85
   network 74
   of subcortical actin 223
   Poiseuille 85
   radial 75
   reactive interpenetrating (RIF) 56, 96
   retrograde actin 81
   tangential 75
   viscous 73
fluid
   incompressible 56
   Newtonian 96
      viscous incompressible 85
   non-Newtonian 56
   reactive interpenetrating 73
   two-phase cytoplasm 74
   viscoelastic 234
   viscous 102, 232
fluorescence microscopy 50
flux
   cytoplasmic 7
   ion 284
foam 284
focal
   adhesion 36
   contacts 8, 101, 118
Fokker-Planck equation 95
force(s)
   attracting 106
   cohesive 59
   constriction 70
   contractile 56, 163
   electrostatic 121
   generation 168
   osmotic 56
   propulsion 95
   pushing 97
   retraction 20
   traction 123
formation
   de novo 243, 245
   furrow 231
   mechanical 215
   organ 171
   osmotic 56
Fourier
   analysis 19, 26, 30
   coefficient 152
   series 152, 177
   transformation 152, 178
frictional resistance 74
frog eggs 28
froth 275
fruiting body 183
functional maturation 56
Fundulus 231
furrow 62
   formation 231

G-actin 28, 76, 91, 94, 101
G-protein 142, 194
   model 195
Galerkin finite element method 61
galvanotaxis 209
gastrocoel 221
gastrula 224
gastrulation 215, 230, 231
Gaussian
   distribution 126
   white noise 189
gel-sol transition 56, 88
gelation 59
gelsolin 20, 22, 101
gene
   activity 215
   expression 284
genotype 244
gibberelline 272
gliding motion 9, 183
global
   directionality 179
   selection pressure 191

gradient
  continuous temporal 147
  diffusion 218
  osmotic 214
  phase 91
  receptor-measured 142
  spatial 115, 133, 139
  steepness 154
  step temporal chemotactic factor (CF) 147
  temporal 115, 142
gravitropic response 242, 262
growth 203, 230, 283
  differential 263, 270
  diffusion limited 112, 205
  factors 164
  field 268
  signals 163
  strategies 206
  tensor 268
  transverse 258
  tumour 166

haptotaxis 115, 154, 209
Hele-Shaw approximation 232
Helianthus 241
helical structure 101
homeostasis 159, 283
Hooke's law 124, 237, 256
Hopf bifurcation 80, 89
humoral immune system 93
hyaline caps 117
hydra 221
hydraulic pressure 10
hydroid polyps 228
hydrostatic pressure 20, 73, 235
hyperbolic
  -elliptic system 96
  mass transport equation 75
hypocotyl 235, 241

image
  analysis 33, 47, 125
  decomposition 51

  enhancement 49
  processing 15
    software 125
immunofluorescence 11
immunofluorescent staining 111
  anti-actin 60
in-plane compression 251
incompressibility 256
incompressible fluid 56
inflammatory reactions 21
inositol-3-phosphate (IP3) 85
insect
  flight muscle 91
  larvae 183
instability
  chemotaxis-driven 202
integro-differential equation 94, 104
  partial 104, 171, 209
integral membrane proteins 68
integrin 20
interacting fields 65
intercalation 222
interference 24
interfilament repulsion 56
intermediate filaments 81
internal
  cAMP 198
  stress 59
intracellular rhythms 28
ion
  channel 284
  flux 284
  pumps 213
isotropy 213
  network 57

jump process 154

Karhunen-Loéve expansion 25, 30
keratinocytes 73, 78, 80
    fish 42
    human epidermal 29
      normal (nHEK) 15
      transfected (trHEK) 15

keratocyte 7, 130
kinematic principle 64
kinematical relations 265
kinetics
    assembly 22, 34, 73, 89, 139
    assembly-disassembly 89, 92
    polymerization 99
klinokinesis 149
klinotaxis 149

lambda phage 244
lamella 7, 15, 29, 101
    tip 74
    dynamics 30
lamellipodia (-pods) 15, 73, 121, 141
Langevin equation 189
Laplace
    transform 158
    -'s equation 232
    -'s law 223
    -Young condition 232
laser scanning microscopy 11
lattice-gas cellular automaton 172, 179
law
    Aboav's 279
    Hooke's 124, 237, 238, 256
    Laplace's 223
    Lewi's 279
    position-dependence 216
leukocyte(s) 73, 132, 183
Lewi's law 279
linear stability analysis 107, 174, 200, 209, 233
Listeria monocytogenes 6, 93
local dry mass density 111
locomotion 7, 73, 119, 130, 133, 139, 141, 167
    amoeboid 33
locomotory machinery 34
longitudinal strain 237
Lymnaea 224
lymphocytic leukemia 47, 119
lysogeny-lysis 244
L1210 cells 47, 53, 119

macrophages 93
many particle system 189, 191
maple 244
Markov process 154, 157
Martiel-Goldbeter (MG) equations 197, 200
maximum entropy 276
maximum likelihood
    bead displacements 131
    image 129
mean-field approximation 180
mechanical
    constraint 168
    coupling 91
    equilibrium 222
    forces 215
    signals 168
    stress (MS) 221, 283
mechanism
    pseudo-spatial 147
    purse-string 230
mechanistic models 115, 168
mechano-receptor 182
medium
    diffusing 183
    elastic 279
    excitable 194, 207
    nondiffusing 183
    porous 85
melanoma 276
membrane
    -associated cell components 67
    -binding proteins 101, 103
    cortex 68
    -cortex interactions 103
    proteins 7
    protrusions 73
    ruffled 28, 101
mesenchyme 214, 229
mesoderm cells 8
messenger
    extracellular 193
    second 58, 60
metabolism 167

metastasis 15, 21
metazoan cells 118
method
    Galerkin finite element method 61
    least squares 37
    shooting 60
    deconvolution 48
    spectroscopy 134
    Yule-Walker 45
    moment's 50
microfilament bundles 230
microplasmodia 88
microscopy
    differential interference contrast (DIC) 33
    digitally recorded interference microscopy with automatic phase-shifting (DRIMAPS) 33
    image classification 47
    confocal laser scan 12
    fluorescence 50
    laser scanning 11
    reflection interference contrast microscopy (RICM) 33
microtubules 10, 80, 112, 222, 224
microvilli 69
migration 83, 160, 162, 165, 167, 171
Mimosa pudica 241
minimal energy 251
minimization of free energy 208
minipodia 121
mint 244
mitosis 230
mitotic
    apparatus 57
    rate 232
models
    basic morphogenetic 77, 80
    bilayer couple 68
    Brownian ratchet 112
    cellular automaton 171, 174, 179, 203
    chemotaxis 115

contact guidance 115
continuum 85
contraction-relaxation 90
cytomechanical 143, 147
discrete 183
discrete-continuum hybrid 172
epithelial 230
G-protein 195
lattice-gas cellular automaton 172, 179
Martiel-Goldbeter 200
mechanistic 115, 168
Physarum droplet 89
physico-mathematical 5
reaction-diffusion 81, 171, 194, 244
reaction-diffusion-advection 174
tissue-interaction 234
two-phase cytoplasm 74
molecular motors 3, 214
mollusk 224
moment's method 50
Monte Carlo techniques 129, 232
morphogenesis 166, 173, 214, 229, 233, 283, 284
    branching 229
    dynamics 215
    field theory 216
    movements 222
motility 59, 81, 123, 165, 167
    amoeboid 65
    bacterial 93
    parameters 133, 149
    spontaneous 17
motion 209, 283
    amoeboid 83
    cellular 284
    curvature-dependent 232
    gliding 9, 183
    self-organized 112
movement 149, 200, 222
muscle
    insect flight 91
mutation 204
myosin 10, 20, 57, 73, 105, 132

I 22
II 22, 76
light-chain kinases 85
myxobacteria 172, 182, 183

Naegleria gruberi 118
Navier-Stokes equations 232
network
  assembly 59
    rate 74
  contractility 56
  flow 74
  isotropy 57
  orthogonal 108
  phase 56
  plasmodial 83
  polymerization rate 56
  viscosity 56, 97
neurula 224
neurulation 215, 230, 231
neutrophils 22, 24, 139, 161
  polymorphonuclear (PMNs) 141
Newtonian
  fluid 96
    viscous incompressible 85
non-local interaction 173
non-Newtonian fluid 56
nondiffusing media 183

Occam's razor 5
opposite-leaved 244
organ
  development (organogenesis) 243
  formation 171
orientational
  singularities 178
  tensor 171, 176, 209
orthogonal networks 108
orthokinesis 149
orthotaxis 149
oscillation 22, 83, 87, 139, 261
  autonomous chemical 91
  torsion 91

oscillatory
  contractile phenomena 83
  system 25
osmotic
  forces 56
  gradient 214
  pressure 256

paired spiral 27
parametrization 29
parasite 93
pattern formation 179, 209, 231, 249
pattern(s) 24, 243
  horizontal stripes 25
  periodic 17, 32, 214
  schlieren-like 24
  spatio-angular 176
  spirals 249
  target 27
  waves 197
periodic solutions 80
periodicity 249
peristaltic contractions 83
persistence 184
perturbation 106, 108
  multiscale analysis 95
phagocytosis 21, 117
phantom cells, 52
phase
  gradient 91
  transition 180
phenotype 244
phospho
  -diesterase 195, 200
  -inositides 22
phosphorylation 194
  -dephosphorylation 91
Physarum polycephalum 5, 81, 85–87, 90
  droplet model 89
physical
  fragmentation 282
  moment 29
physico-mathematical modelling 5

picture frame hypothesis 163
pili 182
pinocytosis 117
plane stress 124
plant development 221
plasma membrane 57, 67, 73, 118
plasmodesmata 255
plasmodial
   network 83
   strand 83
plasmodium 83, 87
plastic extension 236
plate theory 251
Poiseuille flow 85
Poisson ratio 236, 238, 241
polarity 41, 146, 149
   tracks 45
polarization 222
polymerization 20, 56, 79, 94, 103
   kinetics 99
   rate 95
polymorphonuclear neutrophils (PMNs) 141
ponticulin 22
population distribution 173
porous medium 85
position-dependence law 215
positional information 214, 216
   theory (PI) 243
potato chip 244
potential energy 70
prespore cells 208
pressure 97
   hydraulic 10
   osmotic 256
   stress 222
   surface 251
   turgor 213, 221, 255
prestalk cells 208
principal curvatures 67, 71
profilin 20, 22, 94
proliferation 165
propulsion force 95
protein(s)
   bundling 102
   capping 101
   cross-linking 20, 59, 102
   extracellular matrix 147
   membrane 7, 68
   membrane-binding 101, 103
   synthesis 168
   transmembrane linker 20
protoplasmic streaming 83
protrusion 8, 15, 32, 33
   frequency 42, 43, 144
   vectors 39
   -retraction cycle 80
protrusive activity 231
pseudo-spatial mechanism 147
pseudo-steady state approximation 144
pseudocolour coding 43
pseudopodia (-pods) 21, 117, 139
psoriasis 276
pulsation 80
   synchronous and synphasic radial pulsations 83
purse-string mechanism 230
pushing force 97

quiescent centre (QC) 267

radial flow 75
random
   motility tensor 150
   undulation 245
   walk 172, 184, 206
      model 149, 157
      reinforced 185, 186
reaction(s)
   inflammatory 21
   -diffusion
      model 81, 171, 194, 244
      process 27
      wave 28
   -advection model 172, 174
reactive
   fluid 80
   interpenetrating

flow (RIF) 56, 96
fluid 73
receptor 194
  mechano- 182
  -measured gradient 142
reflection interference contrast
  microscopy (RICM) 33
regulation
  chemical 262
relaxation 81, 225
reproduction 213
repulsion 206
  interfilament 56
residual stress 213
resistance
  drag 74
  frictional 74
retraction 9, 15, 33, 77, 147
  force 20
  vectors 39
  -induced spreading 34
retrograde actin flow 81
Reynolds number 85, 232
Reynoutria 242
rheological equation 164
rheology 264
rhythms
  intracellular 28
rhythmic contractions 83
root
  apex 214, 267, 269
  cap 267
  movements 213
rotating waves 30, 88
rotational
  diffusion 104, 175
  velocities 149
ruffle(s) 15, 28, 73, 121
ruffled membrane 28, 101

salivary gland 230
scale linearization 49
schlieren-like patterns 24
scroll waves 27

sea-urchin egg 215
selection equation (of Eigen-Fisher
  type) 191
self
  -assembly 6
  -correcting properties 243
  -organization 243
    spatio-angular 173
    theory 218
  -organized motion 112
severin 22
shape 7, 47, 73
  descriptors 49
  topological 277
shooting method 60
signal(s) 194
  astral 62
  chemical 173
  chemotactic 136
  mechanical 168
  second messenger 58, 60
  temporal 133
  transduction 144, 255
singularities
  orientational 178
slime 192
  mold amoebae 183
  trails 184
  crawling 172
snapdragon 244
sol-gel transformation 22, 88
solid
  elastic 251
solvation 56
spatial gradient 115, 133, 139
spatio-angular
  dynamics 29
  organization 94
  patterns 176
  self-organization 173
spectroscopy 134
spiral
  double 27
    -armed 199

patterns 27, 249
  waves 197
spontaneous motility 17
spreading 35
  retraction-induced 34
stable adhesion zone 120
standing waves 24, 30, 83
starvation 206
stationary pulsations 80
steady state 80
stiffness 213
stimulus fields 154
stochastic
  fluctuation 144
  process 174
Stokes
  elliptic pseudo stationary system 75
  equation 232
strain 168
  longitudinal 237
  rate 232
  -induced activation 85, 89, 91
  tangential-transverse 237
strength of interaction 180
stress 168, 213, 215
  compressive 235
  contractile 59
  internal 59
  lines 225
  mechanical 283
  plane 124
  residual 213
  swelling 97
  tensile 235, 256
    turgor-induced 242
  traction 132
substrate adhesion 8
sunflower 241
surface
  pressure 251
  tension 57, 73, 77, 279
swelling 58, 73, 76, 97
  stress 97
symmetry properties 109

synergetic approach 22

talin 22
tangential
  flow 75
  stretching 222
  -transverse strain 237
target pattern 27, 197
taxis
  chemo- 116, 133, 141, 154, 163, 168,
    183, 193, 198, 205
  tropo- 149
Taylor
  series 174, 177
  transform 178
temporal
  dynamics 173
  gradient 115, 142
  signals 133
tensile stress 235, 256
tension 262
  cortex 80
  surface 57, 73, 77, 279
  visco-contractile 20, 92
tensor
  diffusion 150, 157
  directional persistence 153
  fiber orientation 160
  orientational 171, 176, 209
  random motility 150
theory
  plate 251
  positional information (PI) 243
  reaction-diffusion 244
  self-organization 218
thermodynamics 280
thymosin 20, 22
tissue 213, 282
  cells 33
  deformation 222, 257, 284
  environment 115
  epithelial 275
  -equivalents 167
  formation 159, 171

motion 209
regeneration 164
repair 165
stress 214, 235, 258
viscosity 233
tomato 272
topological
  correlation 282
  energy 279, 280
  shape 277
  transformation 275, 276
torsion oscillation 91
traction 160, 167
  density field 125
  forces 123
  images 123
  stress 132
translational velocities 149
translocation 35
transmembrane linker protein 20
transport-diffusion equation 109
transverse growth 258
travelling wave 24, 91, 92
  circular 80
treadmilling 103
trHEK (transfected human epidermal keratinocytes) 15
tropotaxis 149
tubulin 28
tumour(s) 165, 283
  growth 165
  progression 165
tunica 246, 251
turgor 235
  pressure 213, 221, 255
  -induced tensile stress 242
turning
  probability 103, 154
  rate 175

undulation 245

villin 102
vinculin 120
visco
  -contractile tension 20, 92
  -elastic fluids 234
    Voigt element 85, 88
viscosity 56, 233
  cytoskeletal 59
  network 56, 97
  tensor 74
viscous
  flow 73
  fluid 102, 232
Voigt element 85, 88
Volterra integral equation 259
von Bertalanffy function 261
von Kármán equations 252

Walker carcinosarcoma 43
  cell 42
wave(s)
  circular travelling 80
  excitatory 81
  generation 25
  length 244
  of alignment 178
  of cAMP 208
  reaction-diffusion 28
  rotating 30, 88
  scroll 27
  spiral 197
  standing 24, 83
    pulsating 30
  travelling 24, 91
weighed momental ellipse 18, 30
wound
  contraction 159, 167
    fibroblast-driven dermal 116
  edge 160
  healing 15, 21, 164, 171, 210, 276
  repair 159

Xenopus 7, 231
XTH-2 cells 7

Young's modulus 86, 124
Yule-Walker method 45